THE WORLD ATLAS
OF RIVERS, ESTUARIES,
AND DELTAS

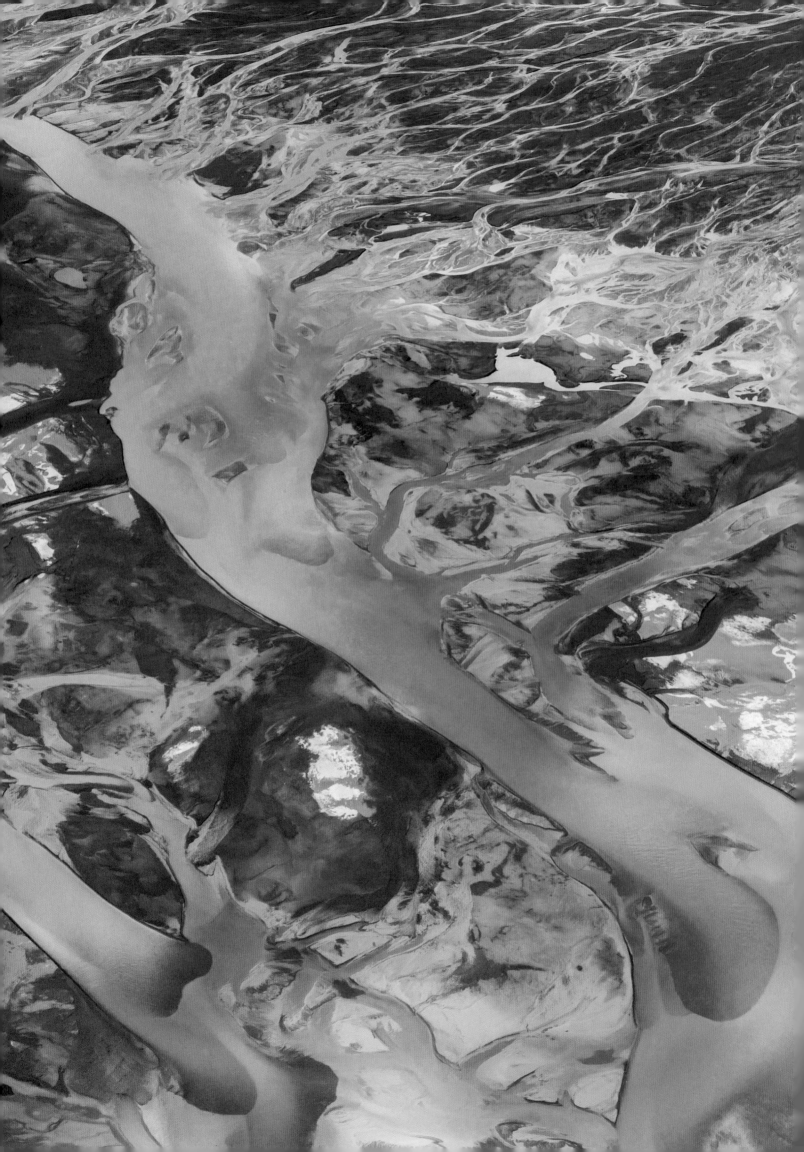

THE WORLD ATLAS OF RIVERS, ESTUARIES, AND DELTAS

JIM BEST · STEPHEN DARBY · LUCIANA ESTEVES · CAROL WILSON

PRINCETON UNIVERSITY PRESS

PRINCETON AND OXFORD

Published by Princeton University Press
41 William Street, Princeton, New Jersey 08540
99 Banbury Road, Oxford OX2 6JX
press.princeton.edu

Library of Congress Control Number 2023943776

ISBN 978-0-691-24483-9
Ebook ISBN 978-0-691-24484-6

Typeset in Ulises and Autor

Printed and bound in Malaysia
10 9 8 7 6 5 4 3 2 1

British Library Cataloging-in-Publication
Data is available

This book was conceived, designed, and
produced by UniPress Books Limited
Publisher: Nigel Browning
Commissioning editor: Kate Shanahan
Art director: Alex Coco
Project manager: David Price-Goodfellow
Designer: Lindsey Johns
Editor: Susi Bailey
Illustrators: Martin Brown, Sarah Skeate and
John Woodcock
Picture researcher: Julia Ruxton

Cover image: © Tom Wagenbrenner
Cover design: Wanda España

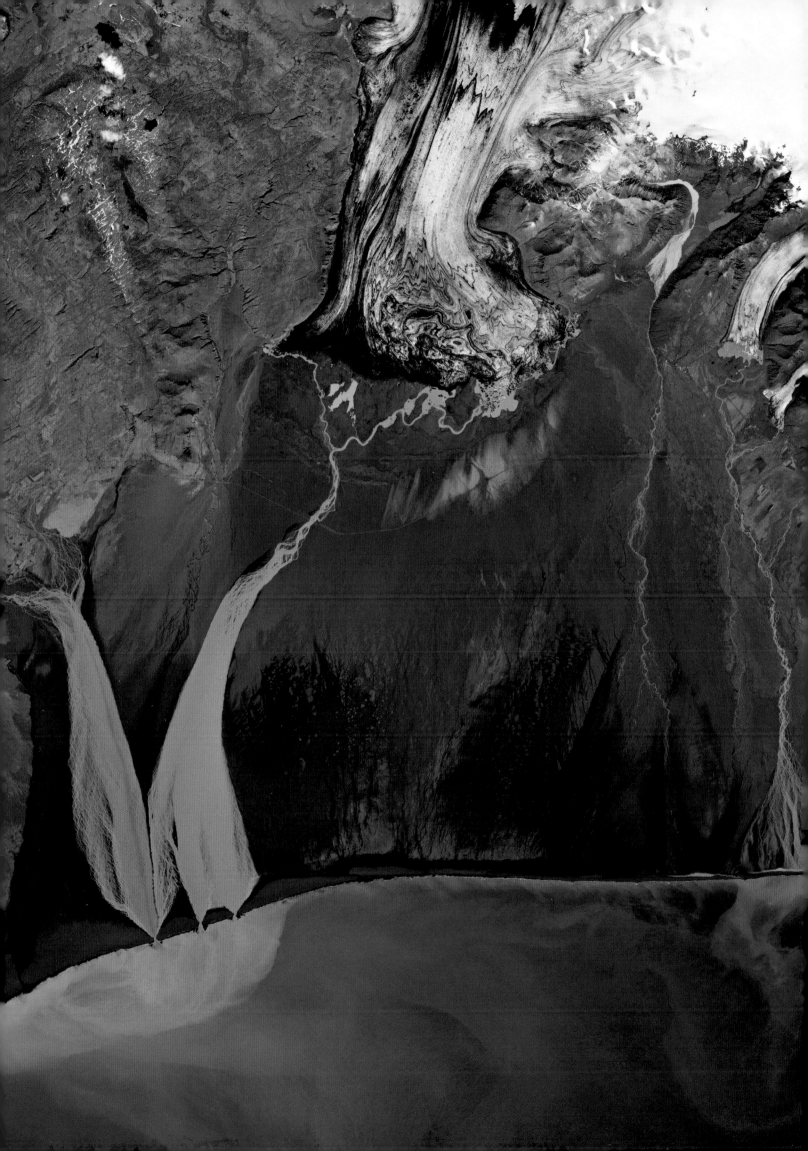

Contents

Introduction

Rivers, estuaries, and deltas comprise some of the most spectacular landscapes on Earth. Having acted as cradles of civilizations and now providing livelihoods for billions of people, they also lie at the very heart of human existence. This atlas examines their processes, form, and ecology, and their links with the human experience. It is a book about rivers, estuaries, and deltas—those elements of the landscape that connect the mountains to the oceans, and on which the world's most diverse ecosystems thrive.

As it makes its epic journey across the landscape to the sea, water forms the unifying thread between the world's rivers, estuaries, and deltas. Rivers are the conduits of water across the landscape, connecting to estuaries and deltas where they empty into lakes, seas, and oceans. Earth is the only known planet to have water in three phases—solid ice, liquid, and water vapor—and exists in a kind of a "sweet spot" of cosmic serendipity due to our distance from the sun. As such, water carves and molds the landscape through the erosion and deposition of sediment, permits the establishment of life, and generates conditions that have, over geological time, led to diverse ecosystems of global importance.

A living mosaic

River, estuary, and delta environments are also home to a teeming variety of life, supporting some of the most diverse ecosystems on our planet and creating a living mosaic of bewildering complexity and beauty. Yet not only are these environments home to great ecological diversity, but their evolution and character are also part of the very reason why such diversity has evolved. We have now come to realize that the ecosystems we live with today, and that provide humans with much natural capital—agriculture, water, food, materials, recreation—have been coevolving with the world's rivers, estuaries, and deltas. The behavior and morphology of rivers, estuaries, and deltas are dependent on the characteristics of the physical, chemical, and biological processes that have been shaping these landscapes for many millions of years.

▶ **Spreading out**
As rivers meet a body of water, they spread out and deposit sediment to form deltas. Here, the sediment-laden Tsiribihina River in Madagascar empties into the Mozambique Channel.

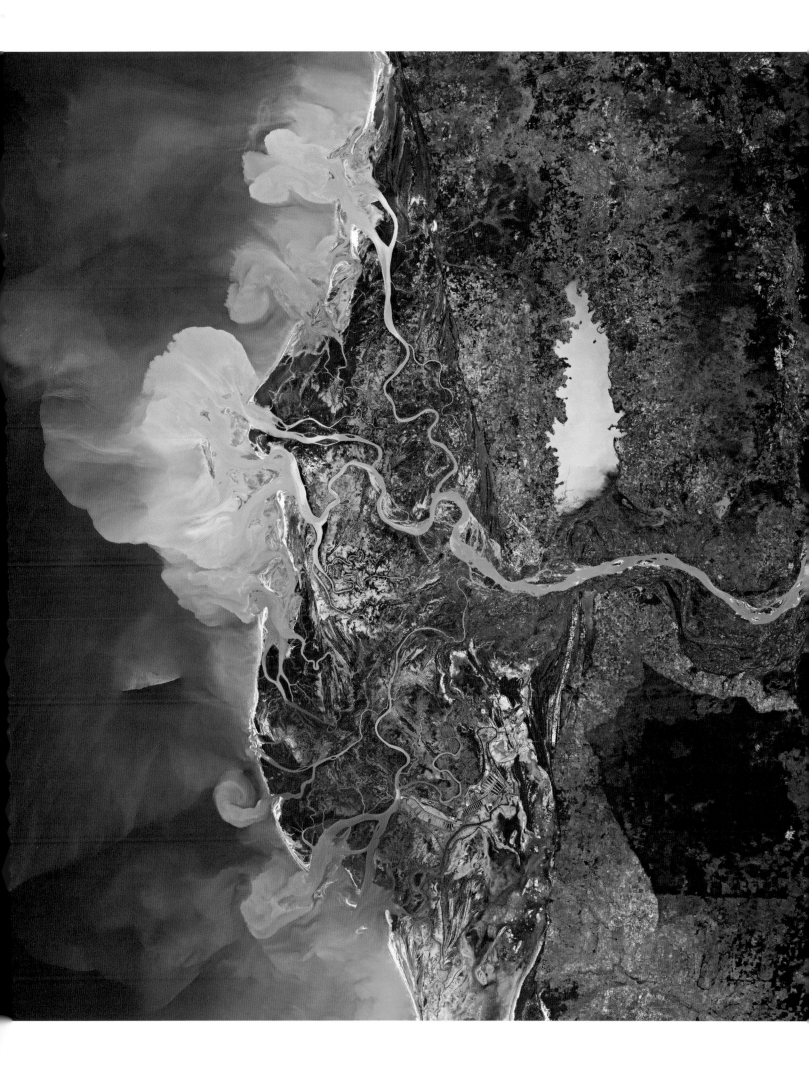

▲ River culture

The Arno River snakes its way through the city of Florence, Italy, a center for the explosion of Renaissance science and art.

The human experience

More than this, however, rivers, estuaries, and deltas sit at the heart of human experience. They have formed routes for human migration across the globe, provided the water and fertile sediments needed to stimulate the growth of agriculture and cities, and acted as the inspiration of much human culture and art.

The natural patterns of rivers, estuaries, and deltas provide shapes and forms that are at once fascinating, reassuring, and thought-provoking. How did these patterns emerge? What controls their characteristics? How long have they existed and how have they changed over space and time in the past? And how is life entwined with the different facets of these landscapes? Philosophers, hunters, farmers, nomads, artists, scientists, poets, and travelers have all posed these same questions over the course of human history, seeking to gain both a better insight into the world as they have perceived it, and to use this understanding to look more closely into their own existence and experiences. Through this book, we will give our own answers to these questions.

Legendary links

Rivers, estuaries, and deltas have also served as objects of admiration and metaphors for much within the human condition throughout our history, holding importance for how we view ourselves and for the human journey through time. Rivers have held mythical significance since the earliest of times. They have, and still do, adopt a religious importance for billions of humans who have come to revere, respect, and worship these elements of the natural landscape.

The Hindu goddess Ganga, the personification of the Ganges River, represents purification, wellness, and benevolence. Legend has it that Ganga came to Earth after hearing cries from people who were dying from drought. The great Hindu deity Lord Shiva divided Ganga into seven streams to flood the Earth, with part remaining in the heavens as the Milky Way. The rest flows through India as the Ganges River, where the goddess lives. The waters of this sacred river thus have religious and cultural significance for millions, who believe that bathing in its waters will bestow powers of protection, forgiveness, and good health. Hindu pilgrims cast the ashes of their kin into the river so that their souls are brought closer to liberation from the cycle of life and death. River, estuary, and delta landscapes are thus far more than the physical, chemical, and biological processes from which they originate.

▲ **Holy river**
Statue of the Hindu goddess Ganga on the Ganges River, Rishikesh, India.

Source of inspiration

Rivers, estuaries, and deltas have captured the attention and fascination of artists for centuries, presenting natural vistas captured on stone, canvas, paper, fabric, and film. Depictions of life in, on, and around these landscapes are portrayed in ancient stone carvings, in paintings, on film, and in words.

These landscapes also offer metaphors for many aspects of human existence. The beautiful poem "The Negro Speaks of Rivers," written in 1920 by American poet Langston Hughes (1901-1967), traces Black history from the earliest moments of human civilization through to the horrors of slavery, and in doing so celebrates the strength and perseverance of Black heritage. Its lines compare the multilayered aspects of rivers and their progression through a landscape to the writer's soul, which is like the most ancient and longest of rivers.

I've known rivers:
I've known rivers ancient as the world and older than the flow
of human blood in human veins.

My soul has grown deep like the rivers.

I bathed in the Euphrates when dawns were young.
I built my hut near the Congo and it lulled me to sleep.
I looked upon the Nile and raised the pyramids above it.
I heard the singing of the Mississippi when Abe Lincoln went down to
New Orleans, and I've seen its muddy bosom turn all golden in the sunset.

I've known rivers:
Ancient, dusky rivers.

My soul has grown deep like the rivers.

The landscapes of rivers, estuaries, and deltas thus lie deep in our souls, being the origin of our evolution, where we have thrived and died, where we have coped with success and tragedy, and where we have seen great civilizations rise and fall.

The structure of this atlas will examine, in turn, rivers, estuaries, and deltas, following the journey of water downstream as it passes through these landscapes. In each landscape, we will examine the processes that control the flow of water and movement of sediment. This will then allow us to look at the shape and morphology of these landscapes—their anatomy—and consider the variety of life that exists in each, creating environments at a variety of scales in time and space. Novel technologies are providing us with fascinating new insights into these landscapes, enabling us to understand better their form and function.

We will then consider how humans live in these landscapes and the intimate links between natural processes and human wellbeing. In our rapidly changing world, river, estuary, and delta landscapes are among those under most pressure from a wide range of factors, from sea-level rise to global climate change, and from land-use change and urbanization to the construction of huge megadams. These anthropogenic stresses are placing huge pressure on our rivers, estuaries, and deltas, and in some cases are even threatening their very existence. It is no exaggeration to say that how we deal with these threats is critical to the future of humanity.

Finally, we will peer into the future worlds of rivers, estuaries, and deltas. We will consider how we can manage these environments, how we can better study and understand them, and what new tools we possess to aid us in this. In this way, we hope to focus our attention on how we can better live with these environments and progress toward a more sustainable future for rivers, estuaries, and deltas, their ecosystems, and ourselves.

▲ **Starry Night Over the Rhône (1888)**
Vincent van Gogh's (1853-1890) stunning depiction of night light over the River Rhône at Arles, France, at the head of the Rhône delta.

The World Atlas of Rivers, Estuaries, and Deltas provides a journey across these landscapes, guiding the reader through the unique conditions on Earth that permit their existence, how they change over time, and how the lives and livelihoods of humans are entwined with the lifeblood that water provides. Using maps, diagrams, and a range of beautiful images, we will guide you through the story of water and life in these landscapes. At the end of the atlas, we provide references to additional reading and several online resources for those who wish to pursue some of the aspects raised in this book further. The longevity of these landscapes reaches beyond human experience and history, and inspires us to view our stewardship of them through a lens that looks back into deep time and forward into the future.

▼ Safe haven

Estuaries such as the Taw-Torridge estuary in England have provided safe harbor for seafarers for centuries.

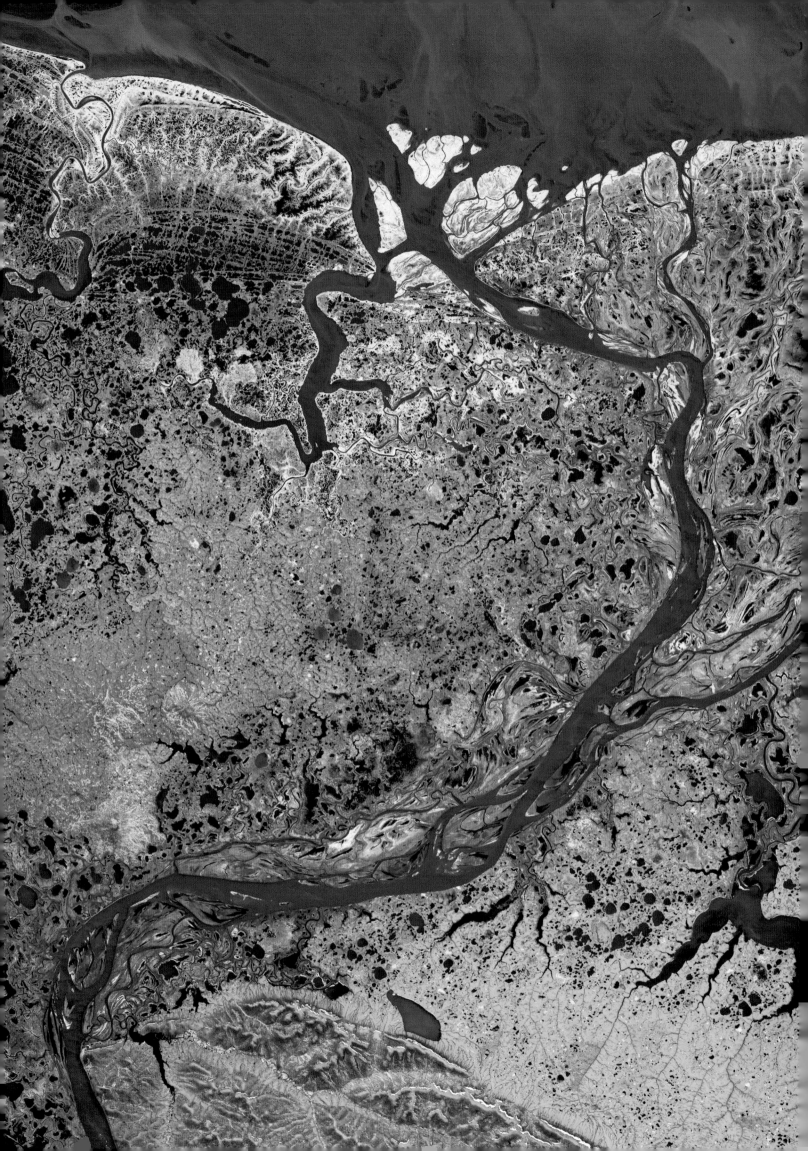

The World of Rivers, Estuaries, and Deltas

A historical context

Rivers, estuaries, and deltas have been regions of especial significance across the entirety of human history. They have been home to growing civilizations, now host modern megacities, and comprise sites of critical historical importance. As such, they form a lens by which we can look to the future but also gaze into the past.

River civilizations

Rivers, estuaries, and deltas provide environments in which human civilization has thrived over millennia. River valleys supply fertile soils and water for irrigation, and an abundant source of protein from fish, forming havens where the transition to agricultural societies could flourish. Deltas also constitute productive land and act as gateways for trade, with estuaries forming safe harbors for ships that ply the world's oceans. Estuaries and deltas thus adopted a central place in the exploration of the world's seas and continents, providing bases from which human expansion could proceed.

RIVER VALLEY CIVILIZATIONS

The ancient civilizations of the Old World were based around river valleys and their deltas, which provided fertile soils, water, and access, and where agriculture, transport, and trade could flourish. These river-valley civilizations possessed very different languages, cultures, religions, and political systems, and were the birthplace of many advances in technology, agricultural practices, societal organization, science, and art.

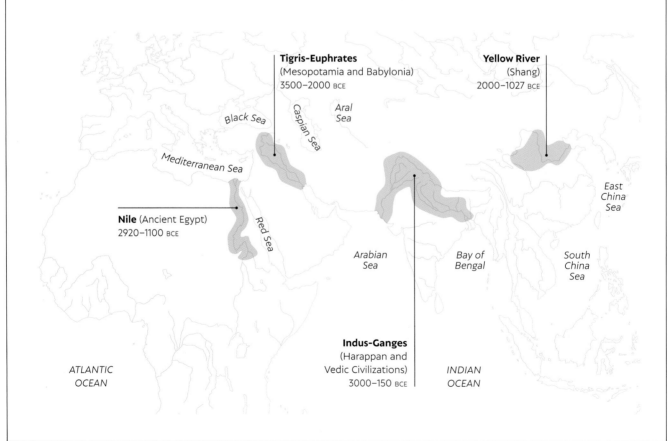

Tigris-Euphrates
(Mesopotamia and Babylonia)
3500–2000 BCE

Yellow River
(Shang)
2000–1027 BCE

Aral Sea

Black Sea

Caspian Sea

Mediterranean Sea

East China Sea

Nile (Ancient Egypt)
2920–1100 BCE

Red Sea

Arabian Sea

Bay of Bengal

South China Sea

Indus-Ganges
(Harappan and Vedic Civilizations)
3000–150 BCE

INDIAN OCEAN

ATLANTIC OCEAN

Some of the earliest civilizations thus lay in the great river valleys of the world: the Tigris and Euphrates rivers (Mesopotamia, meaning "between the rivers"), which were home to the Sumerian, Assyrian, Akkadian, and Babylonian civilizations; the Nile River valley, with its great Egyptian dynasties; the Indus valley, where the Harappan and Vedic societies developed; and the Huang He (Yellow) River in China. Newer civilizations also thrived in river valleys, including those of the Niger River in western Africa, and the Khmer and Funan societies in Southeast Asia.

The Egyptian dynasties used the River Nile to manage and grow their kingdoms, and developed technologies for irrigating the river floodplain that allowed the practice of year-round agriculture. The ancient Egyptians also realized that measuring the height of the river, through gauges termed "nilometers," made it possible for them to ascertain when the annual flood was coming. Not only did this enable them to manage water on the floodplain more effectively, but it also allowed them to predict the likely yield of the yearly crop, including its success or failure, and thus set taxes that could be levied on agricultural production.

Civilizations in the Americas also had an intimate relationship with rivers, in both their growth and, perhaps, their demise. Cahokia, in modern-day Illinois, was the largest city ever built north of Mexico in pre-Columbian times. It consisted of a series of massive square-bottomed, flat-topped earthen pyramids, with the largest—termed Monks Mound—around 31 m (100 ft) high. Agricultural intensification, expansion of settlements, and development of the city were based around the productivity yielded by the river and its floodplain, with a relatively dry climate from 600 CE to 1200 CE perhaps facilitating this growth. There is also debate as to the possible role of increasingly larger floods in aiding the demise of the city after 1200 CE.

▲ **Ancient measurements of river height**

The nilometer at Rhoda Island in Cairo, Egypt, constructed in 861 CE, permitted the level of the River Nile to be measured using the octagonal vertical column, with the stilling chamber being linked to the river by three horizontal tunnels.

Mapping the world

As humans started exploring the globe, the access provided by deltas, estuaries, and rivers into continental interiors proved pivotal. Mapping of river courses and their exit into the oceans has thus charted the progression of human knowledge concerning the Earth's surface.

In depicting the geography of the world, early mapmakers relied on major physical features—oceans, mountains, and rivers—and tried to place them in their spatial context in light of existing knowledge and views of the time, which could include religious, social, and political doctrines. The map of the world according to the Greek philosopher, historian, geographer, and astronomer Posidonius (c. 135-51 BCE), as reconstructed and interpreted by the Flemish cartographer Petrus Bertius (1565-1629), was published in 1630. The map depicts Armenia at the center of the globe, with the Nile (including the great bend of the river), Tigris, Euphrates, and Danube Rivers clearly apparent.

Some of the earliest maps were also engraved in stone; of these, "Yu ji tu," or the "Map of the Tracks of Yu" is the oldest known and is exceptionally preserved. The map depicts the waterways in China, with nearly 80 rivers named.

A country dominated by a river

The Gambia in West Africa (outlined with yellow line), the smallest country of the African mainland, is only 475 km (295 miles) long and 25–50 km (15–30 miles) wide, and is surrounded by the country of Senegal. It is dominated by its river and estuary, with the Gambia River running the whole length of the country into its estuary on the Atlantic coast, where its capital, Banjul (1), is located on an island at the river mouth. The unusual shape of the country is a product of its imperialist colonial past, and agreements between Britain, which controlled the lower Gambia River, and France, which ruled Senegal. Until 2019, the only way to get from one side of the country to the other was by boat, or by making the overland journey through Senegal. Centuries of trading and travel using unreliable ferries ended in 2019 with the opening of the 1.9 km-long (1.2-mile) Senegambia Bridge (2), which now unites the two halves of The Gambia.

The Gambia
A country dominated by its river and estuary. The image depicts the border of The Gambia with Senegal.

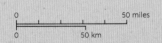

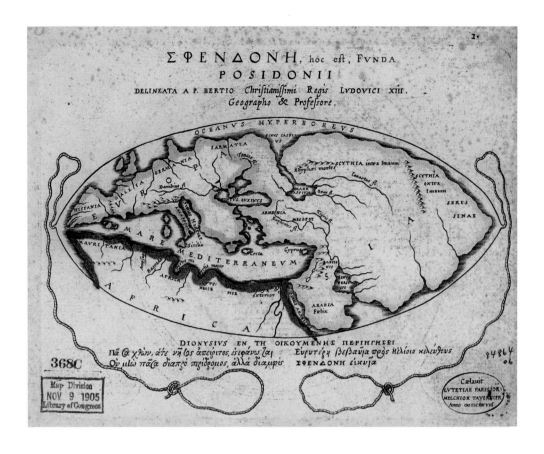

An early world view

The world according to Posidonius (c. 135-51 BCE) as drawn in 1628 by cartographers Petrus Bertius (1565-1629 CE) and Melchior Tavernier (1594-1665 CE). The map is surrounded by bowline knots that highlight the nautical emphasis of the Mediterranean region at the time.

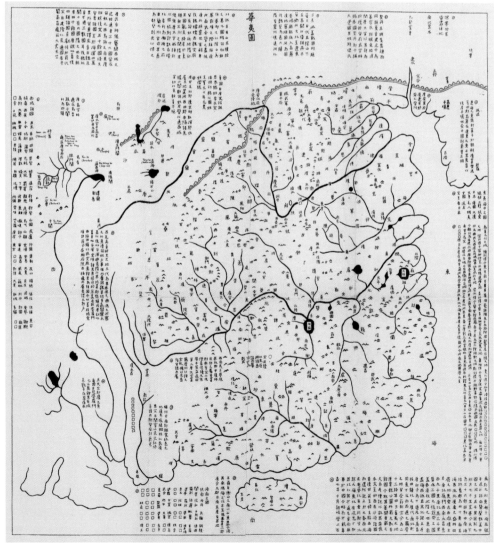

Set in stone

Rubbing made in 1903 of the "Yu ji tu" map of China, which was engraved into the face of an upright stone in 1136 CE. The map depicts the deltas of China's two major rivers (the Huang He/Yellow River and Chang Jiang/Yangtze River), the Great Wall of China, and more than 500 place names. The map is oriented with north at the top.

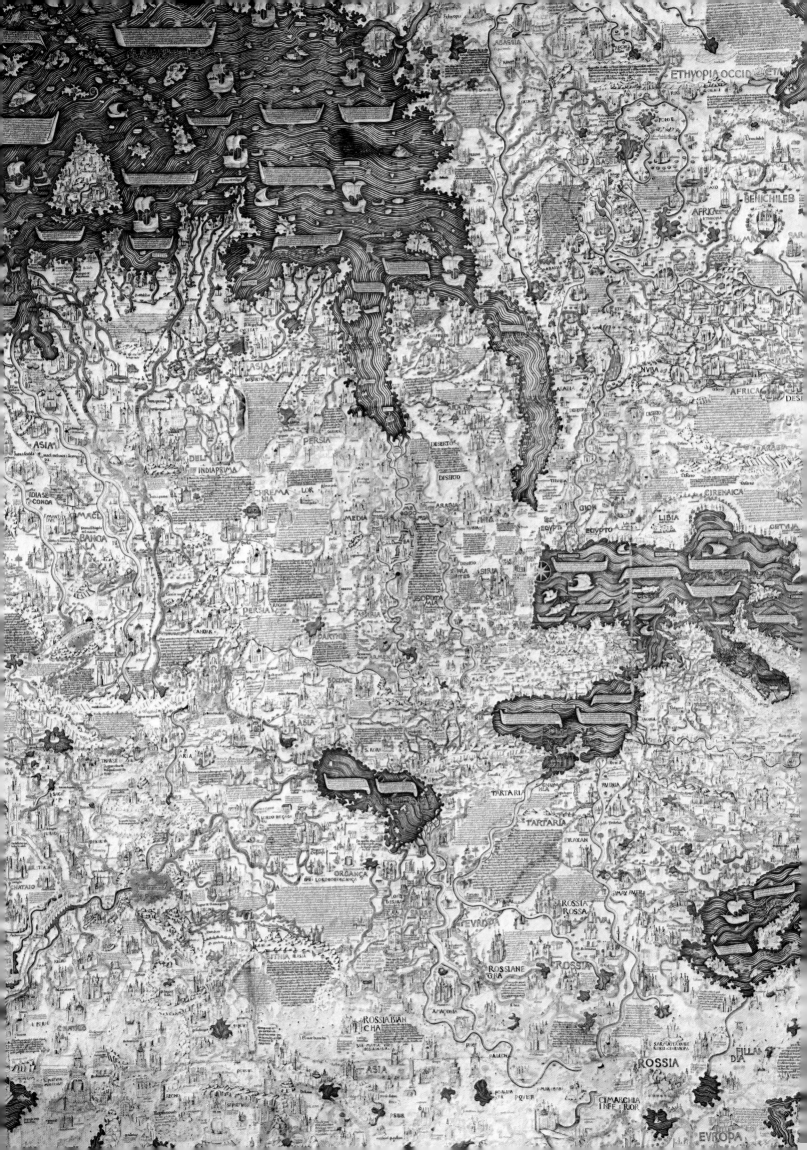

As human exploration expanded, cartographers began drawing maps using information gained from direct accounts of travels. One such pioneer was Fra Mauro, a fifteenth-century Camaldolese monk living on the island of Murano in the Venetian Lagoon, who was commissioned by King Alfonso V of Portugal to produce a map of the world. Using accounts of travelers and traders landing in the port of Venice from all over the world, and integrating this with knowledge from existing maps and books, Fra Mauro drew a map—measuring 2.4 m (7¾ feet) square—that placed cartographic accuracy ahead of religious or traditional beliefs. This astonishing map, first created *c.* 1450 CE, depicts rivers, estuaries, deltas, and mountains, combined with hundreds of detailed illustrations and annotations. Rivers shown on the map, with names often different to their modern-day equivalents, include those in Eurasia (such as the Dnieper, Elbe, Rhine, Rhône, Oder, Po, Saône, Tagus, Tigris-Euphrates, Vistula), Africa (Jordan, Nile, Niger), Asia (Ganges, Helmund, Indus, Ayeyarwady), Russia (Volga) and China (Hong, Huang He/ Yellow, Chang Jiang/Yangtze), with the deltas of the Danube, Ganges, Ayeyarwady, Nile, Niger and Po Rivers also being shown. These maps thus capture landscapes at the time they were drawn, providing records of the world as known at the time.

◀ **Incredible map**
Part of the world map created by Fra Mauro in 1450, oriented with north at the bottom. The Nile, Tigris-Euphrates, and Ganges rivers are all evident on the map.

◤ **Which is longer?**
Chart printed in 1834 by the Society for the Diffusion of Useful Knowledge that visualizes and compares the length, planform shape, and general geography of the world's major rivers, their estuaries, and their deltas. Concentric circles show the general lengths of the rivers, in miles, as the bird flies, with direction being illustrated visually.

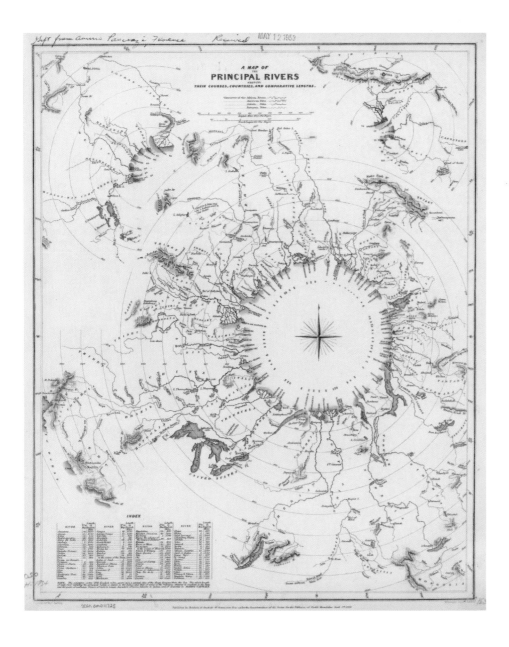

Some watery essentials

Rivers, estuaries, and deltas are controlled by physical, chemical, and biological processes that shape their evolution. To understand their evolution, form, and functioning, we need to examine some common concepts concerning water that set the background for these landscapes.

Water cycle

Due to its distance from the sun (150 million km/91 million miles), Earth is the only planet in the solar system to have water in three states (frozen, liquid, and vapor). The water cycle describes how these three states move across the landscape, from evaporation out of large waterbodies, across the sky as clouds and water vapor, precipitation in either frozen or liquid form (depending on climate), runoff across the landscape due to the force of gravity, and percolation into the soil as groundwater. The water cycle also includes how water is drawn up from the soil by plants and its evaporation from leaves (a process called transpiration), as well as how humans have altered water flow pathways—for example, by placing dams across rivers, diverting water for crop irrigation, and extracting groundwater.

Salinity

▼ **Staying afloat**

The hypersaline conditions in the Dead Sea between Israel and Jordan help keep swimmers afloat, due to the increased water density.

Pure water (H_2O) comprises two molecules of hydrogen bound with one molecule of oxygen. However, several other ions can readily be dissolved in water, including salts such as sodium chloride (NaCl) and potassium chloride (KCl). Salinity describes how much salt (mostly Na+ and Cl- ions) is dissolved in water, and is typically measured in units of parts per thousand (ppt). Measurements start at 0 ppt for pure water, and range up to 33 ppt for seawater and >50 ppt for hypersaline conditions. Fresh water in rivers usually ranges from 0 ppt to 0.5 ppt, and brackish water (0.5-30 ppt) in estuaries is a mix of fresh water and seawater. The oceans have

greater salinity than rivers because of the concentration of ions delivered to them by rivers over millions of years. Hypersaline conditions are reached when river or oceanic water is subjected to intense evaporation, concentrating the salts. Salinity is a major factor influencing the distribution of flora and fauna, as although salt is a necessary nutrient, too much salt disrupts cellular functioning. Only specially adapted organisms thrive under saline and hypersaline conditions.

THE WATER CYCLE

Water moves across the Earth's surface as water vapor evaporated from waterbodies, precipitation as rain or snowfall, surface runoff into rivers, and groundwater percolation and movement.

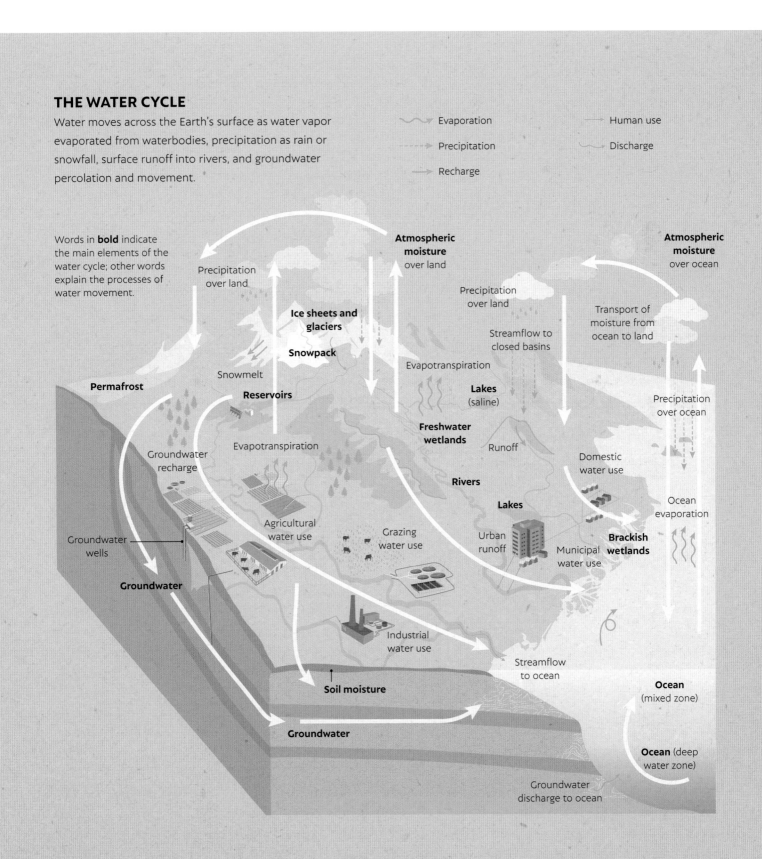

Waves

Waves, the orbital motion of water molecules, are generated when wind blows across the surface of a waterbody. The water molecules move in an orbital (circular) direction, and if you were in the middle of the ocean, they would first move you up, then laterally forward, then down, then laterally back. Due to decreasing energy at greater depth, the size of the orbital motion becomes smaller and smaller. Eventually, the wave motion diminishes to zero at a particular depth, called the "wave base." When waves approach a shoreline, the wave orbitals at depth start to "feel" the bottom, deforming them from circular to elliptical in shape. Friction with the bottom slows the wave, making it taller, and eventually the top of the wave moves faster than the bottom, causing the wave to "break." These spectacular movements of water can transport sediment particles along a beach, either in beautiful linear ripples along the bottom, or up and down and along the beach from currents called littoral drift.

Tides

Tides are a special kind of wave. They result from water in the oceans being affected by the gravitational pull of celestial bodies, primarily the moon and, to a lesser extent, the sun. The moon is a quarter the size of Earth, but it is only 400,000 km (240,000 miles) away. Due to this proximity, water in the ocean "facing" the moon is pulled toward it, forming a bulge. Because of complex gravitational and centripetal forces on Earth, a similar bulge also occurs on the opposite side of the planet, facing away from the moon. These bulges are "high tides," where the water level in the ocean is elevated above the average. At locations in between the bulges, "low tide" occurs. Because the Earth undergoes a complete rotation over a 24-hour time frame (the course of a day) but the moon moves only slightly in its orbit around us, the Earth rotates under this gravitational bulge, causing water levels in the oceans to undergo two high tides and two low tides per day, roughly six hours apart. Each day, the timing of the high and low tides shifts approximately 50 minutes because the moon is orbiting around the Earth over the course of a month (28 days). The difference in elevation between high and low tides is called the tidal range (see page 163), and this varies across the world. Spring and neap tides greatly influence the migration of fish through estuaries and deltas, and where wetland vegetation may take hold on land between high and low tides—the intertidal zone.

▼ **Wave breaking in Queensland, Australia**
Waves break when the top of the wave moves faster than the bottom.

TIDES ON EARTH

Due to its proximity, the moon is the primary gravitational force creating tides, but the sun also affects the magnitude of tides over the course of a month. When the moon is at its full and new phases, the tidal range is at its maximum, called a "spring tide." When the moon is at its first and third quarter phases, the tidal range is at a minimum, called a "neap tide."

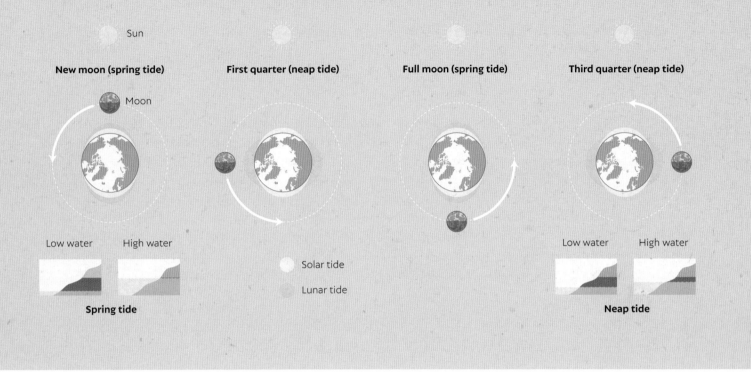

Sun

New moon (spring tide)   **First quarter (neap tide)**   **Full moon (spring tide)**   **Third quarter (neap tide)**

Moon

Low water   High water

Solar tide

Lunar tide

Spring tide

Low water   High water

Neap tide

Sea level

Around the world, water flows from rivers into estuaries and deltas, and eventually the oceans. The water level in the oceans, called "sea level," is not always stationary. Tides can cause fluctuations over the course of a few hours, while winds and storms can push water up against the shoreline over a daily, monthly, or seasonal timescale. In general, scientists consider mean sea level as the average of these daily, monthly, and seasonal water-level fluctuations. Over the course of the past 10,000 years, sea level has been rising as the Earth has warmed, melting glaciers and transferring the water stored in these land-based ice masses into the oceans. Recent additions of greenhouse gases into the atmosphere by humans have accelerated this warming: the International Panel on Climate Change has shown that the Earth's temperature has risen by an average of 0.08 °C (0.14 °F) per decade since 1880, or about 1.1 °C (2 °F) in total. Consequently, mean global sea-level rise (SLR) is now in the order of 3–4 mm ($^1/_8$ in) per year.

▼ **Submerged**

Sea-level rise, together with coastal erosion and land subsidence, has led to the loss of this mosque in Jakarta, Indonesia.

Controls on geographical distribution

Earth is called the "blue and green planet" because water is present across large parts of its surface and on land this enables the growth of vegetation. Rivers, estuaries, and deltas are a central part of this water pathway, and they are present all across the globe, from lush tropical areas to deserts and polar landscapes. What dictates this geographical distribution?

▼ **The Earth's plates**

The theory of plate tectonics shows that the Earth is composed of lithospheric plates that are created, modified, and destroyed along plate boundaries. The continents that we know today—and thus the locations of modern rivers, estuaries, and deltas—were created by these interactions over geological time.

Plate tectonics

The Earth is unique in that its hard rock outer shell (called the lithosphere) is slowly but constantly moving on top of very hot plastic-like rock (called the asthenosphere). It was only in the 1960s that scientists discovered that the creation of mountain chains, volcanoes, and deep-sea trenches all over the world is related to the moving plates of the Earth's outer shell, a process called plate tectonics. The lithosphere is divided into oceanic crust, which is primarily composed of the rock basalt, and continental crust, which is primarily composed of the rock granite. When continental crust collides with continental crust during the geographic wanderings of the Earth's plates, large mountain ranges form—such as the Himalayas. When continental crust collides with oceanic crust, the denser oceanic crust (density = 3.1 g/cm^3; 193 lb/ft^3) sinks and flows under the lighter continental crust (density = 2.8 g/cm^3; 175 lb/ft^3); this process is called subduction. This typically results in the subducted crust melting at depth and gives rise to volcanoes along the margin of the subducting plate (a volcanic arc). Over millions

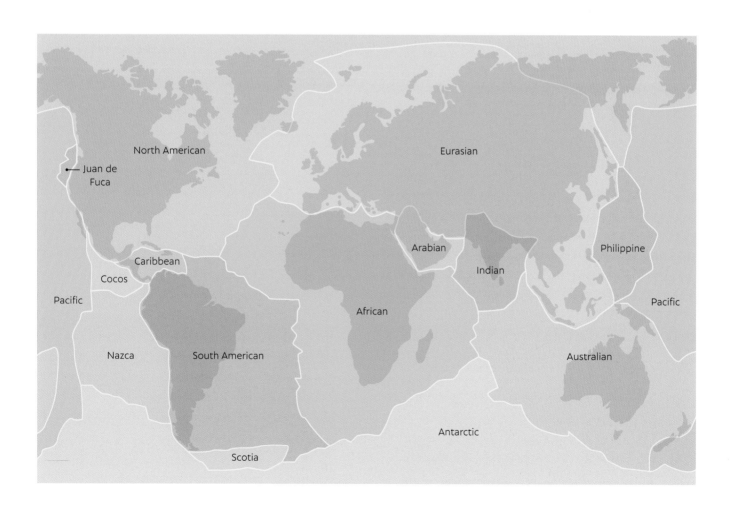

of years, the Earth's plates have wandered substantially, constantly separating and merging. The location of rivers, estuaries, and deltas is therefore also constantly changing due to these interactions over geological time, as the tectonic movements determine the distributions of landmasses and the slopes that water must follow.

Climate zonation on Earth

Earth is divided into several different climate zones that are based on long-term patterns of temperature and precipitation. Near the equator, the energy from the sun is at its maximum, while at higher latitudes incoming radiation diminishes, and the poles receive the most indirect solar energy. As a result, mean temperatures decrease with distance from the equator. Tropical climates are characteristically warm, while moderate and continental mid-latitude climates are more temperate (except over large landmasses in summer and winter) and the polar climate is cold.

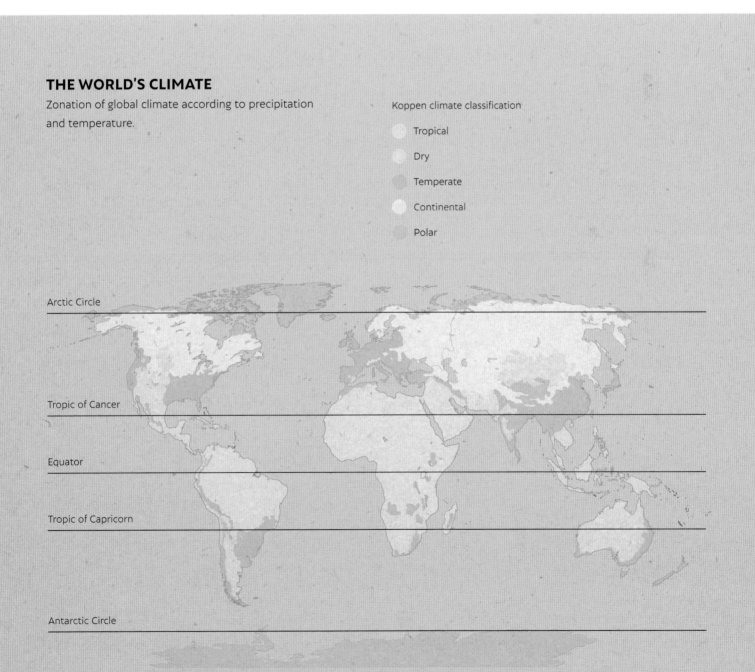

THE WORLD'S CLIMATE

Zonation of global climate according to precipitation and temperature.

Koppen climate classification

- Tropical
- Dry
- Temperate
- Continental
- Polar

Arctic Circle

Tropic of Cancer

Equator

Tropic of Capricorn

Antarctic Circle

These climate zones are further subdivided by how much precipitation they receive: in tropical climates, 100-300 cm (40-120 in) of rain per year is common, while moderate and continental climates receive less annual precipitation (25-100 cm/10-40 in, for example). Polar and desert regions are parched, receiving less than 25 cm (10 in) of precipitation a year. Many deserts around the world are found at the latitudes 30°N and 30°S. This is due to the presence of atmospheric circulation cells, which form due to the redistribution of heat. At the equator, warm air rises (a process called convection), generating clouds as the water in the air condenses. The air is then moved to the north and south of the equator, becoming denser as it cools, and at the same time is steered by the Coriolis force caused by the Earth's rotation. Eventually, the drier, denser air descends at 30° latitude and moves back toward the equator along the Earth's surface,

A WORLD OF RIVERS, ESTUARIES, AND DELTAS

The principal rivers, estuaries, and deltas of the world, together with major cities.

▲   Major delta

■   Major estuary

〰   Rivers and lakes

Cities

o   0.5–2 million

○   2–5 million

○   >5 million

forming the trade winds and regions of drier climate. These circulation cells are called Hadley cells after George Hadley (1685-1768), an amateur meteorologist who first explained the formation of the trade winds. The Sahara Desert in Africa, and the Gibson and Great Sandy Deserts in Australia, are prime examples of these dry climate zones. Beyond these zones and before reaching the cold polar regions, Earth is quite lush, with vegetation thriving from the abundant water supply.

Rivers, estuaries, and deltas occur within all of the Earth's climate zones as long as enough water is able to pass across the surface (rivers), and empty into lakes and ocean basins (estuaries and deltas). Climate not only dictates how much water may be present, but also what kind of flora and fauna thrives there.

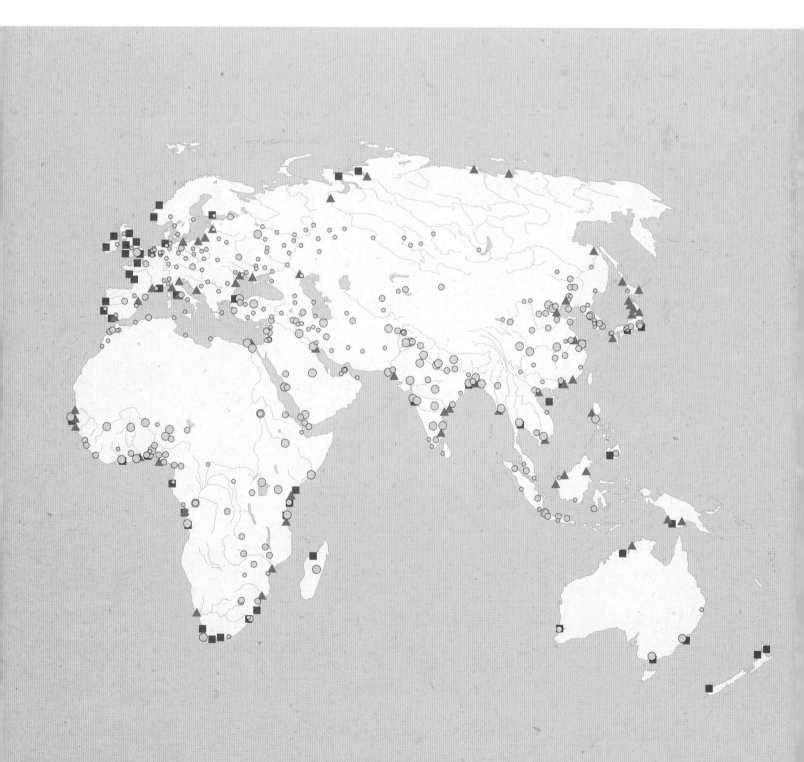

From source to sink

Rivers transport water, sediment, and nutrients from their catchment to the coast and into the sea. Maintaining this flux and connectivity across the catchment-to-coast continuum is crucial to sustaining the range of habitats and conditions required by freshwater, estuarine, and marine life.

A river catchment, or drainage basin, is an area where all surface water and groundwater ultimately drains to the same outlet—a lake, a river, or the sea. Due to gravity, water flows to lower elevations, carrying sediment and other material from upstream landscapes to areas downstream. Therefore, changes in soil conditions, vegetation cover, and human activities in catchment areas can deeply affect the environmental conditions from the river to the deep ocean.

In natural catchments, water flow is controlled by the topography, geology, and presence of vegetation. The vegetation on riverbanks and floodplains plays a key role in regulating water flows along the catchment and producing organic matter that is the source of nutrients to primary producers at the base of freshwater and marine food chains. The volume of water moving down the catchment depends on the amount of precipitation, and how much of this evaporates, is retained

SOURCE TO SINK

Catchments supply water and sediment to downstream areas, and rivers are the main transport pathway on land, while turbidity currents transfer terrestrial material to the deep sea, connecting each landscape from source to sink. Tectonic movements control the dimensions, steepness, and evolution of this connected system.

Source

Sink

Flux of sediment

Submarine canyon

Turbidity current

Catchment area

Fluvial system

Tectonic uplift

Shoreline

Shelf

River sediment plume

Continental slope

Deep-sea fan

by vegetation, or infiltrates into the soil. The soil and vegetation are natural filters, retaining sediment and pollutants and improving water quality. Water that infiltrates permeable soils, or is used by vegetation, will take longer to reach downstream watercourses, which reduces both the spread of pollutants and the risk of flooding.

Nourishing the sea

Rivers transfer massive quantities of nutrients—including carbon, nitrogen, and phosphorus—from land into the sea. Terrestrial and marine plants need nitrogen and phosphorus to grow. Farther from the coast, marine waters tend to be depleted of nutrients, which limits the development of phytoplankton, the microscopic plants at the base of the food chain for aquatic fauna. Rivers with high discharge can form a freshwater plume that flows hundreds of kilometers into the sea, enriching ocean waters with nutrients. The Amazon River, for example, provides 25 percent of the nitrogen input to the North Atlantic Ocean.

Underwater avalanches

Large volumes of sediment transported by rivers accumulate near river mouths, forming a dense layer of mud and sand that, when disturbed, can turn into a fast current moving over the seabed. These are known as turbidity currents and are the main source of sand, mud, nutrients, and, unfortunately, pollution to the deep ocean. Turbidity currents can be triggered by underwater avalanches (which themselves may be caused by earthquakes), spring tides, and storms, creating turbid plumes that later descend into the deep ocean. They may also be triggered by river floods, which feed sediment-rich water to lakes and oceans. When these currents flow within the confines of the steep walls of submarine canyons, they can accelerate and erode material that, in turn, will increase the current density and thus its speed in a feedback loop. They can last many days, losing speed over the gentler gradients of the deep sea. The coarser, heavier sand they carry is deposited first, while the finest particles travel longer distances suspended in water. Turbidity currents form elongated fan-shaped deposits extending over a large area that, through time, can be hundreds of meters thick. These deposits support the development of deep-sea ecosystems and are a sink of terrestrial carbon, but they are also a hazard, eroding the seafloor along their path and breaking submarine network cables that support most intercontinental internet communications.

▲ **Washed out**
Deforestation along the Betsiboka River in northwest Madagascar has increased water runoff and soil erosion, which intensifies after heavy rains, tinging the water red. This image was taken from the International Space Station after the passage of the category 5 Cyclone Gafilo in early March 2004.

DEEP IN THE CONGO

The Congo River has many special characteristics, which mostly derive from tectonic uplifts that have shaped its catchment. The river drains approximately 12 percent of Africa's land surface and its average flow could fill about 16 Olympic-sized swimming pools every second, and up to 30 during peak discharge. Downstream from the wide, calm waters of the Malebo Pool, the large volume of water received from the catchment drains into the narrower, steeper Lower Congo, speeding up the river flow (see also page 86).

Estuaries are often a sediment sink, as only some of the sediment they receive from rivers reaches the sea. In the Congo, a submarine canyon starts 30 km (20 miles) upstream of the estuary mouth and extends through the continental shelf, facilitating the transfer of sediment to the deep ocean via turbidity currents. Changes in depth along the Congo submarine canyon and channel between January and March 2020 indicate that the amount of sediment eroded by turbidity currents represented 19–37 percent of the total suspended sediment flux to the world's oceans from all rivers. Current speeds reached 40 kph (25 mph), breaking underwater cables and affecting the internet in countries from Nigeria to South Africa.

Turbidity currents also play a key role in transferring carbon produced by terrestrial vegetation to the deep ocean. About half of the terrestrial organic carbon that reaches the Lower Congo River settles into the deeper waters of the submarine canyon and eventually is transported to the deep ocean. Deposits formed by the Congo turbidity currents sink about 2 percent of the world's terrestrial organic carbon transported into the deep ocean each year. The low availability of dissolved iron in the ocean is a major constraint for phytoplankton growth. The Congo River reduces this deficit as it supplies more iron to the world's oceans than any other river. The concentration of iron bypassing the Congo estuary represents 40 percent of the total input into the South Atlantic.

Flushed
Rocks in the Congo catchment are rich in iron, and high concentrations of the metal bypass the estuary, helped by the high river discharge and turbidity currents flushing large volumes of water and sediment out into the sea.

1 Democratic Republic of the Congo
2 Angola
3 River plume
4 Congo estuary
5 Head of the submarine canyon

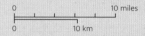

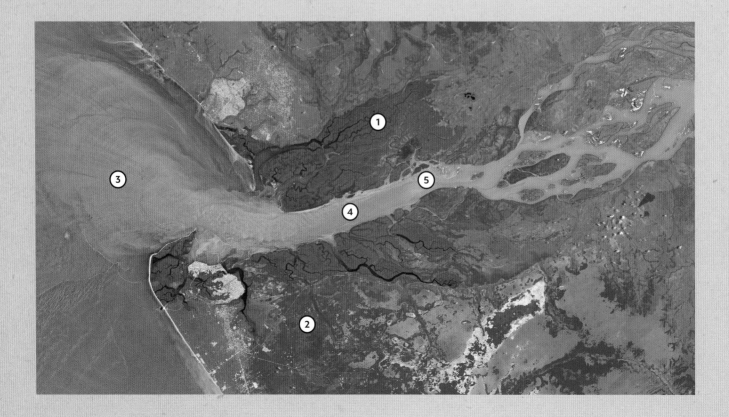

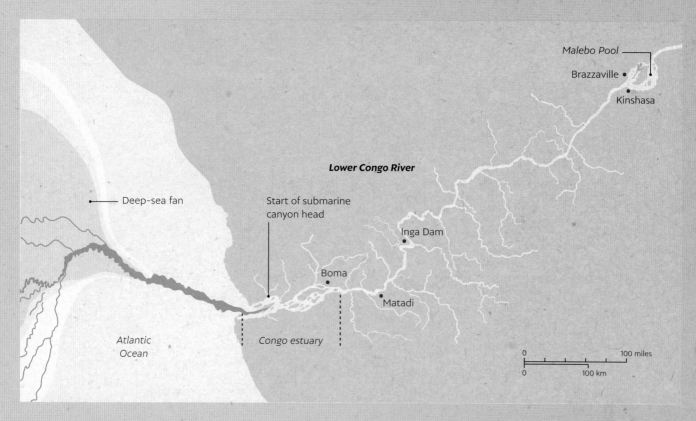

Lower Congo River

Malebo Pool
Brazzaville
Kinshasa
Deep-sea fan
Start of submarine canyon head
Inga Dam
Boma
Matadi
Atlantic Ocean
Congo estuary

0   100 miles
0   100 km

Deep connection

Between Kinshasa and Matadi, the Lower Congo River drops 270 m (900 ft) in 350 km (220 miles), forming a series of rapids along the stretches known as the Livingstone Falls and Inga Falls (see cross-section, right). Turbidity currents travel 1,000 km (600 miles) along the Congo submarine canyon and underwater channel system (see lower graph), reaching depths of 5,000 m (16,400 ft) and forming the longest known flow of sediments into the ocean.

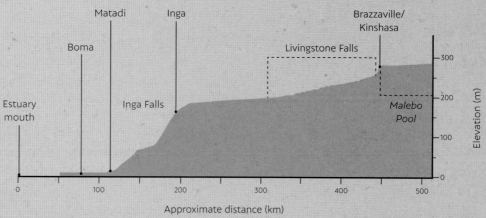

Matadi
Boma
Inga
Brazzaville/ Kinshasa
Livingstone Falls
Estuary mouth
Inga Falls
Malebo Pool
Elevation (m)
Approximate distance (km)

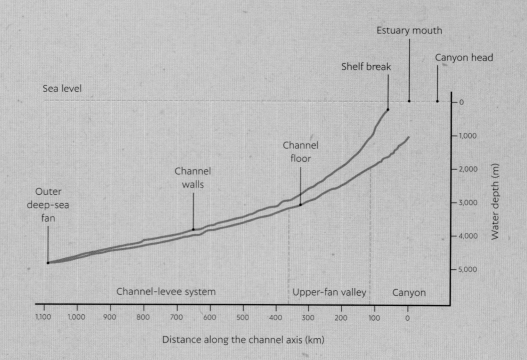

Estuary mouth
Shelf break
Canyon head
Sea level
Channel floor
Channel walls
Outer deep-sea fan
Channel-levee system
Upper-fan valley
Canyon
Water depth (m)
Distance along the channel axis (km)

Rivers, estuaries, deltas, and planetary health

Water and food sourced from rivers, estuaries, and deltas have long supported the development of human societies. However, Earth's waterbodies are constantly changing in response to environmental and human pressures. Protecting our rivers, estuaries, and deltas from climate change and other anthropogenic activities (such as mining, pollution, and dam construction), and understanding how these changes affect human health (such as through the spread of waterborne and infectious diseases) has become a critical concern.

The term "planetary health" is used to capture the concept that there is an interdependent relationship between the healthy functioning of our planet's natural systems and human wellbeing. This relationship is well illustrated by the world's waterbodies. Through their role in providing fresh water for drinking and bathing, and because rivers, estuaries, and deltas supply and store nutrient-rich sediments, these vital environments have underpinned the development of human societies since the adoption of agriculture.

Clean water and food

Water is fundamental to life, but the World Health Organization has estimated that 2 billion people do not have access to safely managed drinking water. With a growing proportion of the world's waterbodies becoming contaminated, it is unsurprising that so many people (an estimated 829,000 per year) die from diarrhea, either as a direct result of drinking unsafe water, or because handwashing is not a priority where water is not readily available. Prevalent waterborne diseases such as cholera, giardiasis, and typhoid are associated with poor-quality water and pollution, mostly due to inadequate sanitation. In 2017, more than 220 million people needed preventative treatment for schistosomiasis, an acute and chronic disease caused by parasitic worms that is contracted through exposure to infested water.

Rivers, estuaries, and deltas also support the plants and animals that are harvested for our food. According to the United Nations Food and Agriculture Organization (FAO), more than 12 million tonnes (13 million tons) of fish were harvested from freshwater fisheries in 2018. The most productive river basins in terms of freshwater fisheries are the Mekong (15 percent of total global catch), Nile and Lake Victoria (9 percent), Ayeyarwady (7.8 percent), Chang Jiang/Yangtze (6.8 percent), Amazon (4.3 percent), and Ganges (3.5 percent). However, global threat maps show that the majority of the world's freshwater fishery catch comes from regions that have moderate to high threat scores of 4–5 (47 percent) or 6–7 (38 percent), while a further 10 percent of catch comes from areas with the highest threat index values.

▶ **Healthy livelihoods**
Vegetables such as morning glory and water mimosa are farmed on floodplain wetlands around Phnom Penh, providing nutritious and locally sourced produce for the markets of Cambodia's capital.

UNDER PRESSURE

Map of the threat to freshwater fisheries developed by the United Nations Food and Agriculture Organization. The highest threat scores are associated with areas of high water abstraction, high population density, intensive land-use change, and pollution.

Threat score

| 0.0-2.0 | 2.1-4.0 | 4.1-6.0 | 6.1-8.0 | 8.1-10.0 |

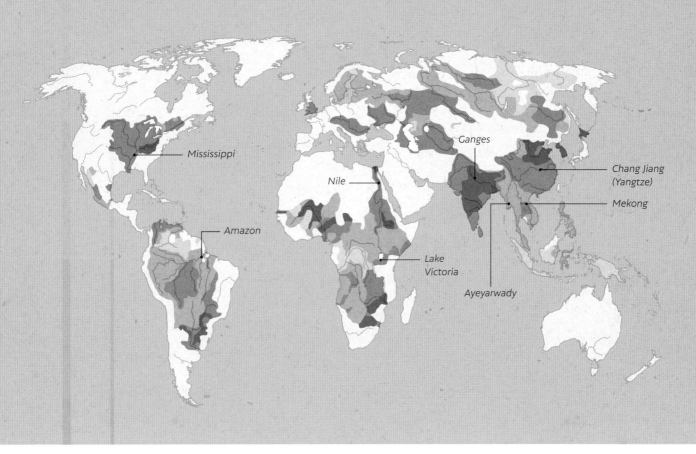

Mississippi

Ganges

Nile

Chang Jiang (Yangtze)

Amazon

Mekong

Lake Victoria

Ayeyarwady

Disrupting the balance

Rivers, estuaries, and deltas are not static; they change constantly in response
to environmental pressures, many of which are induced by humans. At about the
same time that William A. Anders captured the iconic image of the Earth rising
over the moon during the Apollo 8 mission in December 1968, we entered a new
geological epoch—termed the Anthropocene—that is characterized by humanity
becoming the dominant force shaping our planet. For rivers, estuaries, and deltas,
this large human footprint is manifest through the increasing intensity and
frequency of floods and droughts, fragmentation of rivers by damming, water
abstraction to irrigate intensively farmed crops, pollution of estuaries, and the
salinization of deltas as they shrink and sink in response to accelerating sea-level
rise. These pressures are disrupting the functioning of rivers, estuaries, and
deltas in both direct and indirect ways.

River hazards, displacement, and health

Many river hazards, such as flooding and the erosion of land adjacent to rivers, can
drive people to migrate away from their homes. The impacts of forced displacement
on human health are known to be large, with significant associated risks of
malnutrition and epidemic infectious disease. It is also increasingly recognized
that the associated psychological trauma of forced displacement can lead to severe
mental illness. Overall, displacement caused by river hazards is responsible for some
of the largest burdens of disease associated with global environmental change.

Dams and disease

The impacts of dams on fragmenting the natural flow of water, sediment, and nutrients through rivers and into their estuaries and deltas are covered on pages 140-43, but they also have an impact on human health. For example, dams have long been associated with elevated burdens of human schistosomiasis (bilharzia), which is second only to malaria as the parasitic disease most devastating to humans. But what causes this increased burden? In the Senegal River basin, migratory river prawns prey on the snail hosts of schistosomiasis, and so dams that block prawn migrations can lead to increased cases of the disease. River prawn habitats are widespread globally—so much so that dam projects place an extra 400 million people at higher risk of contracting schistosomiasis.

◀ **Protected by prawns**

Restoring prawns to the Senegal River upstream of the Diama Dam reduced snail density and schistosomiasis reinfection rates in people, while also providing new livelihood opportunities for local fishers to market the prawns as a nutritious food.

Mapping and techniques

Monitoring the shape of the Earth's surface and its changes over time has been revolutionized over the past 60 years by new technologies: satellite remote sensing from space, application of light detection and ranging (LiDAR) methods, and new uses of aerial images. Combined, these now provide us with unprecedented capabilities to measure the Earth's surface and how it is changing due to natural and anthropogenic stresses.

Principles of remote sensing

Large-scale mapping of the Earth's surface is possible through the application of various remote-sensing techniques using airborne or space-borne sensors. These sensors work by detecting radiation emitted from the Earth's surface, with different sensors focusing on different regions (wavelengths) of the electromagnetic spectrum (EMS). Since the amount of radiation reflected is influenced by both the properties of the surface and the incoming radiation, different sensors can capture different

SENSING EARTH'S WATERBODIES

Different features on the Earth reflect solar radiation in distinctive ways. Earth observation satellites can measure these reflected signals at varying temporal and spatial scales.

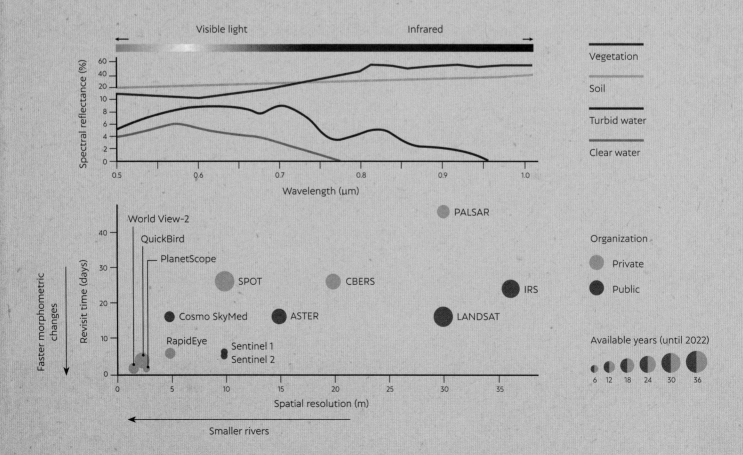

information about surface properties. For example, conventional aerial photography, long used in mapmaking, simply utilizes light in the visible spectrum. Especially pertinent to the mapping of rivers, estuaries, and deltas is the fact that water has a highly distinctive spectral signature compared to adjacent land, making it possible to delimit water features. Compared to land, water more readily absorbs incoming solar radiation in the infrared portion of the EMS (wavelengths of 0.75–3.0 μm), but it absorbs less in the visible range (wavelengths of 0.38–0.75 μm). This means that, so long as corrections are made for the fact that water in waterbodies is rarely perfectly clear (due to the presence of suspended sediment, phytoplankton, and chlorophyll-a), remote detection of water features is possible using optical sensors that have at least one detecting "band" in the infrared spectrum.

Many remote-sensing techniques are based on passive sensors that rely on receiving solar radiation reflected from the Earth's surface. However, active sensors are also employed. For example, synthetic aperture radars (SARs) emit and then receive microwave radiation backscattered by the Earth's surface. The great advantage of SARs is that they can penetrate cloud cover that would otherwise limit the use of optical sensors.

Earth images from early spy satellites

A fascinating glimpse into the shape of the Earth's surface in the 1960s comes from the 1995 release of classified images taken by reconnaissance satellites operated by the United States Central Intelligence Agency (CIA) between 1959 and 1972.

The CORONA project was rapidly advanced after a United States U-2 spy plane was shot down over the Soviet Union on May 1, 1960, and eventually consisted of eight separate satellite missions. On each mission, a space vehicle flew at altitudes of about 185 km (115 miles), capturing images on photographic film using a rotating stereo panoramic camera system. The film was developed on board the vehicle and fed into cassettes in recovery capsules that eventually descended Earthward, where the film was retrieved. Each payload consisted of an astonishing 9,600 m (31,500 ft) of 70 mm film, with the program collecting more than 800,000 images in 12 years. The images have a ground resolution of 8 m (25 ft), increasing to 2 m (6 ft) as technology improved.

Although designed as spy satellites to record details of Russian and Chinese military operations, the CORONA archive—now declassified and available to all—provides an invaluable record of many parts of the Earth's surface in the 1960s, including its rivers, deltas, and estuaries.

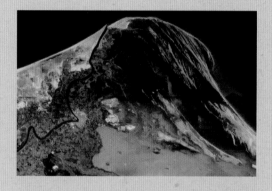

Landscape change in the Nile delta
Images of the downstream terminus of the Damietta branch channel taken in 1968 (upper) by the United States CORONA spy satellite, and in 2022 (lower) by the Planet CubeSat satellite.

0   5 miles

0   5 km

TRACKING RIVER MOTION

Analysis of 20 years of Landsat imagery allows river migration
rates to be estimated along 370,000 km (230,000 miles) of
Earth's river network. The median annual rate of channel
migration is 1.52 m (5 ft), but most rivers are moving across their
floodplains at slower rates. The most rapidly migrating rivers
tend to be located in the Amazon Basin and parts of Asia.

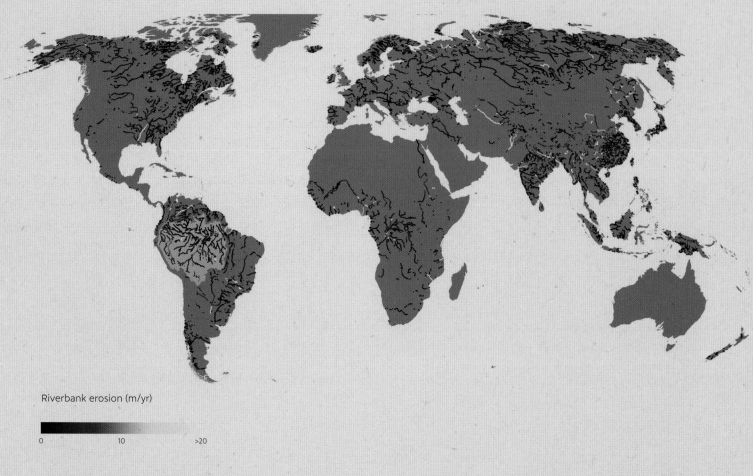

Riverbank erosion (m/yr)

0   10   >20

Amazon River Basin   10 km reaches   200 m nodes

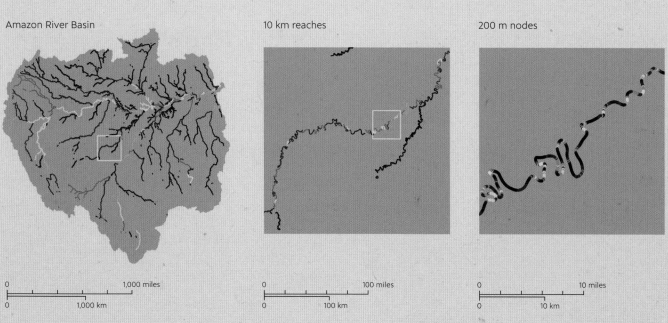

| 0 | 1,000 miles |
| 0 | 1,000 km |

| 0 | 100 miles |
| 0 | 100 km |

| 0 | 10 miles |
| 0 | 10 km |

Landsat: a game-changing development

The Landsat program, funded by the National Aeronautics and Space Administration (NASA), stands out as the most widely used platform for studies of the Earth's surface. The program started with the launch of Landsat 1 in 1972, followed by Landsat 5 (1984), Landsat 7 (1999), Landsat 8 (2013) and, most recently, Landsat 9 (2021). The high spatial resolution (60 m/200 ft for Landsat 4 in 1982, and 30 m/100 ft subsequently) and quality of the Landsat sensors have allowed the expansion of remote sensing—once mainly restricted to land surface changes—to numerous applications.

Three key factors have contributed to Landsat's status as a "game-changer." First, since 1984 the Landsat satellites have employed three different bands in the infrared spectrum: near infrared (NIR), 0.85-0.88 μm; shortwave infrared (SWIR) 1, 1.57-1.65 μm; and SWIR 2, 2.11-2.29 μm. This enables a wide variety of water-detection algorithms and indices to be developed and improved. Second, its longevity (30 m/100 ft spatial resolution imagery available continuously and globally since 1984) has made it possible for researchers to investigate how rivers, estuaries, and deltas evolve over longer timescales, even in inaccessible areas of the globe. Finally, as a government-funded programme, Landsat imagery is free for anyone, anywhere, to use, thereby contributing to an important "democratization" of Earth science.

Global digital elevation models

Maps of the elevation of the Earth's terrestrial surface have been central to advancing our knowledge of the world's landscapes. Several near-global datasets are now available, including that produced by the NASA Shuttle Radar Topography Mission (SRTM), which employed a specially modified radar system onboard the Space Shuttle *Endeavour* during an 11-day mission in February 2000. This produced a digital elevation model (DEM) of the majority of the Earth's surface on a 30 m (100 ft) grid, allowing mapping of many remote regions.

Other global DEMs are now also freely available, including the Japan-USA Advanced Spaceborne Thermal Emission and Reflection Radiometer (ASTER) and European Union Copernicus missions. These datasets, and the various products derived from them, provided an unrivaled quantitative view of the Earth's topography.

▼ **Mapping topography**
Global elevation datasets, such as that from NASA's Shuttle Radar Topography Mission, have revolutionized mapping of the Earth's surface, revealing the intimate connections between topography, rivers, estuaries, and deltas. Here, topography created by mountains, geological faults, and volcanoes shapes the passage of rivers in South Island, Aotearoa New Zealand to coastal plains and the ocean.

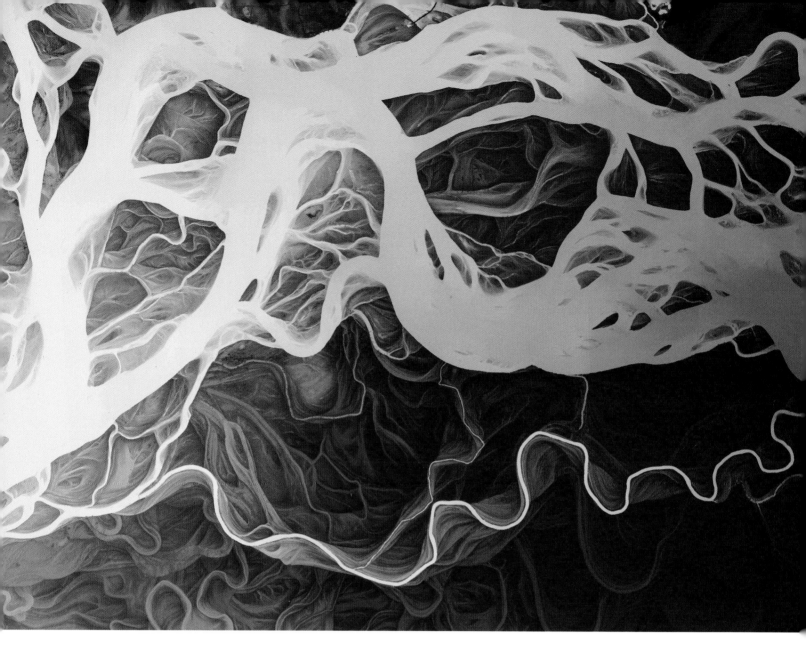

Rivers revealed

LiDAR-derived "bare-earth" surface image of the Yukon River and its floodplain, southeast of Fort Yukon, Alaska.

Using lasers to map the Earth's surface

The widespread adoption of light detection and ranging (LiDAR) has provided the means to map large areas of terrain accurately. LiDAR works by timing the interval required for an emitted laser pulse to be reflected back from the target to the sensor, enabling the distance between the sensor and target to be calculated. By mounting a sensor on a conventional aircraft or drone, whose location is itself accurately measured using a Global Positioning System (GPS), and by having a sensor that emits repeated pulses at extremely high frequency (in some instances up to 100 million times per second), a highly detailed (dense) "point cloud" of accurately geolocated points can be obtained. These can cover very large spatial areas, offering the means to measure surface topography in unprecedented detail.

A key feature of some LiDAR instruments is that the wavelengths of the returned laser pulse can be modified according to surface characteristics, enabling features such as vegetation that obscures the true surface of the Earth to be "removed," revealing the "hidden" detail beneath. Similarly, green-waveform LiDARs (which emit laser pulses with wavelengths in the green part of the visible spectrum) can penetrate shallow water that has good transparency, allowing submerged surfaces to be revealed (topobathymetric LiDAR).

MEASURING THE EARTH'S SURFACE

Three methods for producing high-resolution quantification of the Earth's surface: airborne LiDAR, terrestrial (ground-based) LiDAR, and aerial platform (e.g., drone) structure from motion (SfM). Abbreviations: GPS, Global Positioning System (to measure position); IMU, inertial motion unit (to measure precise three-dimensional movement of the sensor).

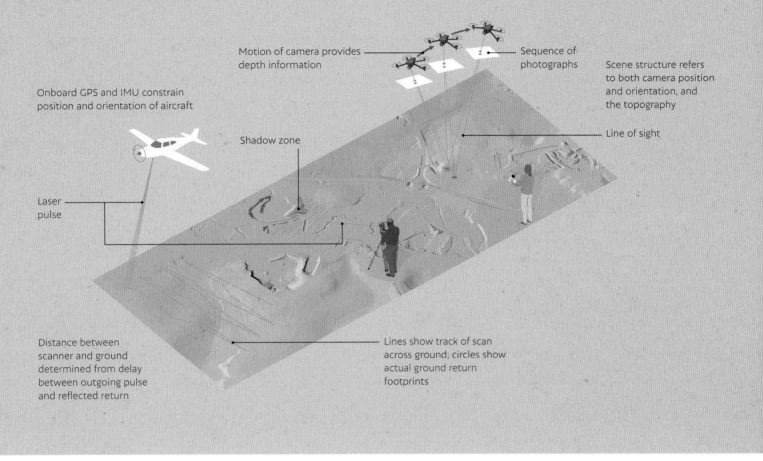

Motion of camera provides depth information

Sequence of photographs

Scene structure refers to both camera position and orientation, and the topography

Onboard GPS and IMU constrain position and orientation of aircraft

Line of sight

Shadow zone

Laser pulse

Distance between scanner and ground determined from delay between outgoing pulse and reflected return

Lines show track of scan across ground; circles show actual ground return footprints

Structure from motion imaging

Stereo photogrammetry, in which overlapping aerial photographs are used to reconstruct three-dimensional models of the terrain based on the shift in apparent position of an object from one image to the other, has long been used to produce detailed terrain maps. Advances in computing power and image-processing algorithms have led to development of the "structure from motion" (SfM) technique, whereby the structure of features is tracked and quantified between multiple overlapping images. Given a known camera position in each image, and/or if the locations of distinct points marked on the ground surface are also known with a high accuracy (so-called ground-control points), features can be matched between images and used to generate a three-dimensional map of the surface. The SfM technique can thus yield a dense cloud of points that depict the height of the surface. Images may be taken from airplanes, but are increasingly being captured using unmanned aerial vehicles (drones). As such, SfM is being rapidly adopted as a low-cost but highly accurate method for measuring the morphology of Earth surface change. Although the SfM point clouds depict the surface imaged, and thus cannot measure the bare-earth surface like LiDAR, the technique is rapid and has been adopted for Earth surface research, forestry, hazard mapping, building inspections, and archeological studies.

IMAGING TOPOGRAPHIC CHANGE

Elevation maps produced from aerial drone imagery and structure from motion (SfM) photogrammetry of the Hester Marsh restoration site, Elkhorn Slough, an estuary in central California. Maps show elevation before and after restoration of the tidal flats, which involved raising the marsh plain and tidal creeks. Quantification of topography can be vital in aiding environmental rehabilitation.

2015 Digital elevation model (DEM)

2018 DEM

Topographic change

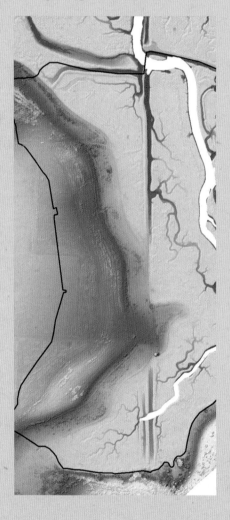
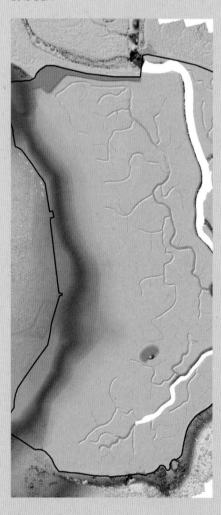
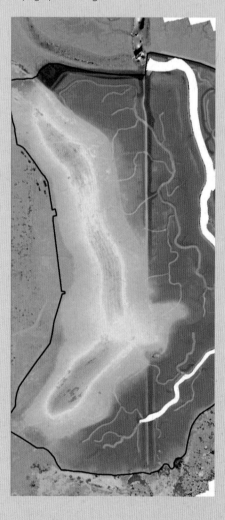

0   500 ft

0   100 m

Elevation (m) with respect to the
North American vertical datum of 1988

High   Low   Elevation change (m)

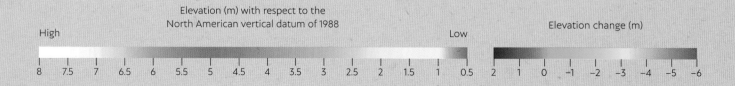

8   7.5   7   6.5   6   5.5   5   4.5   4   3.5   3   2.5   2   1.5   1   0.5   2   1   0   −1   −2   −3   −4   −5   −6

A flock of doves

Technological advances in remote sensing from space have allowed near-daily coverage of most of the Earth's surface using constellations of small CubeSat satellites, termed "doves." These small satellites are the size of a shoebox and less costly than larger satellites. While they have more limited capabilities, they can be deployed in their hundreds, may possess multiple spectral bands, and can yield images with a resolution between 5 m (16 ft) and 0.3 m (1 ft). These flocks of doves provide an unrivaled opportunity to monitor natural events such as floods, hurricanes, and landslides, and human-induced changes such as deforestation and urban expansion. The frequency and high spatial resolution of such images can be of vital use in our response to natural disasters and their management.

▶ **Flood watch**

CubeSat imaging permits the monitoring of flood extent and inundation, as here at low-water (top) and flood (bottom) states on the Sacramento River, California, February 2017.

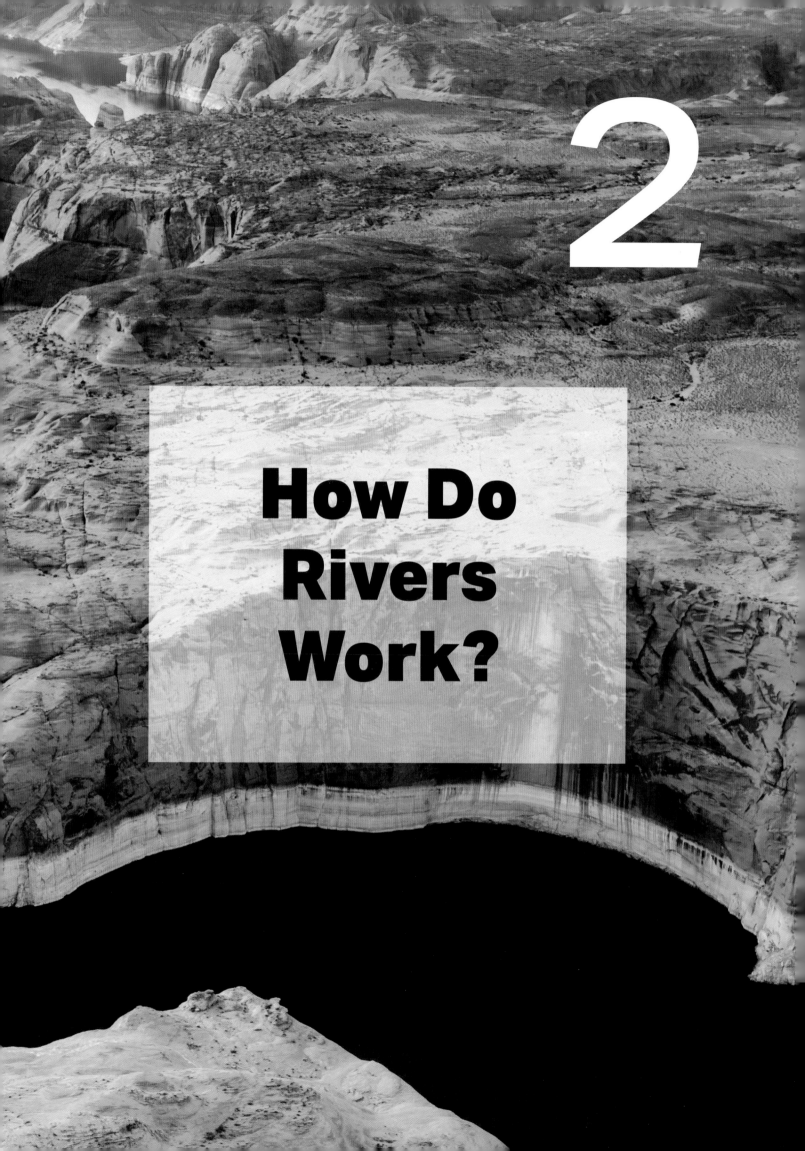

2

How Do Rivers Work?

Why do rivers flow where they do?

Rivers sculpt the Earth's surface through the erosion, transport, and deposition of sediment, creating a rich tapestry of landscapes that support immense ecological diversity and are home to billions of humans. Earth's surface topography, created by tectonics and worn down by erosion, determines the paths that rivers follow.

Follow a drop of water in its natural path along a river and you will always be going downhill—water acts under gravity to trace the gradient of the surface over which it is passing, from steep mountainous headwaters to low-gradient rivers that enter the ocean. This simple movement of water, and the sediment it transports, ties rivers to topography. It also provides a feedback mechanism where the longer-term erosion of the landscape takes place through the action of flowing water. Rivers thus respond to topography, seeking the steepest route, being steered around higher ground, and depositing sediment where flows decelerate.

Dendritic drainage networks

As rivers develop on a fresh surface, their channels etch a pattern, or drainage network, into the landscape that routes water from the land. Upstream channel growth takes place by erosion at the tips of the river network—a process known as headcut erosion. On a relatively uniform sloping surface, or over very long periods of time, this produces a tree-like, or dendritic, drainage network, with the branches of the river tree feeding water and sediment, via channel confluences, into ever-larger channels that eventually form the main trunk channel downstream. Dendritic drainage networks can be found on all scales, from small rivers to some of the world's largest drainage basins.

River networks thus unite channels of different size and provide ecosystem connections—for instance, as paths for the upstream migration of salmon from the ocean to small headwater streams, where they spawn and where individuals of some species then die. Their decomposing carcasses subsequently contribute food for the entire ecosystem. However, human interventions such as damming are making this natural journey far more challenging.

▼ **Intricate networks**
Dendritic networks feed water and sediment into meandering rivers and their floodplains, Amazon River Basin, Brazil. The image shows the elevation from high (black/purple) to low (yellow/blue) and is plotted from the global FABDEM terrain data.

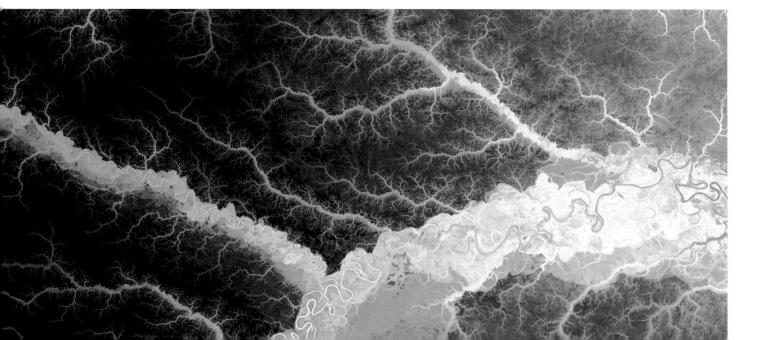

RIVER PATTERNS

The map shows river basins of North America in different colours. The dendritic network of the mighty Mississippi River drains 41 percent of the contiguous United States.

The Colorado and Columbia Rivers each drain around 8 percent of the same area with the Rio Grande collecting water from c. 6 percent of this land area.

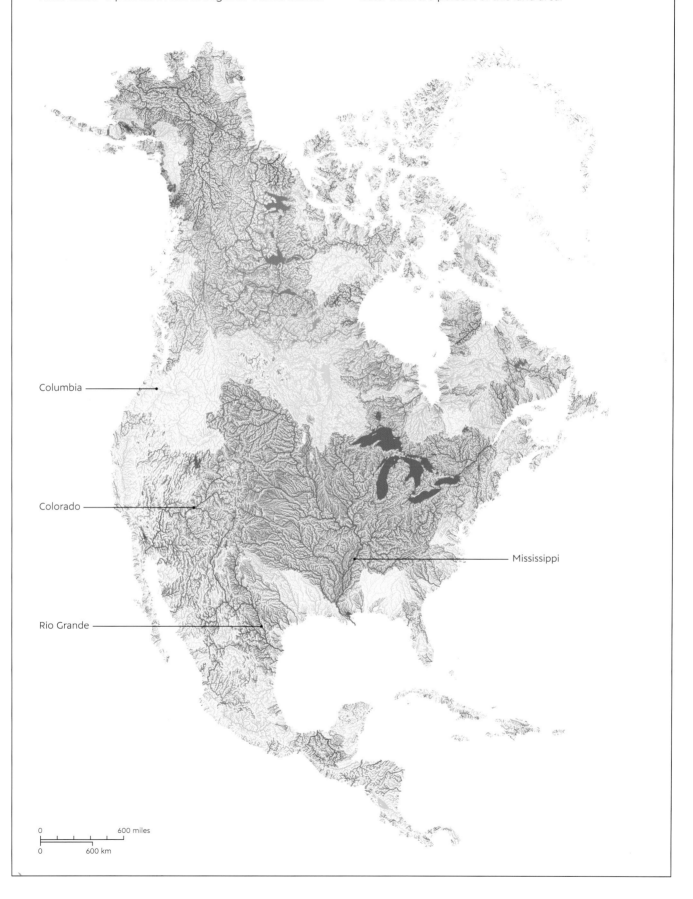

Columbia

Colorado

Rio Grande

Mississippi

0   600 miles

0   600 km

Restoration of the Elwha River in Washington State is the largest dam-removal project in US history. It saw the removal of Glines Canyon Dam between September 2011 and August 2014. The Glines was the upper of two dams in the project; the downstream Elwha Dam was dismantled by March 2012.

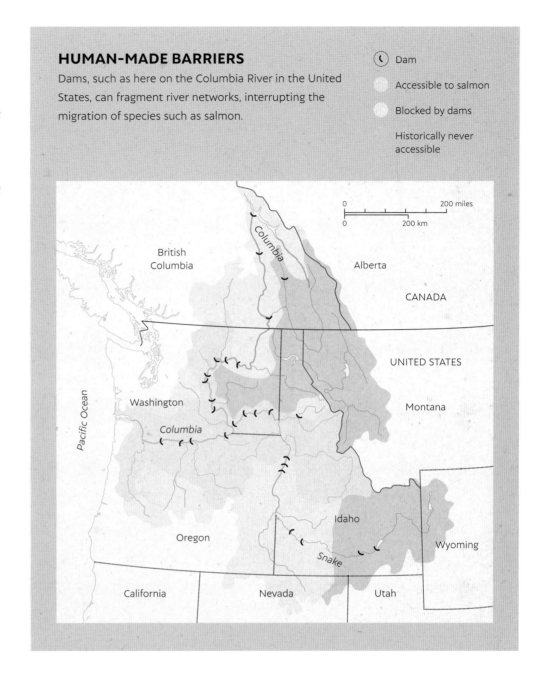

HUMAN-MADE BARRIERS

Dams, such as here on the Columbia River in the United States, can fragment river networks, interrupting the migration of species such as salmon.

☾ Dam

Accessible to salmon

Blocked by dams

Historically never accessible

▶ **Upstream migration**

After a long journey upstream through a river network, a female Sockeye Salmon (Oncorhynchus nerka) uses her tail to excavate a pit, or redd, in an Alaskan riverbed to lay her eggs.

River networks and controls

The development of river networks is controlled by surface topography, climate, and the level of the waterbody into which they flow. Over tens of millions of years, the uplift of mountains and movement of continents provides the template on which rivers develop.

The rivers of the world act over long periods of time to erode and sculpt the landscapes over which they flow as water travels downhill in its journey from the uplands to the ocean (although some rivers terminate in inland deserts and swamps, never reaching the sea). Because rivers act as conduits for the flow of water, sediment, carbon, nutrients, and contaminants over long periods, they are also indicators of some of the long-term controls that shape our continental landscapes.

Three of the largest-scale controls are those of plate tectonics, climate, and relative sea level, which may change radically over periods ranging from thousands to hundreds of millions of years. Deciphering the pattern of river drainage networks on the Earth's surface thus helps us interpret the changes in these broadscale controls in geological deep time. In addition, the tectonic makeup and climatic character of the Earth's surface changes greatly across the globe, and the world's rivers reflect these changing attributes in their course and the pattern of their drainage networks.

▶ **Stepping down strata**

A river responding to geology and topography in this false-color Landsat image: the Ugab River, Namibia. The river steps its way along, and across, topography created by folded rock strata. This topography forces rivers to run parallel to the ridges in some places, until they incise and cut across the strata, forming steps in the river planform. The river paths also take advantage of geological faults that can be seen as straight lines running northwest to southeast in some parts of the image.

REVERSAL OF THE AMAZON

Uplift of the Andes Mountains since c. 50 million years ago has caused the mighty Amazon River to reverse its course. It once flowed to the northwest and north, but uplift of the Andes resulted in a gradual change in surface topography, first forming the extensive Pebas wetlands and eventually causing the Amazon River to flow eastward into the Atlantic Ocean.

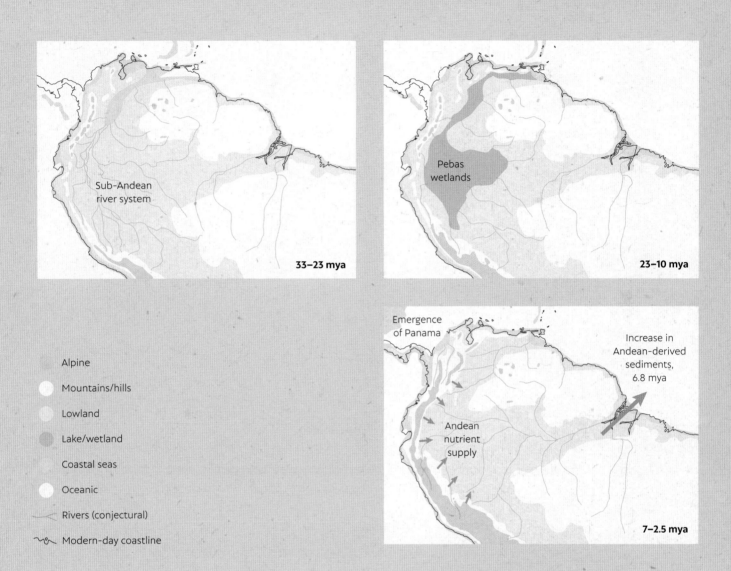

Tectonic controls

Because water follows the gradient of the land surface, the routes and patterns of riverine drainage networks reflect large-scale plate tectonics and evolution of the Earth's crust. Water is diverted by topography, such as that created by tectonic folds and faults, or incises into bedrock that is rising over millions of years due to tectonic uplift. Tectonics provide the fabric of the canvas on which the world's rivers are painted, and act as a principal control on how their drainage networks are organized. Some river networks develop rectangular or "trellis" planform patterns, whereas those draining from domed topography—for example, around volcanoes—generate river channels that run radially from the highest point. Drainage networks thus provide a sensitive record of crustal deformation, and as plate tectonics alter surface gradients through geological time, so the surface routing of rivers responds accordingly.

Climate controls

The weathering of rocks and transport of sediment are strongly influenced by climate, which helps determine the type of weathering as well as the volume of water supplied to river networks. Climate thus exerts a first-order control on the dynamics of rivers. Climate changes both spatially across the globe and temporally through geological time as the Earth has warmed and cooled.

RIVER CLASSIFICATION

Classification of the world's rivers based on characteristics of their water flow (hydrology, shown by line thickness) and their physical and climatic controls (indicated by different-colored lines).

Physio-climatic zone

Cold, high moisture

Cold, low and medium moisture

Warm, medium moisture

Warm, high moisture

Warm and hot, low moisture

Very hot, low moisture

Hot, high moisture

Very hot, high moisture

Cold and warm, high elevation

Hot and very hot, high elevation

Climate has also changed as the Earth's crust has evolved through time. For example, uplift of the Himalayas, which began around 50-60 million years ago and is caused by the northward drift of the Indian tectonic plate and its collision with the Eurasian plate, has resulted in dramatic shifts in the Earth's climate, changing global circulation in the last 10 million years and leading to the onset of the Indian summer monsoons. The Himalayas are still rising today—by about 1 cm (⅔ in) per year or 10 km (6 miles) in the last million years—signifying the intense weathering and erosion that is feeding sediment to the great rivers flowing from the high mountains. Himalayan uplift and associated climate change has thus shaped both the paths of the many rivers in this region and the climate. In turn, the rivers have fed immense quantities of sediment to the oceans, forming some of the world's greatest estuaries and deltas.

Sea-level controls

Rivers flow to lower elevations, many eventually forming estuaries and deltas as they enter the ocean. As a consequence, sea level acts as the key control on the distal end of a river system—the baselevel to which the end of the river must adjust. However, baselevel changes as sea level rises or falls—for example, absolute changes in sea level caused by melting of the ice caps or relative changes due to more local processes such as regional tectonics or land subsidence. Land subsidence in many of the world's present-day deltas (for example, caused by groundwater extraction) is exacerbating the effect of sea-level rise due to a warming climate.

OUT OF SUNDALAND

Schematic representation of the rivers of Sundaland, a vast area south of the present-day China Sea, at minimum sea level at the height of the last glacial maximum some 18,000 years ago. Rivers extended further across the continental shelf, and likely provided a home for early humans who transited across this landmass. The inset shows sea level over the last 21,000 years.

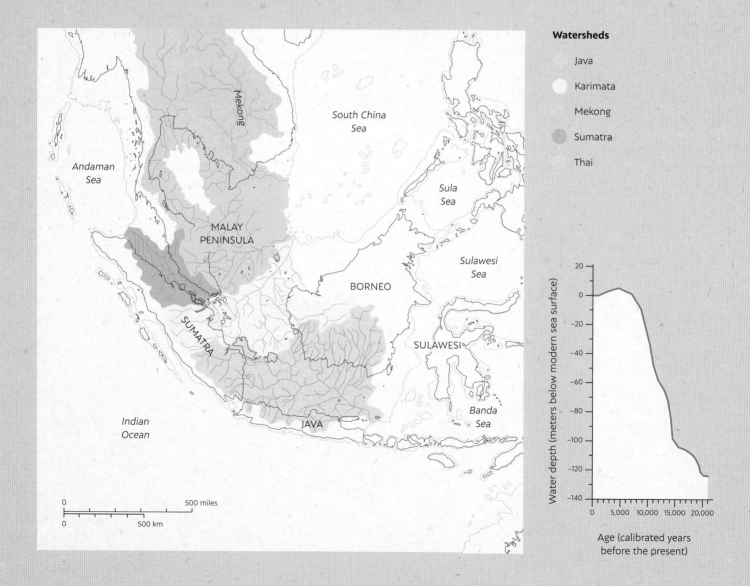

Today, we live in a world with a sea level that is relatively high in relation to the last few million years of Earth's history—as shown by the existence of many estuaries along our contemporary coastlines that represent river valleys drowned by high sea levels. Yet, at the height of the last glacial maximum around 18,000 years ago, when much water was locked up in extensive continental ice sheets, the global sea level was, on average, some 120 m (400 ft) below where it is today. As such, the distal ends of some rivers extended many tens or hundreds of kilometers further across the continental shelves and have subsequently been drowned, and buried with sediment, as sea levels have risen. Such expanded land areas at low sea level also provided routes for human dispersal and migration across the globe. Scientists speculate that one such landmass, Sundaland (between present-day Thailand, Borneo, Java, and Sumatra), once formed a region that fostered the spread of humans across the globe and was perhaps the cradle of civilization.

Multiple controls: river terraces

As one might expect, rivers are often shaped by a combination of controls that act over different scales in both space and time. One example is found in the common presence of river terraces within alluvial valleys. These are flat-topped benches that represent the surfaces of river floodplains that have become abandoned due to vertical incision (or aggradation) by the river. Multiple terraces are separated by steeper steps, with the oldest terraces usually at a higher elevation and more fragmented due to longer periods of incision and erosion.

River terraces may be formed by erosion into existing sediments (alluvial terraces) or bedrock (strath terraces). Incision is driven either by external factors—such as tectonic uplift, falls in baselevel, or changes in climate, which may enable the river to erode more—or due to the inherent lateral migration of river channels across their floodplains. Some terraces can be formed by sediment deposition as the bed of the river valley aggrades through time, perhaps burying older terrace fragments. These flat-topped benches may extend for long distances along a river valley, and they may occur in pairs on either side of the valley but with the top of each pair being at the same elevation. The number and elevation of terraces can thus inform us about the longer-term evolution of the river and help reveal the controls on how the alluvial valley has been sculpted through time.

TERRACE FORMATION

River terraces formed by incision of a channel into older alluvial sediments or bedrock, with successively younger terraces (T1–T5) forming as incision continues.

Weathering and erosion

For sediment to be supplied to a river, the source rocks in its upstream catchment must be weathered to yield sediment particles of many different sizes that can be transported downstream. Rock weathering occurs through a range of physical, chemical, and biological processes, all of which are strongly modulated by climate.

▼ **The power of ice**
Freeze-thaw weathering can split boulders apart, eventually yielding smaller fragments of rock to the river system.

Types of weathering

Physical weathering can involve the thermal expansion and contraction of rock surfaces (insolation), caused by daytime solar heating and nighttime cooling, producing stresses that fragment the surface. If water is present in fractures and cracks, its freezing and thawing can also split rock surfaces apart.

The presence of water in rock fissures can also create chemical change, such as the breakdown of rock by acidic water (hydrolysis) to produce clays and soluble salts, the dissolution of soluble materials such as calcium carbonate, and the oxidation of iron-rich minerals. Chemical weathering becomes more important in warmer, wetter climates, where chemical reactions can take place more easily and rapidly, and where unstable minerals such as feldspars (a very common aluminosilicate mineral) can be weathered to yield smaller grains and alteration products such as clays.

Rock weathering is also significantly influenced by biological activity. This includes the action of plant roots and, especially, that of algae and fungi, which produce organic acids that aid rock disintegration.

Weathering gives rise to the formation of soils, which constitute an important weathering rind on the Earth's surface that supports life. In tropical environments, weathering to a depth of up to around 30 m (100 ft) produces brick-red lateritic soils that are rich in iron and aluminum and have formed the building blocks of many architectural wonders.

Erosion

Over time, weathering thus rots unstable minerals, generating rock fragments and mineral grains for transport by water, yielding a range of elements to groundwater, producing alteration products such as clays, and progressively leaving behind the more stable minerals such as quartz (silicon dioxide) and hard, heavy minerals. Upland, mountainous terrains feed sediment into river channels by a mixture of processes, such as direct rockfalls, the slower downslope creep of material, and transport by flows of debris, mud, or water. Fan-shaped accumulations of sediment, known as alluvial fans, are deposited at breaks in slope between uplands and a lower-gradient valley floor. These can provide abundant sources of loose, unconsolidated sediment that are then eroded by the river and transported downstream. Thus begins the long journey of sediment from its erosive source to its eventual depositional 'sink', whether that be within an alluvial valley, estuary, delta, beach, or deep-sea environment.

Fanning out
The weathering of mountains feeds sediment to alluvial fans and then into river networks.

Royal building material
The ultimate product of tropical weathering— red iron-rich soils, called laterites—can be used as building stones, such as in the Pre Rup temple, Angkor Wat, Cambodia, which was dedicated to the Khmer King Rajendravarman in 961 or 962 CE.

Sediment transport

Water flowing over a bed of sediment exerts forces that drive different types of grain movement. Measuring such sediment transport is fundamental to understanding how rivers shift this material and estimating the changing quantities of sediment supplied to the world's estuaries and deltas.

▶ **Mobile sediment**

The mobile bed of the River Markarfljot, Iceland, is sculpted into barforms that are indicative of active sediment transport.

Types of sediment transport

The solid sediment particles moved along the course of a river are transported in two distinctive modes. The bedload is the portion of sediment that is transported by intermittent rolling, sliding, or "hopping" (termed saltation) of grains along or near the riverbed, typically at average speeds much less than that of the flowing water. In contrast, the suspended load comprises particles that are held aloft by turbulent eddies in the main body of the water, which move at roughly the speed of the flow and are carried along without significant contact with the riverbed.

With an abundant availability of sediment, the overall rate of sediment transport increases rapidly as flow velocity increases. The proportional split between bedload and suspended load fractions is controlled by a balance between the size of the sediment particles being transported and the rate at which sediment is mixed upward into the flow by turbulence. This means that larger particles tend to move more as bedload.

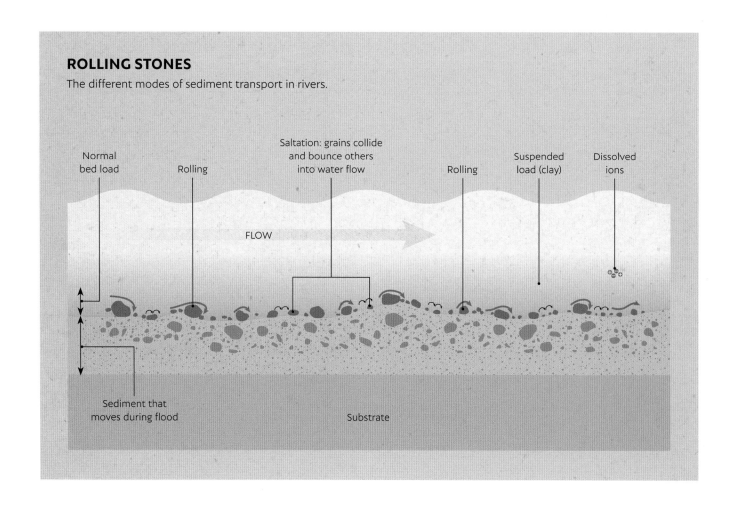

ROLLING STONES

The different modes of sediment transport in rivers.

Normal bed load

Rolling

Saltation: grains collide and bounce others into water flow

Rolling

Suspended load (clay)

Dissolved ions

FLOW

Sediment that moves during flood

Substrate

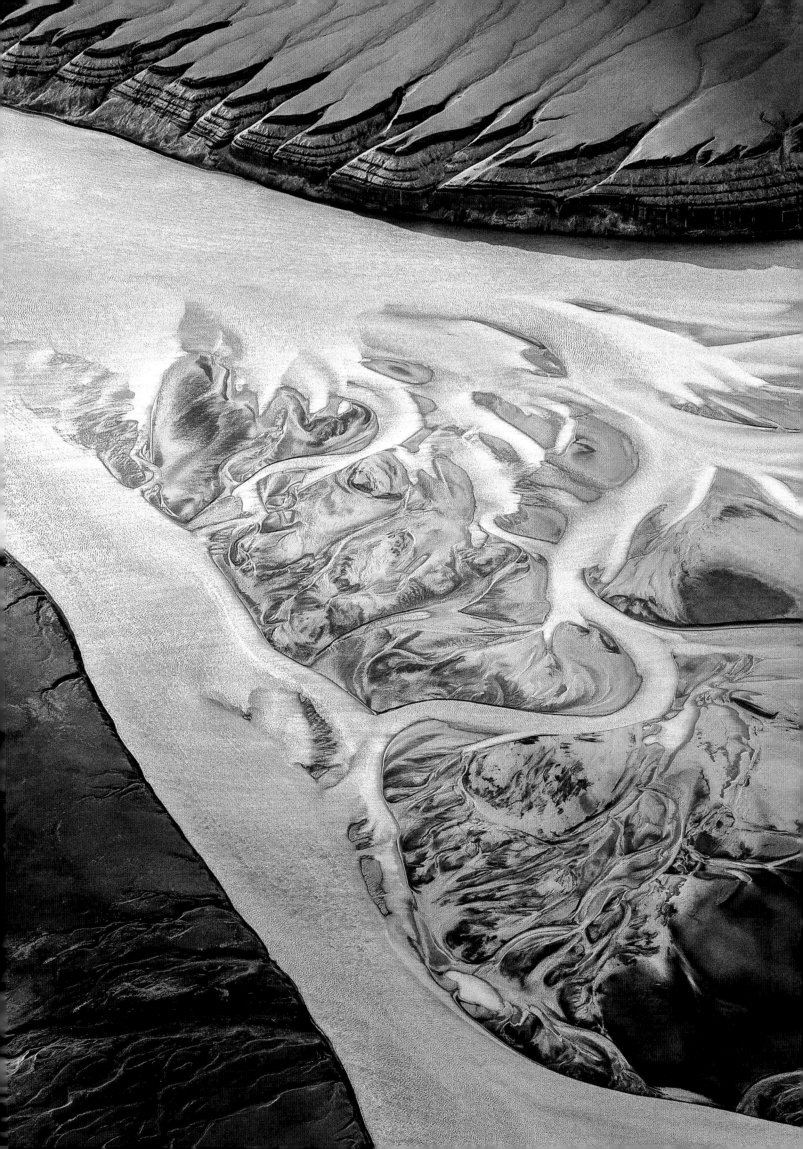

Traditional techniques to measure sediment transport

Most sediment is therefore transported during floods, which means that measuring sediment transport can be very challenging and even dangerous. Traditional "intrusive" measurement methods typically rely on deploying sampling devices of varying designs into the flow to capture the mass, or concentration (per unit volume of water), of moving sediment; this sometimes requires major infrastructure to trap moving sediments. However, unless such sediment samplers are very carefully designed and deployed, they can alter the flow, or become rapidly infilled or fouled, meaning that they may not always provide a representative picture of the actual rate of sediment transport. Moreover, such techniques may be difficult, or impossible, to use in large or remote rivers.

New techniques for measuring sediment transport

To overcome these limitations, modern measurement of sediment transport uses passive, non-intrusive techniques that rely on the principles of hydroacoustics. For example, hydrophones (underwater microphones) can monitor the noise made as grains collide and then calibrate this as a proxy measure of the intensity of bedload transport. Alternatively, sonar mapping can be used to create detailed maps of the riverbed through time, allowing the migration of sandbars to be tracked. This information can then be used to determine the rate of bulk bedload sediment motion.

The development of acoustic Doppler current profilers (ADCPs) has revolutionized the measurement of flow velocities and suspended sediment concentrations in the world's rivers, lakes, estuaries, deltas, and oceans. These instruments transmit acoustic energy into the water column from an array of transducers, with the energy backscattered to the instrument from different depths being proportional to the concentration of suspended sediment in the water column (accounting for the loss in energy due to the spread of the acoustic beams). The frequency of the backscattered sound changes due to interaction

▶ **Traditional measurement**

Measuring suspended load in the shallow North Fork Toutle River, Mount St. Helens, Washington State, using sampling bottles.

USING SOUND TO MEASURE FLOW VELOCITY

Flow velocities measured by an acoustic Doppler current profiler (ADCP) in a 90 m-deep (300 ft) cross section of the Lower Congo River. Note that negative primary velocities represent localized instances of upstream-directed flow. Four acoustic beams are emitted from the ADCP on the survey boat, and the signal detected is used to calculate flow velocities. Here, high velocities are shown in red and lower velocities in blue, with the arrows indicating the nature of large-scale circulations in the river flow. Such measurements were impossible before the advent of ADCP technology.

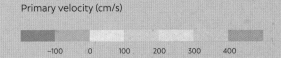

Primary velocity (cm/s)

−100   0   100   200   300   400

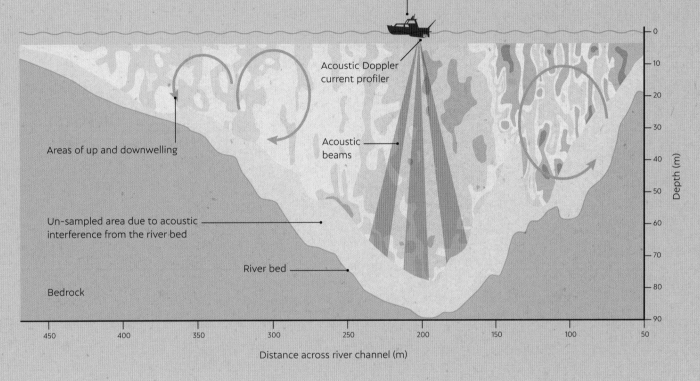

Global positioning system (GPS) receiver

Acoustic Doppler current profiler

Areas of up and downwelling

Acoustic beams

Un-sampled area due to acoustic interference from the river bed

River bed

Bedrock

Depth (m)

450   400   350   300   250   200   150   100   50

Distance across river channel (m)

with particles in the water—the so-called Doppler shift, a phenomenon we also notice when the sound of a train changes as it approaches and then recedes from us. This frequency shift is proportional to the velocity of these particles, which are assumed to be moving at the same as the velocity of the water. In this way, ADCPs measure water velocity throughout most of the water depth. Such hydroacoustic monitoring allows measurement of both the speed of the water and the concentration of suspended sediment. We can then multiply these two measurements to calculate the flux of material being carried in suspension. Both sonars and ADCPs can be deployed from moving boats or on the riverbed, and can be used in shallow channels decimeters deep up to the world's biggest river channels. They can also be used during floods when other methods cannot be employed. Perhaps more than any other instrument, ADCPs have lifted the veil on the many secrets of water flow and sediment transport in global rivers.

◀ **Colliding rivers**
Near Manaus, Brazil, the dark (blackwater) Rio Negro joins the pale sandy-colored (whitewater) Amazon River (referred to as the Solimões River in Brazil upriver of this confluence).

SPATIAL VARIABILITY

The difference in suspended sediment load in the world's large river systems (drainage areas >40,000 km² or 15,400 square miles, and water discharge >30 m³/s or 1,060 ft³/s). Data represent average values for the time period 1960–2010.

Suspended sediment (million tonnes/yr)

0–0.1

0.2–1

1.1–10

10.1–100

100.1–1,000

>1,000

The challenges of measurement

Whether intrusive or non-intrusive methods are used, measuring sediment transport is both time-consuming and expensive, and on average globally there is just one suspended sediment monitoring station per 10,000 km (6,200 miles) of river length. Thus, while it is estimated that about 19 billion tonnes (21 billion tons) of sediment are discharged from the world's rivers into the oceans each year, the sparse coverage of the global sediment transport monitoring network means that this estimate is very uncertain—by around ±50 percent. Moreover, there is a very large spatial variability in the rates of suspended sediment transport through the world's rivers, with the highest values located in zones of intense rainfall that coincide with steep, tectonically active mountain regions with erodible rocks, such as in the rivers that drain the Andes, the Himalayas, and Southeast Asia.

SEDIMENT CONCENTRATIONS
FROM SATELLITE

Suspended sediment concentration (SSC) for part of the
Amazon River for September averaged over the period
2000–2016, as measured from satellite imagery. The map
illustrates differences in sediment supply from tributary
channels, patterns of mixing between river flows, and
higher suspended sediment concentrations in some
floodplain lakes.

1 Rio Solimões
2 Rio Negro
3 Rio Madeira
4 Rio Tapajós
5 Manaus

0   50 miles

0   50 km

SSC, mg/L:

| 0 | 20 | 20 | 30 | 40 | 50 | 60 | 70 | 75 |

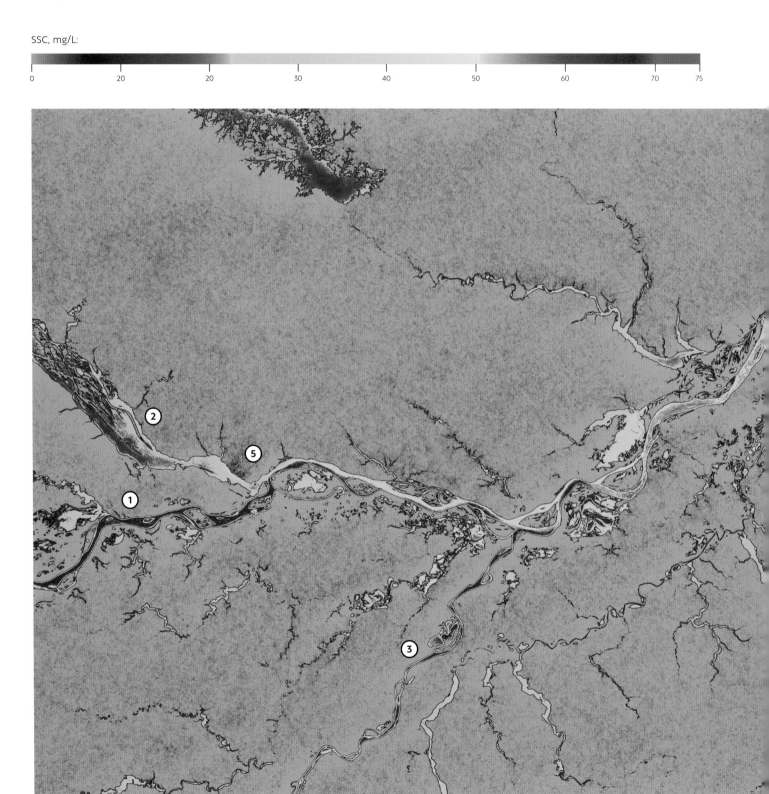

So, how can these challenges of monitoring sediment movement be overcome in the future? One way is to use remote sensing from satellites. The concentration of suspended sediment particles at the water's surface is known to affect the reflectance of incoming solar radiation. By using satellite sensors to measure this surface reflectance, it is now possible to estimate suspended sediment concentrations from space by calibrating this with known river samples. Such satellite-based remote sensing is offering exciting new opportunities to monitor sediment transport across large spatial scales, and on a routine basis, affording new insights into the processes by which erosion and sedimentation reshape some of Earth's most dramatic and inhospitable environments.

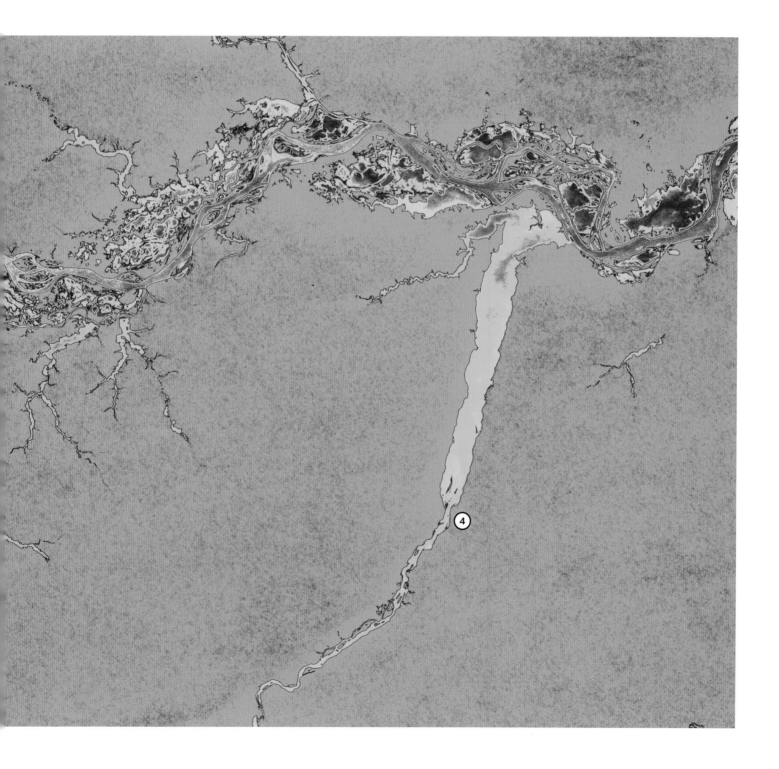

Carbon and nutrient flows through rivers

Besides transporting sediment particles, rivers are responsible for moving huge quantities of nutrients and carbon to the oceans. In addition, rivers and their floodplains provide regions for abundant vegetation growth, and carbon, stored as organic material, can become buried and locked into alluvial sediments.

Supplying nutrients to the oceans

Rivers are vital conveyors of nutrients and carbon from the land to the sea. Nutrients are critical to the development of plant and animal life, and are required for the growth of algae and cyanobacteria that lie at the base of the riverine food web, forming a food source for many small invertebrates and fish. However, healthy aquatic ecosystems require only modest concentrations of nutrients and, as we shall see later, pristine nutrient conditions are essentially non-existent in many rivers, estuaries, and deltas due to a range of human-sourced contaminants. The excessive enrichment of nutrients such as nitrogen and phosphorus—in a process termed eutrophication—is perhaps the Earth's most widespread issue for water quality.

Natural nutrients enter rivers through a range of processes, such as direct input from the weathering of rocks and nutrient release from soils, as well as the decomposition of riverside vegetation and aquatic organisms. Dissolved inorganic nitrogen (DIN) and phosphorus (DIP) are two of the most common and vital nutrients in rivers. Around three-quarters of these riverine nutrients reach the open ocean each year—globally, around 17 million tonnes (19 million tons) of DIN and 1.2 million tonnes (1.3 million tons) of DIP are delivered to the sea to support marine aquatic ecosystems. Some of these nutrients may become processed and changed due to biological activity in estuaries and along coasts, which act as biogeochemical buffers between rivers and the ocean.

▶ **Flood debris**

Mountain stream in flood, with large boulders, trees overhanging the channels, and fallen trees creating logjams.

FLOODPLAIN CARBON

The nutrients and minerals transported from the Andes Mountains and deposited on floodplains give rise to the massive productivity of the Amazon Basin, which is a major source, transporter, and store for carbon. This balance of organic carbon in the mainstream Amazon floodplain shows that large quantities of organic matter return to the river channel to fuel in-channel respiration. All quantities given are for total organic carbon (TOC) and in Tg per year, except for dissolved organic carbon (DOC) and particulate organic carbon (POC). Note: 1 Tg = 1 million tonnes (1.1 million tons), equivalent to the annual emissions from about 217,000 typical passenger vehicles in the United States.

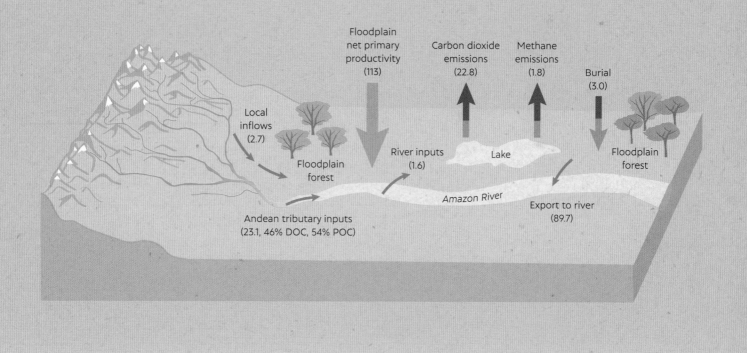

Fertile floodplains

Given a favorable climate, the action of weathering, nutrients, water, and microbes will lead to the creation of fertile ground on which vegetation can grow and thrive. Riverine corridors are thus major producers, transporters, and storers of carbon, and play a central role in the global carbon cycle and budget. The type of organic matter, and its storage, changes from upland mountainous rivers to extensive lowland floodplains, where sediment and carbon can be sequestered for tens of thousands of years. Floodplains also act as dynamic bioreactors, changing the flow of water, sediment, nutrients, and carbon over a wide range of timescales, and adopt a vital role in creating biodiversity. Today, it is estimated that rivers supply approximately 1,000 million tonnes (1 petagram, or 1.1 billion tons) of carbon to the oceans every year, and the global store of carbon in soils is about 2,500 times this value.

However, human-induced changes to floodplains, such as deforestation, changes in land use, interruptions to the natural flood pulse, and the removal of logjams and large woody objects, are causing substantial reductions in the amount of carbon stored in riverine valley corridors and yielding it to the atmosphere. Accounting for the role of carbon storage and release from river floodplains—whether in the world's great tropical rivers or in the melting permafrost floodplains of the high Arctic—will become increasingly important for our ability to tackle global warming.

Sediments deposited in rivers

The gravels, sands, and silts transported by rivers become deposited on the riverbed, where they form a wide array of morphological features, termed bedforms and barforms. These features are found in all the world's rivers, can create their own flow fields, and generate valuable ecological niches. They are also fundamental building blocks for river channel morphology.

Bedforms

Bedforms are generated by the action of flowing water moving unconsolidated sediment grains, and include familiar features such as sand ripples and dunes. At higher flow velocities, or bed shear stresses, these bedforms give way to a far flatter sediment surface over which there is intense sediment transport. At even greater flow velocities, the river surface becomes distorted into a series of waves, which can break upstream and are known to whitewater kayakers as 'stoppers'. These intensely turbulent waves are caused by water motion over so-called supercritical flow bedforms beneath, but can also be generated by motion over obstacles, such as boulders or steps in the bedrock, forming rapids. Where permanent, rapids can fragment the river into distinct segments that may limit, or control, the migration of fish and aquatic animals, providing a link between river hydraulics and both species distributions and evolution.

Bedforms can also form in coarser-grained gravels and boulders. However, the bed shear stresses needed to move these particles are greater than for sands, and hence dunes formed from much larger grains give indications of high-magnitude and potentially catastrophic flows in the past.

Barforms

Bedforms can become stacked and superimposed to generate larger barforms—for example, in regions where the flow expands and decelerates, encouraging sediment deposition. Such barforms adopt a range of sizes and myriad forms, and can become emergent at low flows, thereby promoting the growth of vegetation and potential development of soils and more permanent river islands. River bars may have a length several times the channel width, and can form in a range of locations—in mid-channel, adjacent to the channel banks, downstream of abrupt planform curvature, or associated with sediment deposition on the inside bends of meandering rivers.

Steering the flow

Both bedforms and barforms exert an influence on the river, steering water around their topography, creating their own flow field, and providing a resistance to flow. As such, knowledge of the presence and type of bedforms and barforms in a river is important to understanding and predicting the quantity of sediment it transports during floods and its water levels at such times. In addition, bank erosion may be controlled significantly by the diversion of river flow around barforms that are evolving in the center of the main channel, and thus riverine engineering interventions need to consider the changing riverbed morphology. As discussed above, river barforms owe their origin to episodes of erosion and deposition, often linked to flooding events. The depositional nature of bedforms and barforms in the sedimentary record thus enables geologists to decipher the characteristics of ancient rivers.

▼ **Bedforms, barforms, and river islands**
Waves of sediment revealed in an aerial image of the 400 m-wide (1,300 ft) South Saskatchewan River in Canada, showing the presence of an array of barforms and vegetated river islands, and revealing the riverbed to be covered by smaller sand dunes. The flow is from left to right, and the water is relatively clear due to trapping of fine suspended sediment in an upstream reservoir.

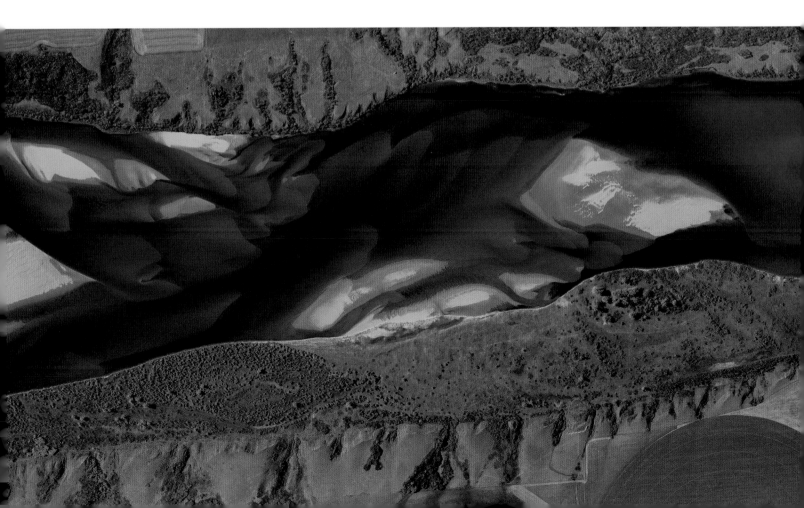

Floodplains

The area adjacent to river channels, the river floodplain, is home to some of the most diverse and biologically rich environments on Earth. Floodplains have also provided the land on which human civilizations have flourished, and their sustainable management is vital for future agriculture, habitation, and climate change.

What are floodplains?

Floodplains are the relatively flat, but far from featureless, areas of land next to most rivers. They are fundamentally transitional environments that lie between the main river channel (or channels) and the surrounding valley sides, and that become partially or fully inundated during periods of high flow. They are therefore complex, dynamic environments that are not separate landscape features but integral parts of rivers and vital to their ecological makeup.

Mapping, frequently relying on satellite-based remote sensing, has only recently revealed the rich array of floodplain shapes and sizes. In total, floodplains have been estimated to cover a land area of around 13.4 million km^2 (5.2 million square miles), or around 10 percent of global land outside of Antarctica. Many rivers are affected by a seasonal "flood pulse" as part of their annual hydrological cycle. These flood pulses bring nutrient-rich floodwaters and sediments out of the river and onto the floodplain, where the flooded vegetation can provide sheltered "nurseries" for young fish. As the flood wave recedes, nutrients from decomposed floodplain vegetation flow back into the river system. In these ways, the flood pulse is integral to sustaining the rich biodiversity and ecological productivity typical of the world's unmodified floodplains.

A fundamental tension

Yet floodplains have a disproportionate importance relative to their spatial extent. As discussed in Chapter 4, many floodplains—at least in their natural, undisturbed, state—are ecologically rich, productive environments. They may cover only 10 percent of the Earth's land surface, but it is estimated that around 70 percent of all terrestrial plant and animal species live on, or utilize, floodplains during their life cycle. The combination of relatively flat land, access to water, and fertile soils means that floodplains have also long been attractive places for human settlement. Whether the Egyptian civilization along the Nile valley, the Harappan culture of the Indus River, or the Mesopotamian settlements on the broad plains between the Tigris and Euphrates Rivers, floodplains were the cradle of our most ancient civilizations.

Today, floodplains are home to some of the world's most dense human populations, with more than a billion people living in areas where the annual risk of flooding exceeds 1 percent (the so-called 100-year flood event). Yet, therein lies a fundamental tension. On the one hand, floodplains are merely parts of rivers and are, without intervention, inundated frequently. But on the other hand, they are attractive environments for humans to live and work on. The price that is paid for this conflict is that flooding is today the world's deadliest and most costly natural hazard. Flood events impact more than 300 million people each year, with average annual financial losses exceeding US$60 billion. As a result of climate change, which is driving the increasing frequency of extreme precipitation events, as well as

GLOBAL FLOODPLAINS

Four of the world's major river floodplains, mapped at
250 m (820 ft) spatial resolution, based on topographic
data acquired using radar aboard the Space Shuttle
Endeavour in February 2000. The mapping reveals the
large extent and complex structure of the floodplains
of the world's biggest rivers.

Mississippi

Niger

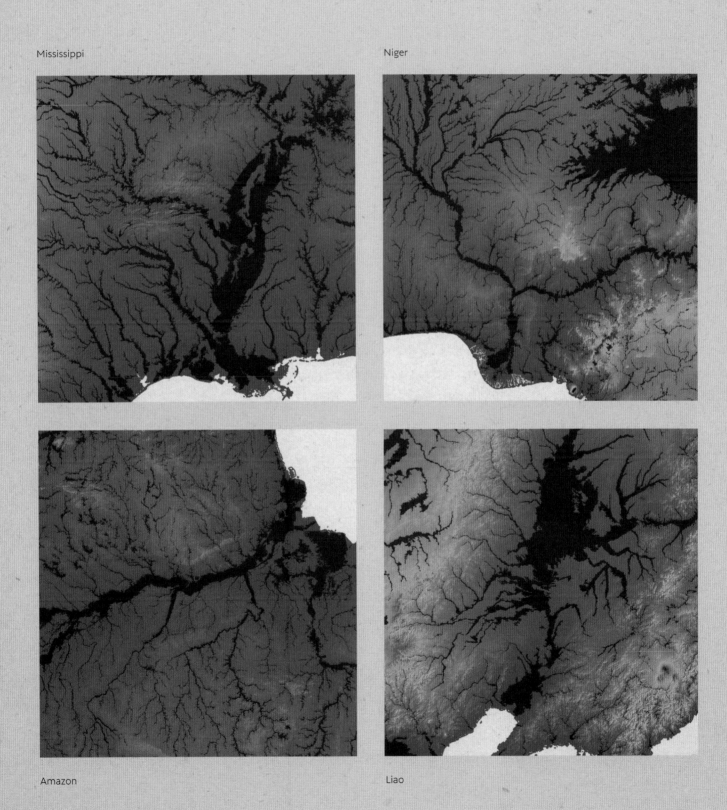

Amazon

Liao

demographic changes and economic development on floodplains, this flood risk is set to double by the year 2050. The growing likelihood of flooding is prompting investment in flood defense infrastructure, which protects people in the short term but has the effect of disconnecting rivers from their floodplains, destroying the seasonal inundation and associated processes that created them and their ecological richness.

Floodplain processes

The processes that shape floodplain morphology reflect a balance between erosional and depositional forces. For example, the migration of meandering rivers results in sediments building up as point bars on the inner bend of the meander but causing erosion at the outer bank. This continually reworks and overturns floodplain sediments, creating new land surfaces even as older areas are destroyed. Sometimes, these processes of lateral channel migration—especially in meandering rivers—cause the meander to extend and bend back on itself, creating an oxbow lake as the neck of the bend is eventually cut off by the river's remorseless movement. In times of flood, flow over the floodplain may be gentle, promoting vertical accretion of the floodplain surface as sediments settle out of the flow, or rapid, scouring new channels into the floodplain surface.

These complex, interacting processes mean that although floodplains may be relatively flat, they are far from featureless environments. Rather, they comprise complex mosaics, with areas of high and low elevation, lakes that may be fully or partially (dis)connected to the river, and networks of active or partially abandoned channels, all of which may be dissected by networks of tributary and distributary channels. It is this topographic complexity, layered onto the regular wetting and drying of floodplains during and after floods, that creates the wide range of ever-changing physical habitats that underpin their ecological richness.

Flooding events

Just as the morphologies of "flat" floodplains are more complex than they seemingly appear, so too are the mechanisms by which floodplains are inundated. When the media report on flood disasters, it is common to hear that rivers have "burst their banks," but this expression can be misleading and is only one of a number of flooding mechanisms. On many rivers, floodplains may gradually fill by diffusive flow as the banks are overtopped. Elsewhere, they may be inundated directly by intense rainfall, even if water levels in the main channel are lower than the riverbank crests. Alternatively, flow from tributaries may spill out onto the floodplain, especially if the drainage of those tributaries is itself impeded by high water levels in the main channel.

Breaking the banks
Satellite image of a partially flooded crevasse splay on the Tigris River, Iraq, December 1988.

Flooded crevasse-splay lobe

Inactive crevasse-splay lobes

At the core of the Amazon biosphere, the richest terrestrial ecosystem on our planet, is the river and its floodplain. It is estimated that 20 percent of the Amazon lowland basin is covered by permanently or seasonally flooded wetlands, as seen here in the igapó forest (see page 120) on the floodplain of the Rio Jutaí, Brazil (upper image). However, many floodplains in populated or intensively cultivated areas, such as in this floodplain in Uruguay (lower image), are protected by flood embankments, reducing the frequency of flooding and clearing the natural riparian vegetation, leading to significant habitat loss.

In floodplains such as the Amazon River, with its complex networks of channels, the water that inundates the floodplain may have been conveyed from distant sources rather than the adjacent river itself. And, of course, rivers do sometimes "burst their banks," such as through spectacular crevasse splays that break through levees. Here, the main river disgorges onto the floodplain, rapidly depositing large volumes of material and forming intricate fans of sediment on the floodplain surface. The precise mechanisms of flooding are vital as they influence whether, when, and where pathways of water, sediments, nutrients, and pollutants are moved from the main river onto the floodplain surface.

THE FLOOD PULSE:
NATURE'S WATER HEARTBEAT

Diagram illustrating the concept of the flood pulse and its
important impacts on the life cycle of many fish.

Flood-tolerant trees

Terrestrial shrubs

Annual terrestrial grasses

Low flow season

Most river-spawning fish
start to breed

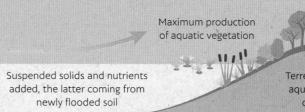

Maximum production of aquatic vegetation

Onset of flood season

Lake and river fish spawn; young
fish and predators follow the moving
shore edge; fish and invertebrate
production is high

Suspended solids and nutrients
added, the latter coming from
newly flooded soil

Terrestrial and older
aquatic vegetation
decomposes

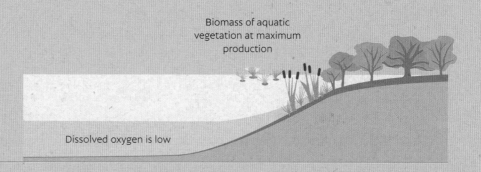

Biomass of aquatic vegetation at maximum production

Peak flood

Adult and young fish disperse,
adequate levels of dissolved
oxygen permitting

Dissolved oxygen is low

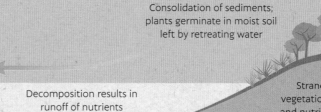

Consolidation of sediments;
plants germinate in moist soil
left by retreating water

Flood begins to recede

Many fish respond to decreases
in water depth by moving to
deeper water

Decomposition results in
runoff of nutrients

Stranded aquatic
vegetation decomposes
and nutrients mineralize

Terrestrial grasses and
shrubs start to regrow

Decomposition results in
runoff and concentration
of nutrients

Consolidation of
sediments

Most remaining
vegetation decomposes

Flood recession almost complete

Fish migrate to main channel,
permanent lakes, or tributaries

Floodplain (transition zone of aquatic
and terrestrial vegetation)

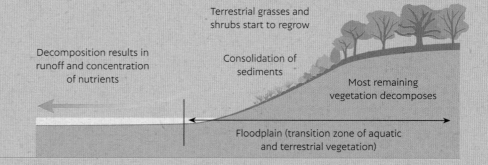

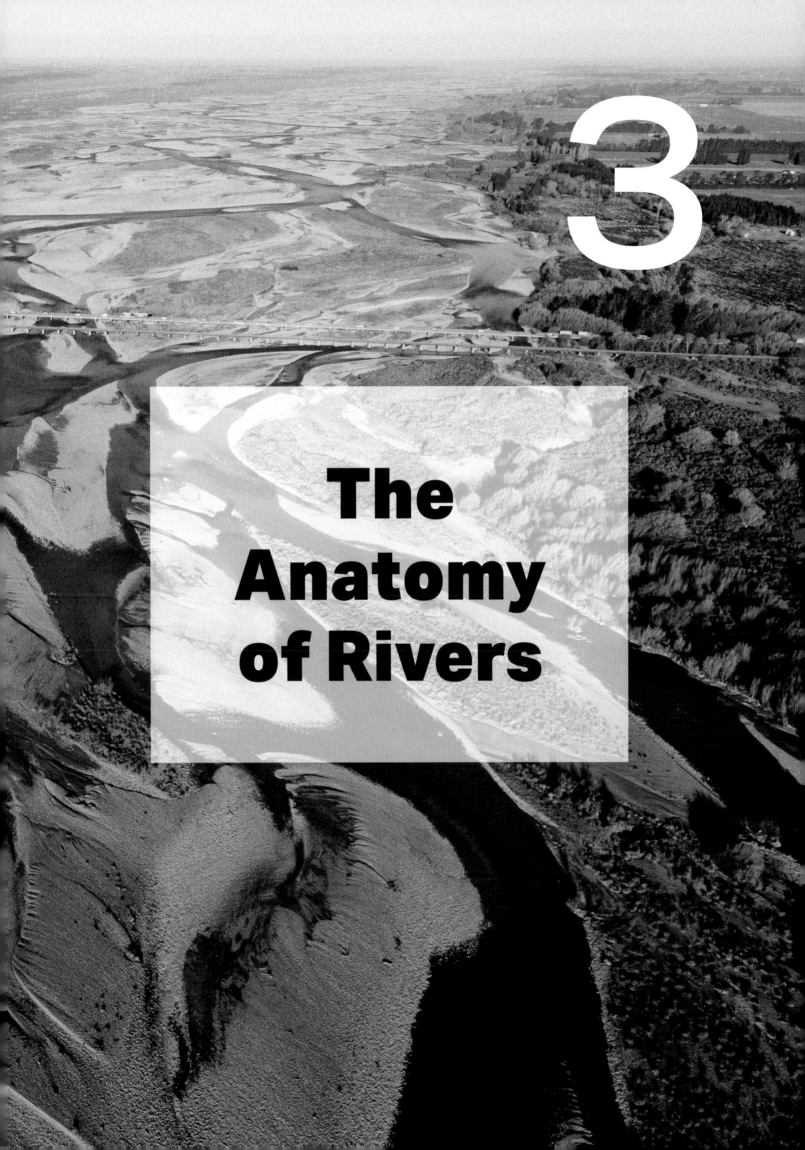

The Anatomy of Rivers

3

Winding down to the sea

The long profile of a river describes how the elevation of its bed declines with distance downstream from the river's source to its mouth. The shape of the long profile not only reflects the outcomes of long-term erosion and sedimentation, but also fundamentally affects contemporary processes.

The long profile is a fundamental morphological attribute of a river, representing the topographic expression of the ways in which climate, tectonics, geology, and human impact have interacted over time to shape the evolution of its drainage basin. Moreover, a feedback mechanism is involved: the river slope exerts a key control on processes of erosion and sedimentation. This is because the local river slope strongly controls flow velocity, and thereby the mechanisms and rate of

GLOBAL RIVER STEEPNESS

A map of global river long profiles produced at a spatial resolution of 6 arc-minutes (around 11 km/7 miles at the equator) derived from the HydroSHEDS global digital elevation model (DEM). The dark blue colors indicate very shallow slopes, while reds are very steep.

Slope (m/m)

0.28

0.00000285

sediment transport. Similarly, slope also controls the gravitational potential exerted on sediment and rock, and thus their susceptibility to erosion and transportation.

Determining long profiles

Long profiles are readily determined either by direct field measurement of the fall in riverbed elevation with distance along the river course, or by manually extracting these parameters from topographic maps. These traditional methods provide accurate representations, but they are time consuming and, until very recently, this has made it challenging to systematically map river long profiles across the world. However, the advent of global topographic data sets, such as global digital elevation models (DEMs) derived from satellite-based radar sensors, alongside the development of computer algorithms to analyze terrain data, has now enabled river scientists to undertake global-scale analyses of river profiles and river slope variability.

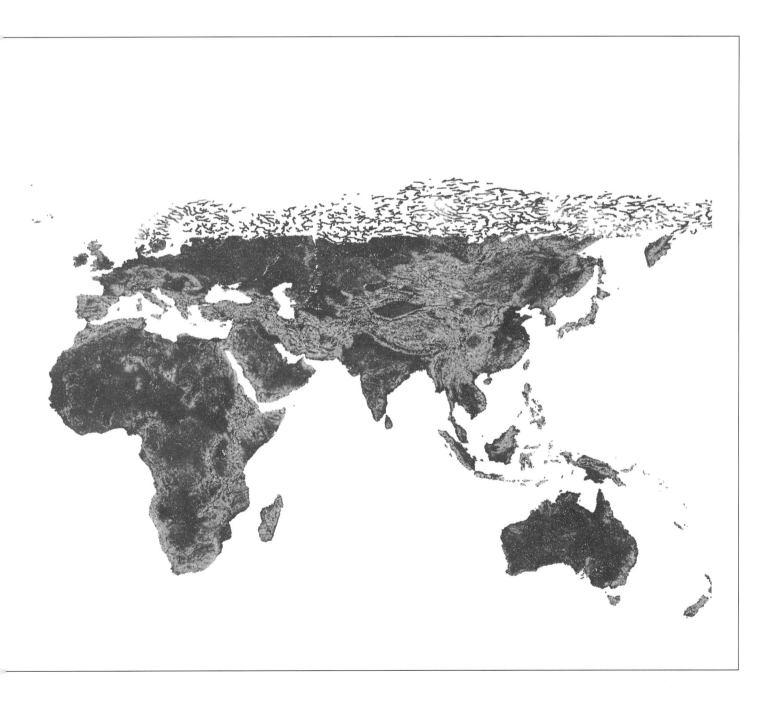

New insights and implications

These large-scale digital mapping analyses provide new insights that are challenging previous conceptions of the global geography of river slopes. For example, the global mean river slope is now known to be 0.0026 m/m, or a fall of 2.6 m in elevation for every 1 km of the river's course (14 ft per mile). However, it is striking that many of the world's rivers have very low gradients: more than half of the world's river segments have slopes less than 0.0006 m/m (a fall of just 60 cm in every 1 km, or 3 ft per mile), and the total length of ultra-low-gradient rivers, with slopes less than 0.00005 m/m (a fall of just 1 cm in every 1 km, or $^2/_3$ in per mile), is around 276,000 km (171,500 miles).

Most ultra-low-gradient rivers flow across the world's coastal plains and their very low slopes have important implications for the sensitivity of these regions to future rises in global mean sea level. This is because the "backwater zone"— the length of river that is hydraulically slowed as it nears the ocean—is controlled by the ratio between flow depth and channel gradient. Thus, a 1 m (3 ft) increase in global mean sea level (which is in line with Intergovernmental Panel on Climate Change projections to 2100), and hence flow depth, would increase the landward extent of the backwater zone in each ultra-low-gradient river by more than 100 km (60 miles), potentially increasing the risk of flooding or shifts in channel position.

▼ **Tumbling down**
The long profiles of rivers are not always smooth; sometimes they are interrupted by sharp steps as illustrated by the Gullfoss Waterfall here in Iceland.

CLIMATE-DRIVEN RIVER SHAPES

Long-profile shapes of the world's major rivers illustrating how these vary across the climatic zones through which the river flows. The normalized concavity index is a measure of long-profile shape where values less and greater than zero represent concave and convex profiles, respectively. Note the prevalence of convex profiles in arid regions.

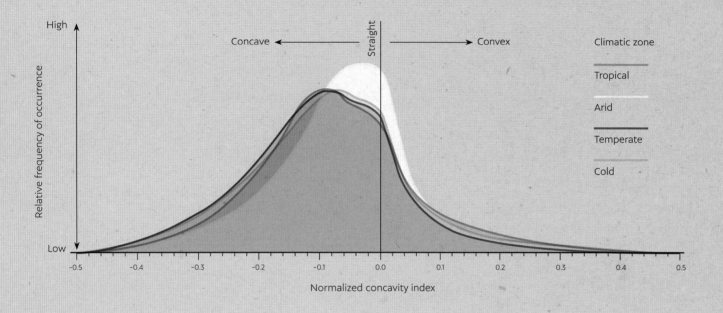

Factors influencing profile slope and shape

Global-scale data sets can also be analyzed to derive the average slope for each of the world's largest rivers. These average-slope maps show that there is a key link between mean river slope and tectonic setting. Three Asian rivers—the Indus, Ganges, and Chang Jiang (Yangtze) —stand out as having some of the largest mean river slopes. It is no coincidence that these rivers are also among the world's most active in terms of their rates of erosion, sediment transport, and sedimentation. In contrast, some of the flattest, least active rivers—such as the Murray-Darling river system in Australia and the St. Lawrence in North America—are located in tectonically stable environments.

The detailed shape of the river long profile is also critical. Traditionally, it was thought that the vast majority of river long profiles have a generally concave-up shape (i.e., the slope becomes progressively flatter with distance downstream), producing a smooth profile that is sometimes interrupted by sharp breaks associated with waterfalls. However, although most river long profiles are indeed concave, we now know that globally they become systematically less concave with increasing aridity (*see* figure above), and also that a substantial minority even become convex in form (their slope becomes steeper with distance downstream). These patterns in long-profile shape are explained by systematic variations in the rate at which streamflow changes with distance downstream. For example, river profiles are increasingly concave in humid environments, because streamflow increases more rapidly with distance downstream than in arid regions.

Channel shapes

The width and depth of rivers adjust to accommodate the flows of water and sediment supplied from upstream, but are also influenced by the resistance of the riverbed and riverbank materials, and the presence or absence of vegetation and other biota. In turn, the cross section of the river channel influences sediment transport processes and the aquatic ecology.

As a typical river flows from its source to the sea, its width and depth both tend to increase to accommodate the greater volume of water. For most rivers, the rate at which channel width increases with increasing flow is greater than that for river depth, meaning that the ratio of width to depth (a key measure of channel shape) tends to increase toward the river mouth—rivers become relatively wider with respect to their depth. However, while flow is undoubtedly the dominant factor influencing river width and depth, changes in the sediment load supplied to the river also are important, and can sometimes drive very abrupt changes in channel morphology.

Other factors also play a role in modulating the width or depth of a river. In particular, the resistance of riverbank materials (including the protective effects of any vegetation) is important, as banks that are weak tend to result in wider (and hence shallower) rivers than would otherwise be the case.

The deepest river

Parts of the Congo River are believed to be the deepest of all the world's rivers. Around 300 km (200 miles) from the ocean, the Congo drops into a series of steep gorges that are constrained by bedrock (see also pages 34–35). Here, the massive flow discharge—on average about 41,000 m³ (1.4 million ft³) every second—and steep gradient lead the river to scour pools that reach 220 m (720 ft) below the surface, so deep that no light can penetrate. Together with the power of the river currents, which makes it difficult for species to move between different parts of the river, these conditions have generated a unique ecosystem.

One denizen of the Congo's deeps is known locally as the mondeli bureau, or "white man in an office." The fish is pale and eyeless, and has a characteristic elongated form, suggesting its evolutionary adaptations may be driven by living in the river's great depths and fast currents.

Denizen of the deep
A blind fish *Lamprologus lethops* from the lower Congo River.

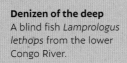

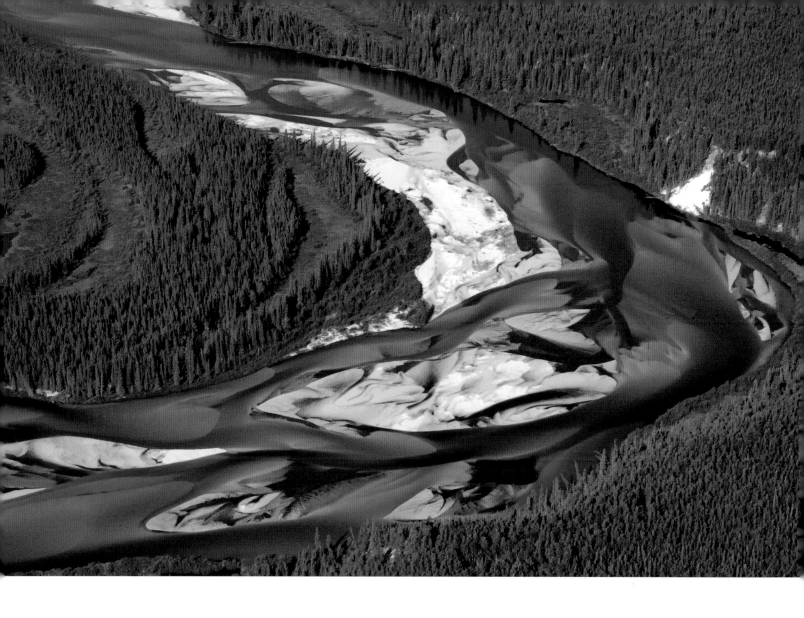

▲ Transformed by sand

A relatively narrow and deep single-channel stream as it flows northward to Lake Athabasca, the lower William River picks up a 40-fold increase in bedload sediment over just 27 km (17 miles) as it passes through the Athabasca Sand Dunes Provincial Park in Saskatchewan, Canada. As a result, the channel abruptly becomes braided while undergoing a five-fold increase in width and a tenfold increase in width/depth ratio.

▶ Unstable riverbanks

Hippopotamus (Hippopotamus amphibius) on a heavily eroded bank of the Luangwa River, North Luangwa National Park, Zambia.

Shifting sands

As rivers transport sediment, the features created—including dunes, bars, and islands of different types—erode, accrete, and migrate. These riverbed features are the building blocks of overall channel and floodplain morphology, but are constantly shifting.

The continuous movement of sediment downstream causes river channels to change their form constantly through time, although the majority of work is done at high flows when the larger volumes of water move most sediment. If you stand in a shallow river and look down at your feet, you will see this constant flux of moving grains along the riverbed. But how can we see what such change looks like on a larger scale? Comparison of maps, charts of river bathymetry, and aerial and satellite images of rivers at different times allows us to measure how riverbeds change on timescales ranging from minutes to years and perhaps even centuries. Such maps show the nature of shifting sands and gravels, revealing the flux of sediment movement and how it constantly reshapes the riverbed.

A changing landscape

Shifting sands are also vital for the ecology of rivers and the humans who live around these waterways and on their floodplains. As river channels evolve, their in-channel and floodplain ecosystems are also subject to change, in turn affecting human use of these rich resources. For example, river bars deflect flows, driving the erosion of riverbanks and of floodplain forests and agriculture. This can displace humans, but also forms new land on which habitations can be built. In addition, the growth of bars and river islands allows plants to establish new vegetation successions, providing a mechanism for the river ecology to renew itself. River valleys are thus constantly changing, although river engineering has often sought to constrain and fix rivers in place in a quest to make floodplains places for agriculture, urban growth, and the development of civilizations. This is a process that has gone on for many thousands of years.

Shifting sands
Changes in the topography of the braided South Saskatchewan River, Canada, over a period of one year (2015–2016). Red–yellow indicates areas of erosion and blue–green regions of deposition, revealing the downstream migration of sandbars, which move at an average rate of around 1 m (3 ft) per day.

Flow
⟶

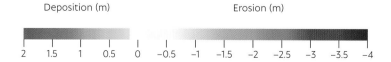

VEGETATION SUCCESSION

As river sandbars form, new vegetation colonizes them,
leading to a succession of plant growth over time.

Early succession
0–60 years

Mid-succession
60–100 years

Late succession/mature
forest over 100 years

River planforms

A river's planform is its shape when viewed from above. The planforms that rivers adopt on their downstream paths are set by interactions between the erodibility of the channel and floodplain sediments, and the processes that help construct bars, islands, and other features. These complex interactions lead to a mosaic of planform states across Earth's river networks.

River planforms are determined by the assemblage of morphological units (bedforms, bars, and islands) deposited in the active river channel. Straight rivers are rare in nature, with most either taking on meandering, braided, or anabranching forms (see illustration). The very term "meander" is derived from the winding Büyük Menderes River, south of Izmir in modern-day Türkiye; to the ancient Greeks it was known as Maiandros, which led to the Latin name Maeander. However, many natural river planforms have been disrupted by human activity, and straightened rivers are now a common indicator of the human imprint on river landscapes.

▼ **Nature abhors straight lines**
Natural rivers with straight planforms are very rare and instead are usually associated with human modification, as is the case here with the straightened Vistula River near its mouth at the Baltic Sea.

Anabranching, anastomosing, and braiding

Anabranching and anastomosing rivers (see illustration) comprise systems of multiple channels that divide areas of a floodplain (i.e., vegetated or otherwise stable alluvial islands that rise above the normal levels of annual flooding). These channel patterns are distinct from braided rivers, which are divided by low-elevation bars that are frequently overtopped during floods. Anabranching frequently occurs concurrently with other patterns, such that individual branches within multithreaded rivers can take on braided, meandering, or straight forms. Anabranching rivers are especially common in large rivers—by length, 90 percent of the alluvial reaches of the world's ten largest rivers comprise this planform.

◄ Meanders

Landsat 8 image of the meandering Tsiribihina River, Madagascar. The yellow-orange colors of the water indicate that it is transporting a high load of suspended sediments, as is common for sinuous, single-thread rivers.

▼ Interweaving channels

Braided rivers tend to be associated with high supplies of sediment and/or highly erodible banks. These conditions are met on the braided Rakaia River, New Zealand, where glacial outwash feeds abundant sediment loads and the cool temperate environment limits stabilizing vegetation growth on the banks.

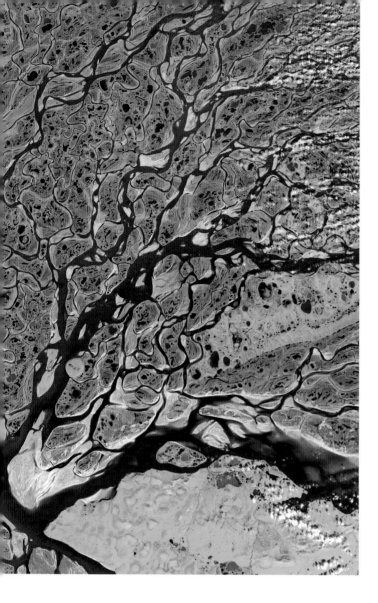

Factors affecting planform shape

So, for those rivers that have not been directly modified by humans, what determines their planform shape? Three factors interact to determine whether a river is likely to be single-threaded or multithreaded. The first of these is the rate of floodplain development. Where sediments build up rapidly, vertical accretion (rather than downstream migration) of sand and silt is favored on mid-channel and lateral bars, which promotes the development of larger, more stable islands. As a result, dynamic braided channels are associated with slower rates of floodplain development. Second, the resistance of riverbanks to erosion is important. More resistant bank materials, such as fine-grained silts and clays, or riverbanks with dense assemblages of bank vegetation, promote lower channel width and fewer channel branches. Third, the nature of the flow hydraulics, and specifically the way in which the flow controls the mechanism of sediment transport, is important.

This last factor illustrates an important property of rivers: large-scale order (here, the planform structure of rivers in their floodplains) often emerges from interactions between water and sediment flows operating at much smaller scales, in this case individual sediment grains. These grain-scale interactions influence the mechanism of sediment transport and hence planform evolution because, as grain size increases, a greater fraction of sediment is transported as bedload (in contact with the riverbed) rather than as suspended load. This is important because sediment transported as bedload is deflected by gravity in the direction of the local bed slope, whereas sand suspended above the bed is not. Thus, when sediment is transported as bedload it may be deflected around local topographic high points, thereby reducing vertical bar growth, slowing the conversion of bars to vegetated islands and floodplains, and widening the channel. In contrast, when more sediment is transported in suspension (for example, due to a reduction in grain size or an increase in flow velocity) there is less potential for it to be deflected around local features such as bars. This promotes vertical bar growth and conversion to floodplain, which in turn drives the formation of narrower, more sinuous channels. This leads to a reduction in the number of channel branches and/or a transition from braiding or anabranching to meandering.

While river planforms are set by predictable interactions between these three driving factors, the factors vary substantially from location to location and so there is no clear emergent geography of river planforms. Instead, Earth's river network comprises a complex mosaic of planform shapes that are no less beautiful for their variable structure.

▲ **Split channels**

Anabranching rivers, such as the River Ob in Russia, occupy a wide range of environments, from low to high energy, and across varied climatic settings. They are particularly widespread in lowland regions.

FROM GRAIN TO FLOODPLAIN

The planforms of the world's largest rivers adjust in response to flow-grain interactions that change the relative proportion of sediment transported as bedload versus suspended load. As the proportion of bedload increases from lowest (Madeira River) to highest (Negro River), channel planforms become more complex.

Madeira River, Brazil
4.1°S 59.3°W

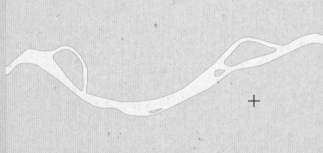

Middle Paraná River, Argentina
2.8°S 69.9°W

Ica River, Peru
32.0°S 60.6°W

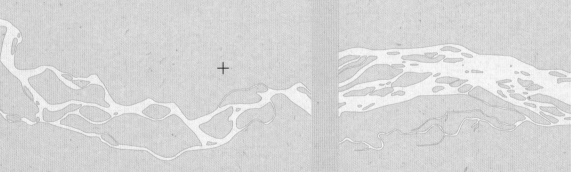

Negro River, Brazil
0.6°S 63.5°W

Flow

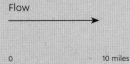

Rivers on the move

Rivers migrate laterally across their floodplains as a result of erosion on one riverbank and deposition on the other. Simple feedbacks between channel curvature and these processes create beautiful, complex landforms as the channel moves, continually reworking and re-creating the floodplain environment. This dynamism maintains rich, healthy ecosystems, but may conflict with human use.

Unless they are locked in place by natural geological constraints, such as bedrock constrictions, or by human structures, river channels are in perpetual motion, migrating backward and forward across their floodplains, while also translating down—or even up—along the axis of the valley. This migration happens wherever river erosion and sedimentation work in tandem to develop coherently structured patterns that are segregated in space. For example, if, rather than being deposited in the center of the channel, a bar instead becomes attached to one of the riverbanks, it may "push" the flow toward the opposite bank. In turn, this may drive bank erosion there, leading to a net shift in the position of the channel.

▶ **Snaking along**

Rivers that flow down a shallow gradient tend to meander, looping back and forth over the terrain. These motions, which take place over decades and centuries, are preserved here in patterns visible in the floodplain of the Selemdzha River in Russia.

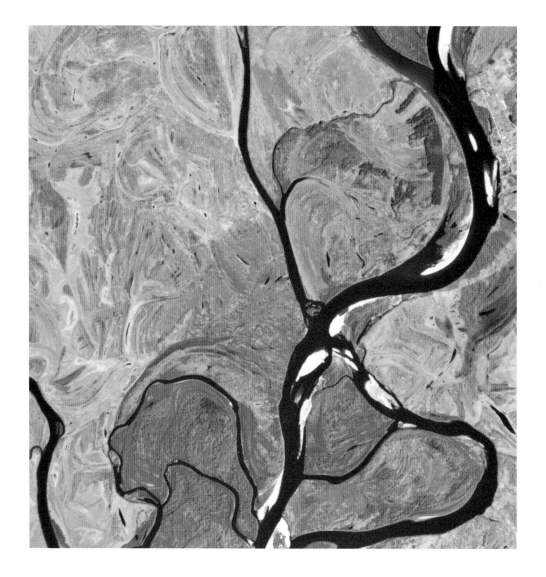

The dynamics of channel migration

In fact, there are strong positive feedbacks between a river's planform geometry (and specifically its local curvature) and the flow hydraulics that drive erosion and sedimentation; it is these feedbacks that create the coherent patterns of erosion and sedimentation that drive channel migration. Just as a passenger in a car moving quickly around a bend will experience an outwards-directed centrifugal force, so too does any river that flows around a curve. This centrifugal force—which increases as either, or both, the water velocity or channel curvature increase—causes fast-moving flow to move preferentially toward the outer bank, favoring erosion there and deposition from the slower-moving water along the inner bank.

Importantly, there is a difference in the location of maximum curvature and the position at which maximum erosion occurs, the latter being slightly downstream of the former. This "spatial lag" is critical in the feedback process because it means that channel migration is directed both outward (driving bend expansion) and along the valley (driving up- or down-valley translation). This has the effect of amplifying channel curvature, eventually causing the river to migrate back on itself and cut off, forming a classic oxbow lake, locally restraightening the channel, and restarting the cycle.

MEANDERS ON THE MOVE

Channel curvature drives riverbank erosion and hence lateral meander migration by directing fast flow to the outer bank, focusing erosion there, while deposition is favored from the slower flow along the inner bank. Critically, the location of maximum migration lags a short distance downstream of the point of maximum curvature. This means that meanders migrate both outwards (driving expansion of the meander belt) and along the valley axis (driving meander belt translation).

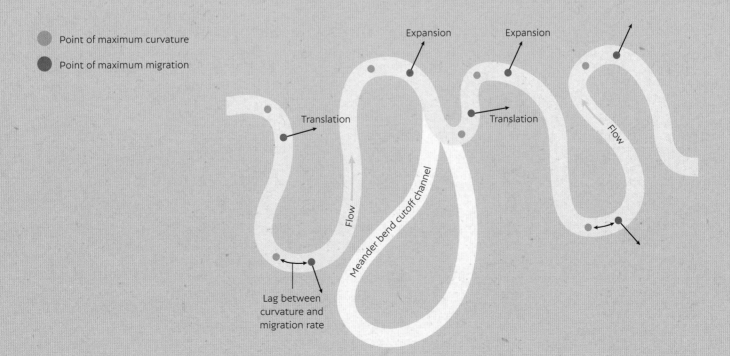

- Point of maximum curvature
- Point of maximum migration

Expansion

Expansion

Translation

Translation

Flow

Flow

Meander bend cutoff channel

Lag between curvature and migration rate

Measuring floodplain changes

Over time, as rivers migrate across their floodplains, the feedbacks between flow, channel curvature, and erosion and sedimentation create intricate mosaics of landforms that are preserved in the floodplain. These features can be revealed by comparing the course of the river on sequences of old maps, and through the subtle variations in local floodplain elevations that can be measured by new surveying techniques such as airborne light detection and ranging (LiDAR). These features include point bars, formed by deposition along the inner part of a river bend as it expands; cut banks, steep cliffs that form along the outside of a bend as a result of the channel eroding laterally into the floodplain; and counterpoint bars, which form along the outer bank of a river when meanders migrate rapidly through downstream translation.

River scientists have long measured the patterns and rates of channel movement by carefully comparing maps of different ages, but the availability of global satellite imagery has enabled a much clearer picture of rivers as they migrate across and along their floodplains. Based on these analyses, we know that migration rates tend to increase as sediment loads supplied from upstream increase, as flow regimes become "flashier," and in weak, unconsolidated, or sparsely vegetated sediments. However, the most influential factor affecting rates of movement is the curvature of the river, emphasizing again the importance of the key feedback mechanism driving river motion.

The continual reworking of floodplain sediments through river migration, and the complex patchwork of landforms created, result in a high degree of physical diversity that sustains high biodiversity. However, rapid channel migration can cause problems for landowners and threaten infrastructure, creating tensions between human use and the preservation of ecosystem services.

▶ **Ecological richness**

Migrating rivers are essential for creating biodiverse floodplains.

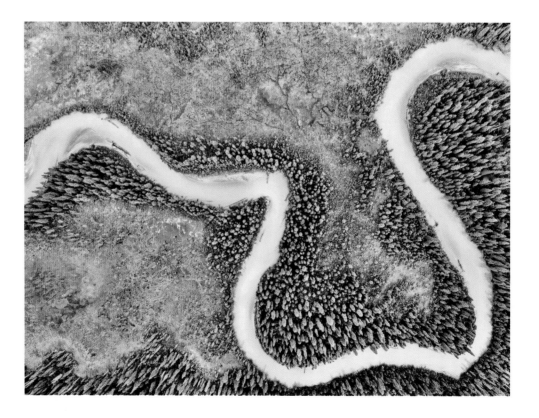

PERPETUAL MOTION

A time-lapse series of tracings of Landsat satellite images showing the dynamic movements of the Mamoré River in Bolivia (a major tributary of the Amazon) between 1990 and 2018. Rates of channel movement of several tens of meters per year are common along this river. The changing landforms are indicated by the colors in subplots A to F (blue colors show low elevations and reds are high elevations).

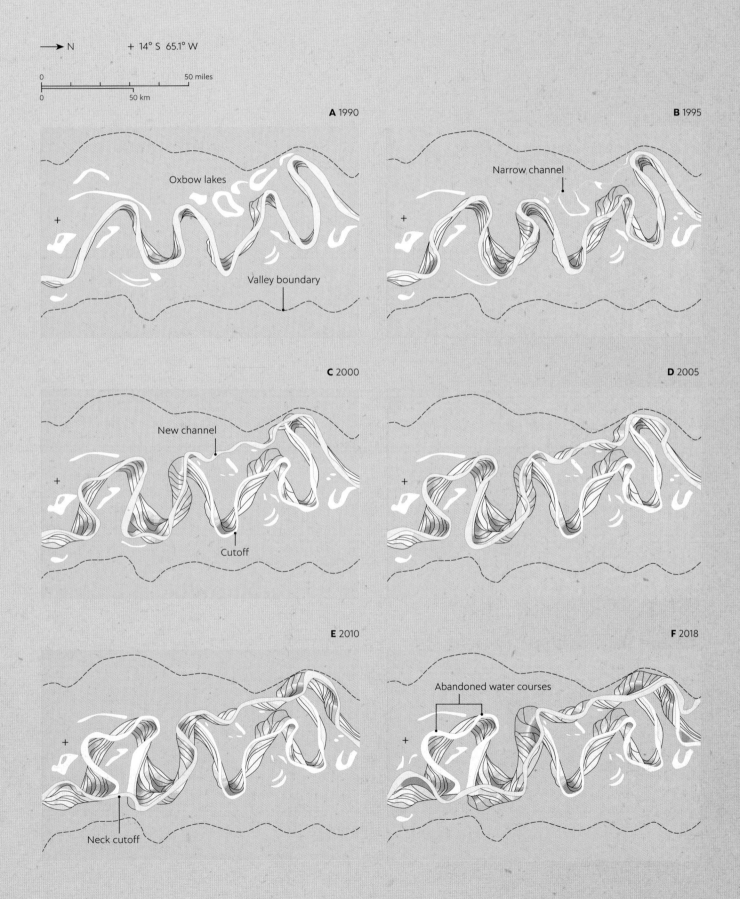

N   + 14° S 65.1° W

0   50 miles
0   50 km

A 1990
Oxbow lakes
Valley boundary

B 1995
Narrow channel

C 2000
New channel
Cutoff

D 2005

E 2010
Neck cutoff

F 2018
Abandoned water courses

Catastrophic channel change

Rivers often undergo gradual changes in their morphology. At times, however, they can experience rapid and large-scale changes—for example, as a result of huge floods—when they adopt a completely new path across their floodplain.

Rivers carve their paths following lines of least resistance as they flow down the steepest path of descent. Where a river deposits its sediments rapidly—for example, in locations where sediment loads are very high and/or areas where the gradient changes abruptly, the riverbed can become higher in elevation locally compared to the surrounding terrain. In such instances, the river's path can change rapidly as it seeks out a new route, and the channel can jump out of its old course by large distances.

▶ **The shifting Brahmaputra 1**

Large-scale avulsion of the Jamuna (Brahmaputra) River in the period 1765-1860 is revealed by comparing the 1776 map of British surveyor Major James Rennell (1742-1830), which shows the river in its old course, and a recent satellite image (opposite).

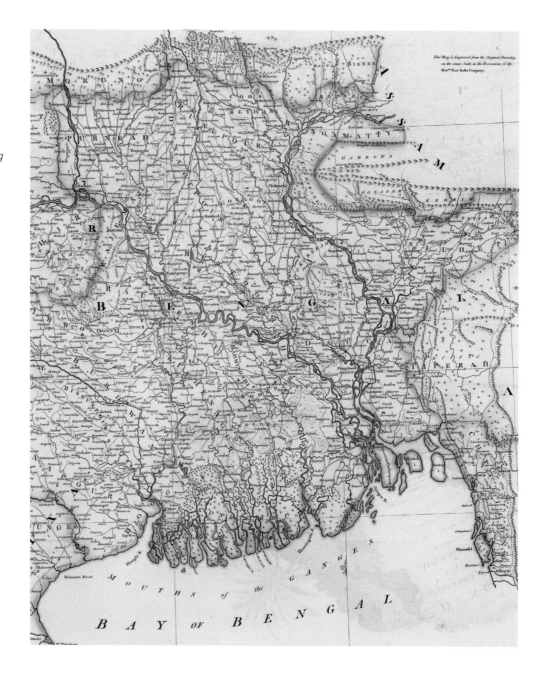

This process, known as river avulsion, occurs frequently on alluvial fans, across lower-gradient river floodplains, and in river deltas as the channels switch around as they enter a lake or ocean. Such movement of channels occurs as a river exploits local slope advantages, shifting laterally to follow a path along a steeper gradient. However, in time, this new channel undergoes sedimentation and so spawns another avulsion. Although avulsion is a natural process occurring in many rivers, human interventions, such as channelization and fixing rivers in place, can create the potential for sudden, catastrophic change, as the river breaks out and follows the path of the steepest gradient. As such, natural and human-assisted river avulsions can pose a threat to the humans who live in these areas—previously habitable land may suddenly become flooded or even be eroded to form part of the new river channel.

◀ **The shifting Brahmaputra 2**

The map opposite and this satellite image indicate gradual abandonment of the old channel (now called the Old Brahmaputra) over a period of around 100 years, likely due to erosion of the riverbank until the channel intersected the course of a smaller channel that flowed to the south. After its shift 80 km (50 miles) to the west, the new river channel underwent significant deepening and widening as it adapted to its new increased flow. The occurrence of such river avulsions has significant implications for human population displacement, loss of life and livelihoods, and ecological change.

Megafloods

The largest known river floods on Earth were caused by the breaching of lakes dammed by glaciers or alluvium that, when they failed, released huge torrents of water into the downstream river valley. Such floods can be massive, with water depths up to 200 m (650 ft), water discharges up to 20 million m³/s (706 million ft³/s) (about 100 times greater than the peak flow of the Amazon River), and flow velocities up to 60 m/s (200 ft/s), and can last from a few days to several weeks.

Historical megaflood events have been documented in many regions, including catastrophic floods that swept down the Brahmaputra valley from ice-dammed lakes in the Himalayas between 9,000 and 30,000 years ago, and torrents that flowed from the melting North American ice sheet 18,000 years ago. Glacier lake outburst floods (GLOFs) occasionally occur today—such as glacial lakes breaching under ice due to volcanic activity in Iceland (termed *jökulhlaups*, meaning "glacial run"). The growth of glacial lakes in a warming climate represents a growing hazard from GLOFs in future.

Fossilized dunes

At West Bar, near Trinidad in Washington State, a LiDAR image and photo reveal giant dunes, up to 9 m (30 ft) high and 100 m (300 ft) long, which were formed around 14,000 years ago by a megaflood from glacial Lake Columbia and are now fossilized in the Columbia River valley. Potholes scoured into the basalt bedrock are preserved on Babcock Bench, some 200 m (650 ft) above these giant dunes.

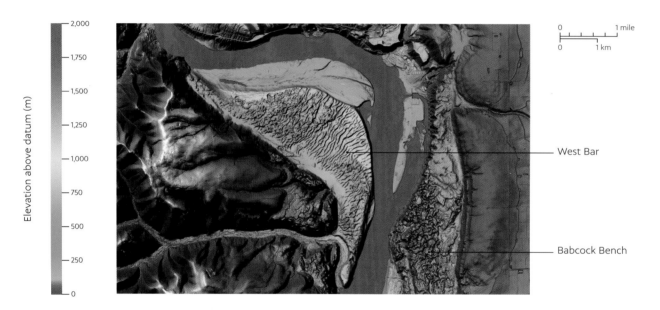

MEGAFLOOD PATHS AND CATASTROPHISM

The Pacific Northwest of the United States underwent giant Ice Age floods from three ice-dammed lakes, with Lake Missoula generating more than 100 repeat breakout floods. These huge flows sculpted the valleys through which they passed, leaving scars on the landscape, witnesses of these massive water flows.

Unveiling the story of these catastrophic floods started with the pioneering work of geologist J. Harlen Bretz (1882–1981). In the 1920s, Bretz proposed the origin of the channeled scablands of eastern Washington State lay in catastrophic floods unlike any normal river flows. His work was met with scepticism and disdain by his peers, but with the source of the floods later revealed by geologist Joseph Thomas Pardee (1871–1960) to be the collapse of ice-dammed lakes, Bretz's extraordinary ideas eventually became accepted. Bretz's work and experiences prompted his famous quote "Ideas without precedent are generally looked upon with disfavor and men are shocked if their conceptions of an orderly world challenged," as well as stimulating the growth of ideas on catastrophism—that Earth's history has been punctuated by catastrophic events that altered the way the Earth and its life have developed.

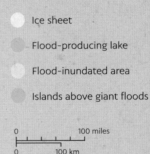

Ice sheet

Flood-producing lake

Flood-inundated area

Islands above giant floods

Old waterfalls
The Pothole Coulee Cataracts, Washington State, with scoured channels leading into them, bear witness to massive erosion generated during the Lake Missoula outburst floods.

Ancient rivers and ancient worlds

River sediments may be preserved within the geological record, providing evidence of the ancient environmental conditions under which they were deposited. These sediments not only provide us with a window through which to view the past, but also hold vital resources such as water, gas, building aggregates, coal, oil, and minerals such as gold and diamonds.

As river deposits become buried by later sediments, and under conditions where they can be preserved, they become incorporated into the geological sedimentary record. The recognition and interpretation of ancient river sediments can be used to provide clues as to what the Earth's terrestrial surface looked like in deep time. These sediments also give hints of conditions that may have been very different to those of the present day, such as before the first land plants evolved some 500 million years ago. Ancient river sediments can be studied in rock outcrops at the Earth's surface, from rock cores obtained from boreholes, and by using a range of geophysical techniques that can image the subsurface. Importantly, our knowledge of contemporary rivers and their sediments provides a modern-day comparison with which to interpret ancient rivers.

Reconstructing ancient landscapes

Geologists have sought to create a series of templates, or models, of the different types of rivers (e.g., meandering, braided, and anastomosing channel planforms), which can also be applied to ancient rivers. Detailed observations can then be used to quantify the various structures in the sediments of ancient rivers, both within their channels (e.g., preserved sand dunes) and their floodplains (e.g., soils and accumulations of organic matter that form coals). Interpreting the size of ancient channels can also give clues as to the depth and width of the original rivers, the amount of water and sediment they transported, and the direction in which they were flowing. All of this information can then be used to develop reconstructions of the land surface and alluvial channel morphology, as well as the changing controls on sedimentation in geological time, such as tectonics, climate, and baselevel.

ANCIENT BENDS

A schematic model of sedimentation in a meandering channel (A), alongside the pattern of bend migration shown by a LiDAR image of a bend in the Mississippi River (B). Such floodplains and their meandering channels can be preserved in ancient sediments (see C for section X–X'), as in the 70 m/230 ft-high face of a coal mine in 260-million-year-old sediments in New South Wales, Australia (D). The mound-shaped sandstone in the coal mine corresponds to section X–X' in the schematic model (A, C), with the sandstones of the meander point bar both underlain and overlain by thick black coals formed from dead plant matter deposited in the floodplain of a large ancient river.

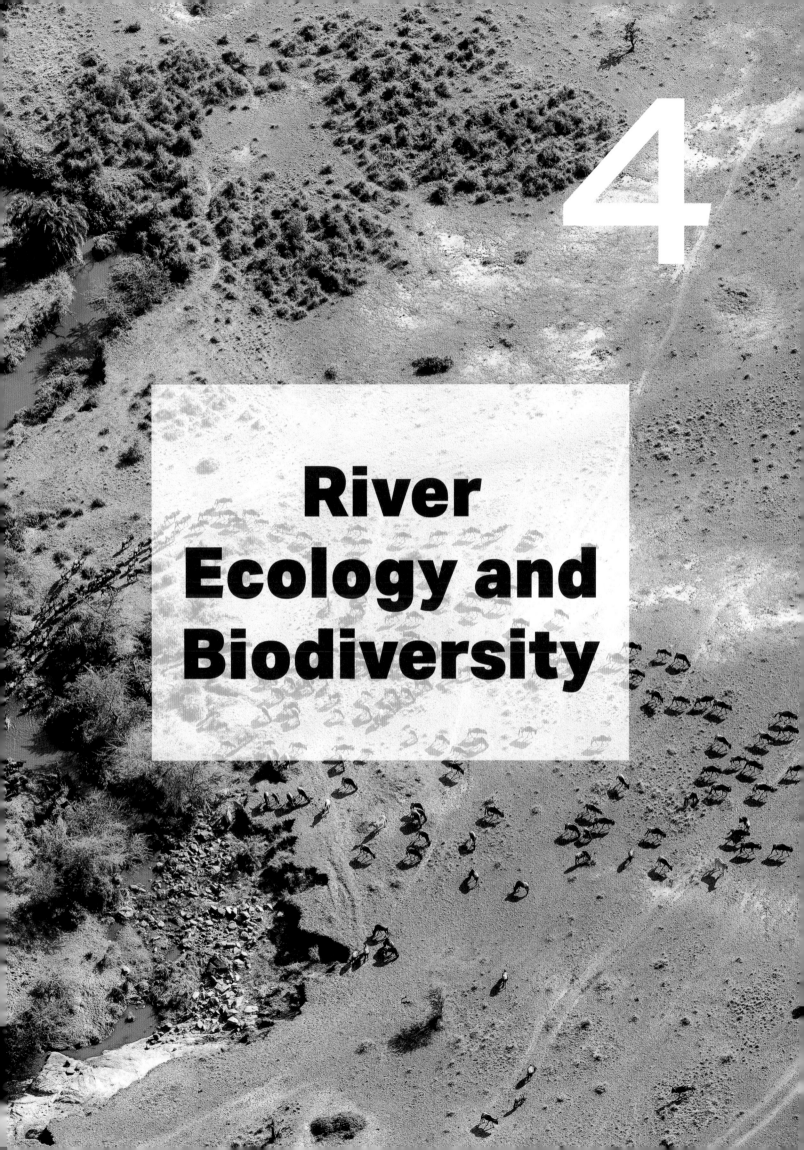

River Ecology and Biodiversity

4

Biodiversity and rivers

Rivers are home to the richest diversity of flora and fauna of all of Earth's terrestrial ecosystems. This huge variety of life reflects the wide diversity of flow conditions, substrates, and channel and floodplain morphologies that form their physical habitat. Some of the most biodiverse river basins are located in large tropical river basins like the Amazon, Congo, and Mekong. However, the organisms that make the world's rivers their home are increasingly threatened by a suite of human pressures.

FISHY HABITATS

The diversity of fish species in the world's rivers shows considerable variations across the globe. The diversity of species reflects fish evolution in response to factors that are affected by the river environment, such as geology, climate, and geomorphology, and increasingly the influence of humans.

Species richness

Low   High

Angolan Tilapia
This freshwater fish (*Oreochromis angolensis*) is endemic to the Middle East and Africa, and is commonly used in aquaculture.

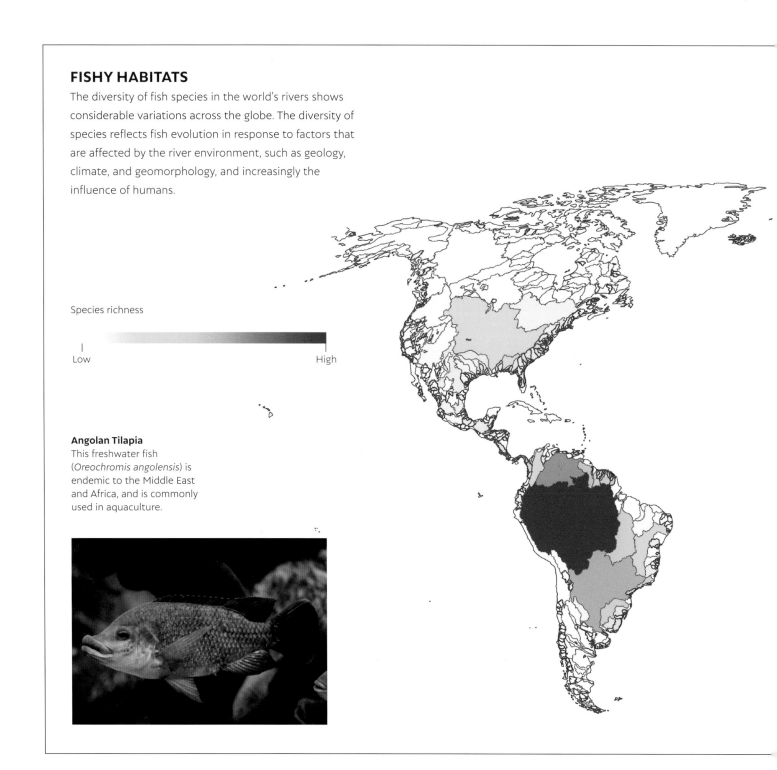

Biodiversity refers to the species, genetic, and ecosystem diversity in a defined area, and is often expressed as an overall species richness—the absolute number of species living there. Rivers are among the most ancient waterbodies in the world, and as such have provided time for many species to evolve adaptations to the unique conditions found in their waters. This long evolutionary "window," combined with the rich and varied physical and chemical conditions found in rivers, means that the running fresh waters of our planet are zones of extraordinarily high biological diversity. Rivers cover only around 0.01 percent of Earth's surface area but host more than 126,000 (7 percent) of all described species, while major tropical river systems such as the Amazon and Mekong are among the most biodiverse ecosystems on the planet.

Floating
Giant water lilies,
Pantanal, Brazil.

◄ **A remarkable**
mammal
The Duck-billed Platypus
(Ornithorhynchus
anatinus) lives in the
rivers of southeastern
Australia. Its distinctive
duck-like snout is
covered in receptors
that help it detect
and scoop up its prey—
insects and larvae—
from the sediments on
the river bed.

Knowledge gap

Although scientists know that the world's rivers are highly biodiverse, the precise spatial distribution of river species is not well understood and until very recently there was no global audit of freshwater biodiversity. This lack of information is problematic because, although there are fears that biodiversity in rivers is declining rapidly, the lack of a sound baseline makes it much harder to direct conservation efforts to the most severely impacted areas. However, in recent years new data, much derived from remote sensing, has emerged and our understanding of the threats to, and status and importance of, freshwater habitats has started to increase.

Chalk rivers

Rivers that emerge from chalk (a form of limestone) aquifers are very pure in quality and rich in minerals, and retain a fairly constant temperature. These conditions allow a diverse range of aquatic plants to grow, which in turn sustain a very large number of invertebrate and fish species. There are only around 200 chalk rivers globally, of which 85 percent are found in southern and eastern England, where they form a quintessential part of idyllic rural landscapes. Indeed, rivers such as the Itchen and Test are world famous for their salmon and trout fishing, and they host a range of plants, including Water-crowfoot, and animals, including Lamprey and White-clawed Crayfish (*Austropotamobius pallipes*). Despite their international importance, chalk rivers are under threat. Some have been drying up during the summer months as a result of climate change and overabstraction of water, while the introduction of untreated sewage poses another threat.

Marsh-marigold
Marsh-marigold (*Caltha palustris*) seen below the rippled water surface of the River Itchen in Hampshire, England. The plant is one of the UK's most ancient, having grown here since the last Ice Age.

In-channel ecologies

Rivers host rich and varied biological communities, but with distinctive characteristics compared to other ecosystems. Here, the downstream flow of water dominates every facet of life.

Ecosystems are areas where organisms and their physical environment interact to form an interconnected web of life. River ecosystems are unique because the organisms that live here must be able to cope with the energetic costs of maintaining their position against the flow of water. On the one hand, flow sustains ecosystem productivity by supplying nutrients from upstream, but high flows can destroy or reconfigure the landforms and substrates on which organisms live. In-channel ecosystems thus experience continuous physical change.

Feeding strategies

The major variations in river ecosystems are determined by the river slope and, therefore, the flow velocity, as well as the width of the channel and valley morphology. In upstream reaches, rivers tend to be narrow and steep, so bankside vegetation limits the penetration of light. In these environments, respiration (consumption) outpaces primary production, and the ecosystem is dominated by invertebrates (so-called "shredders") that rely on breaking down the coarse particulate organic material (CPOM) supplied from riverbank vegetation, as well as those (so-called "collectors") that rely on the fine particulate organic matter (FPOM) produced by the shredders. Fish that feed on these invertebrates need the high concentrations of dissolved oxygen produced by the turbulent flow.

▼ **Underwater worlds**
River ecosystems are rich and varied, and are closely connected with their surrounding environments.

THE RIVER CONTINUUM

As rivers wind from source to sea, in-channel ecologies respond to changing environmental conditions.

Collectors   Grazers

Shredders   Predators

Small headwater streams
Most inputs of energy to headwater streams are supplied from the coarse particulate organic matter (CPOM; leaves, twigs, etc.) shed from adjacent plants. The invertebrates in headwater streams rely on "shredding" this CPOM or collecting fine particulate organic matter (FPOM).

Middle reaches
Inputs of energy on these streams come from the FPOM supplied from the headwaters upstream and the algae and plants that grow in the rivers. Here the dominant invertebrates feed on the plants, and collect the FPOM supplied from upstream.

Large rivers
Phytoplankton can be an important source of energy in large rivers, but FPOM from upstream is usually the largest energy source. Here the dominant invertebrates collect FPOM while the FPOM can also support zooplankton populations.

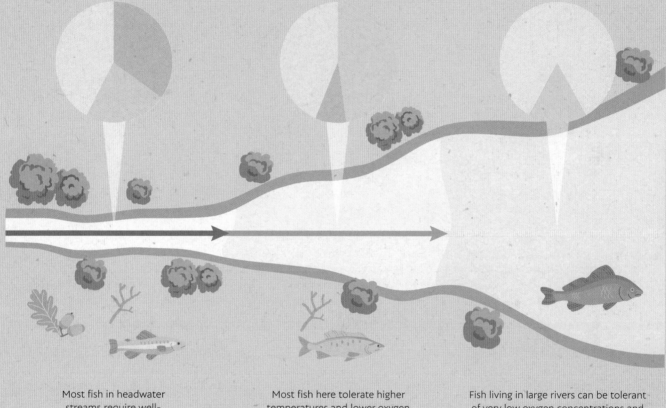

Most fish in headwater streams require well-oxygenated water

Most fish here tolerate higher temperatures and lower oxygen concentrations

Fish living in large rivers can be tolerant of very low oxygen concentrations and warmer water temperatures

In a river's middle reaches, its channel widens and the valley floor broadens, while the bed slope decreases. Here, photosynthesis is more effective and primary productivity more important, so a wider range of feeding strategies flourish.

Further downstream, the high-flow discharge carries a larger supply of FPOM and increased sediment loads make the water more turbid, limiting photosynthesis. Here, invertebrate communities are structured more around collectors and organisms that filter-feed from the water column, as well as fish adapted to the lower oxygen concentrations.

▲ **Microhabitats**

Boulders in the Roaring Fork River, Great Smoky Mountains National Park, Tennessee, create diverse microhabitats to sustain a wide range of river organisms.

▶ **Ocean bounty**

Each year, millions of Pacific salmon return to their natal rivers to spawn. Here at Brooks Falls in Alaska's Katmai National Park and Preserve, Alaskan Brown Bears (Ursus arctos middendorffi) feast on returning Sockeye Salmon (Oncorhynchus nerka).

However, within this continuum of changing downstream feeding strategies, substantial local variations in physical habitat drive further diversity. Such microhabitats arise from local variations in flow velocity, substrate type, and channel and valley morphology, driven by features such as boulders, fallen logs, depositional bars, scour pools, and tributary inputs. This huge variability in physical habitat creates specialized ecological niches that sustain high biodiversity.

In-channel ecosystems are not self-contained, and nutrient exchanges take place between rivers and their surrounding landscapes. For example, mayfly hatches can move biomass large distances from rivers. In some rivers, anadromous fish (species that migrate upriver from the sea to spawn) such as salmon are the dominant supply of nutrients, sustaining a wide array of terrestrial life.

Life in the flow

Organisms living in rivers have to cope with, and adapt to, changing conditions, with the volume, quality, and velocity of flow determining what species live where. Life in the flow is partly determined by flow type, and alterations to this have ramifications for riverine ecology.

Life in a river responds to the quantity and quality of flow, and whether the water is flowing or still—termed lotic and lentic waters, respectively. Fresh water in river valleys can be viewed as comprising running water and standing bodies of water, such as floodplain ponds, larger lakes, and wetlands. Semi-lotic environments may also occur, where flow occurs only at certain times of the year, such as during floods.

How do flow types influence the ecology? Although fish are found in all water types, species such as salmon and trout need fast-flowing rivers, whereas catfish often feed on the beds of slower, muddier rivers and lakes. Some mammals and reptiles live mainly in water but also need access to a riverbank to feed, reproduce, and make their home. Frogs and toads thrive in ponds and swamps, while beavers and otters live both in the water and on land. Insects often prefer quieter lentic waters near the animals and plants on which they depend for food, generally favoring locations with water but also some dry land. Birds such as ducks and geese like water that isn't too fast-flowing, whereas others rely on fishing from moving flows.

By filtering out fine particulates and providing time for the transformation of solutes and contaminants, the storage zones offered by lentic waters are vital for water retention during floods and for riverine ecological health. However, too much storage time may favor growth of algal blooms that lead to hypoxic (low oxygen) conditions. Achieving the right balance of lotic and lentic flows is thus vital.

Reservoirs, lakes, and millponds

Although many lentic habitats are natural, such as floodplain lakes or regions of slower velocity created by beaver dams, the number of bodies of standing water has grown due to human interventions, such as the creation of reservoirs, lakes, and millponds. Mapping lentic waterbodies across the coterminous United States shows the role of its 34,000-plus reservoirs, but also the significant influence of 1.7 million smaller ponds. The latter are more concentrated in the headwaters of rivers, where the principal inputs of water and chemicals may enter the network.

Shortened rivers

Some meandering rivers have been subject to major engineering works to decrease their length and therefore the distance needed to transport goods by boat, as well as to combat bank erosion and channel migration. This has been achieved by straightening the river course and creating artificial bend cutoffs. Although such straightening has provided important economic benefits, it may lead to the disconnection of rivers from their floodplains, altering the balance of lotic–lentic conditions. This can create lentic habitats in artificial cutoffs that become stagnant because they prevent the development of semi-lotic conditions.

Agile fishers

The Common Kingfisher (Alcedo atthis) lives along riverbanks in Eurasia and North Africa, and hunts mainly fish from flowing river waters.

Unwinding a river

Artificial meander cutoffs along the 90 m-wide (300 ft) Kaskaskia River, Illinois. This false-color satellite image reveals the higher concentrations of suspended sediment in the river compared to most of the lentic water in the artificial meander-bend cutoffs. Recent engineering has sought to reconnect some of these cutoffs to the main river and therefore re-establish semi-lotic flows.

Life on the floodplain

Floodplains are distinctive and diverse ecosystems that play a key role in reducing flood risk. The natural functioning of many of the world's floodplains has been disrupted by human activity, but river management agencies are now seeking to restore healthy floodplains to preserve nature and increase resilience to climate change.

▲ **Cooling off**
African Wild Dogs (Lycaon pictus) *cooling down on the floodplains of the Zambezi River.*

Floodplains are productive ecosystems that are delicately adapted to a unique combination of hydrological, geomorphological, and biological processes. In their natural states, these interactions create a rich variety of landforms, generating a high level of habitat heterogeneity that supports a host of flora and fauna.

In the headwater reaches of a river, floodplains can be closely confined by narrow valleys, constraining low-sinuosity channels with coarse sediments. These deposits are often shifted by the high-energy flows of the river, so floodplains here experience frequent physical disturbance; floodplain vegetation is therefore typically dominated by pioneer species such as willows. Further downstream, valleys widen and the river gradient reduces, so floodplains form around larger channels that may be divided into multiple branches. Here, floodplains have a more complex mosaic of disturbed and stable environments, so that colonizing species as well as more mature, late-successional floodplain forests can develop. Trees that die as they age, or that are downed by episodic physical disturbance, add further habitat complexity by diverting floodplain flows and sculpting intricate forms into the floodplain surface.

Adapted to inundation

Many species that live on floodplains require regular flooding and are adapted to specific flood frequencies, durations, and depths. For example, in arid parts of the United States native cottonwoods require regular flooding to ensure they have enough moisture to establish themselves. Many riverine fishes migrate seasonally between river channels and floodplains, while others are adapted specifically, and confined, to floodplain waters. This means that connectivity between the river and its floodplain is critical to sustaining healthy ecosystem functions. When floodplains are no longer inundated, major species shifts can occur, often resulting in an overall reduction in biodiversity.

With increasing human encroachment onto the world's floodplains, and as extreme weather events increase due to climate change, attention in recent years has become more focused on the urgent need to restore natural floodplain function. This is necessary not only to preserve habitats for wildlife, but also to promote natural floodwater storage, which increases resilience to flooding.

FLOODPLAIN TRANSITIONS

Reductions in the frequency of flooding on the Dolores River, Utah, due to upstream damming have driven shifts in floodplain vegetation from trees such as cottonwoods to shrubs such as sagebrush. Climate change will amplify such effects in the future, as the graph demonstrates.

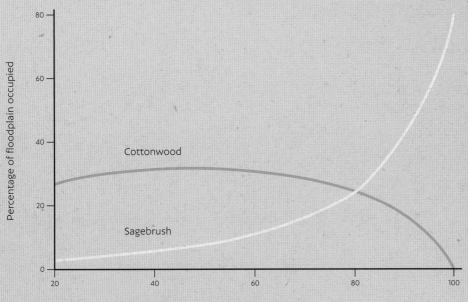

Percentage of floodplain occupied

Cottonwood

Sagebrush

Percentage of years with drought over the next 200 years

How river life evolves

Sculpting of river valleys has occurred over many millions of years, reflecting a range of tectonic, climatic, and sea-level changes. As the physical landscape evolves, life on it also has the chance to adapt and develop, thereby linking the evolution of species to the geomorphological evolution of rivers.

The growth of life in river basins must be considered over a wide range of spatial scales, from that of a sediment bar to that of the entire river basin, and over a wide range of time frames, from months to tens of millions of years. River erosion and deposition provide a continual short-term mechanism for the growth of new vegetation and the establishment of new riverine habitats.

Evolution over geological timescales

On a geological time frame, the interactions between tectonism, climate, and landscape evolution exert a primary control on the evolving ecology. For example, in the Amazon Basin evolution of the river network (see also Chapter 2) has progressively reduced the area of sedimentation derived from the older rocks of the Guiana and Brazilian Shields, and expanded the deposits derived from the Andes, affecting the availability of nutrients. Consequently, the surface geomorphology and biotic habitats have changed. This has given rise to an easterly expansion

▼ **Flooded forest**

Várzea flooded forest (see page 120), Amazon River, Brazil.

RIVER EVOLUTION AND FOREST CHANGE

Numerical modeling shows the evolution of the Amazon River network over a 24-million-year time period, revealing its influence on the type of forest cover (see page 120 for definitions of forest types).

 Várzea forest

 Igapó forest

Terra firma forest

Grasslands with periodic flooding

Mountains above 1,500 m (4,920 ft)

27 mya

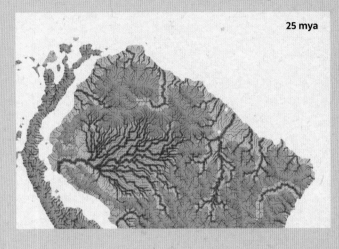

25 mya

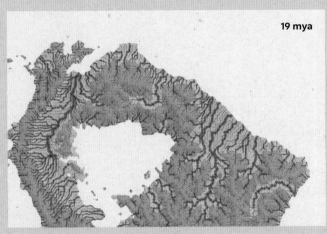

19 mya

17 mya

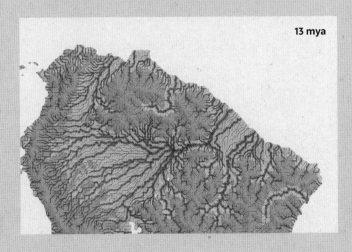

13 mya

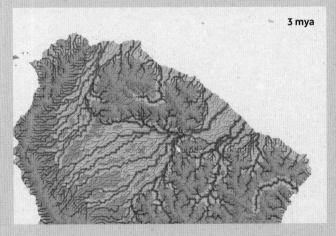

3 mya

of the terra firma forest (often comprising higher river terraces, with a tall, species-rich forest occupying well-drained soils that are very rich in nutrients) and várzea forest (forest that floods annually from inundation by sediment-laden whitewater rivers). This expansion was accompanied by a reduction in the area of igapó forest (forest flooded seasonally by sediment-poor, organic-rich blackwater rivers) and millennial-scale changes to the mosaic of habitats in the lowland river floodplain. The dynamic interaction between tectonics, climate, and surface geomorphology, together with changes in sea level, has thus provided the template for patterns of speciation in the most biodiverse biome on Earth.

Shocking evolution

The electric eels of the Amazon River are not actually true eels, but rather knifefish that have more in common with carp and catfish. Electric eels can grow up to 2.75 m (9 ft) long and weigh up to 23 kg (50 lb), and were the natural inspiration behind development of the first battery by Italian physicist Alessandro Volta (1745–1827) to provide a reliable source of electric current, as well as the recent development of hydrogel batteries for use in medical implants. These eels prefer to live in slow, sluggish waters, and often inhabit small streams and floodplain ponds. They have three electric organs containing cells (electrocytes) that create an electrical current, which in one species (*Electrophorus voltai*) can release up to an astonishing 860 V. To place this shock in context, heart defibrillators commonly apply 200–1,700 V to heart-failure patients.

WHERE ARE AMAZONIAN EELS?

Distribution of the three species of *Electrophorus* electric eel in the Amazon River Basin. Colors depict the distributions of each species as predicted from ecological niche models based on climate and geomorphological variables, with the black dots showing species that were recorded during field surveys.

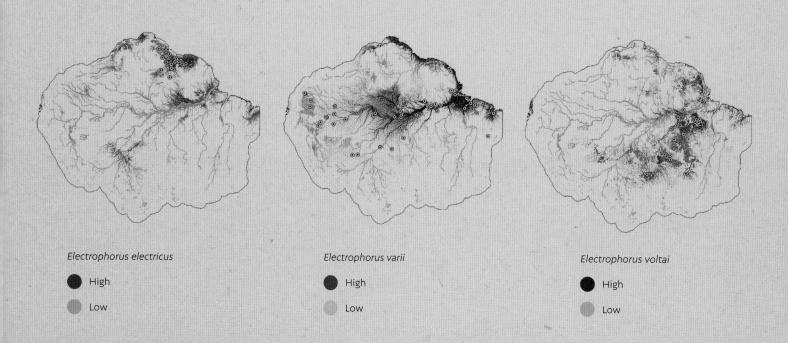

Electrophorus electricus

● High

● Low

Electrophorus varii

● High

● Low

Electrophorus voltai

● High

● Low

It was originally thought there was just one species of electric eel in the Amazon, *Electrophorus electricus*, but research in 2019 revealed two others—the high-voltage *E. voltai* and *E. varii*. Divergence of *E. voltai*, which lives in rivers on the Brazilian Shield, from *E. electricus*, which inhabits waters of the Guiana Shield, is speculated to have begun some 7 million years ago, when barriers to dispersal were created by growth of the high-conductivity river floodplain system of the modern Amazon River. This formed a barrier to the two eel populations, which are both restricted to low-conductivity waters, and allowed them to evolve separately. The subsequent geographic ranges of the three species are controlled largely by altitude, soil and water geochemistry, and mean water temperature and flow discharge, the latter both strongly linked to the river's annual flood pulse.

Our closest relative

Bonobos (*Pan paniscus*) are apes that live in one part of the Congo Basin, to the south of the main river, and together with Chimpanzees (*P. troglodytes*) are our closest living relative—we share 98.7 percent of our genome with them. But Bonobos and Chimpanzees differ from each other in both their physique and their social structure. Bonobos display little aggression and do not hunt cooperatively, use tools, or exhibit lethal aggression, and their societies are dominated by females. In contrast, Chimpanzees—which thrive north of the Congo River—are male dominant, display intense aggression that can be lethal between different groups, use tools and cooperatively hunt monkeys, even eating the young of other Chimpanzee groups. How did these two closely related species come to be so different?

▼ **Shocking eel**
The Amazonian electric eel (Electrophorus electricus) *stuns its prey by generating electricity. These eels are nocturnal and have poor vision, and use electroreceptors to help locate their prey.*

▼ A family outing

Bonobo apes (Pan paniscus) have a matriarchal social structure, with the group hierarchy dominated by a coalition of more senior and experienced females and males. This hierarchy is typically led by an older, experienced matriarch, who takes on the role of decision-maker.

Recent research has shown that Bonobos and Chimpanzees began to form different groups around 1–1.7 million years ago. It is thought that a much-reduced discharge in the Congo River at this time allowed ancestors of the Bonobo to cross the main channel and begin to develop in a forest refugia on the south side of the river. Changes in hydrology raised the river flow again and subsequently isolated the populations, which began to evolve separately, forming two distinct species— Bonobos and Chimpanzees. Over time, their differing environmental situations, and perhaps also their differing competition with other apes, gave rise to the discrete social character of the two species, and also their contrasting physical forms, with Bonobos being slightly shorter and more slender than Chimpanzees. Separation of Chimpanzee species to the north by the physical barrier of the Ubangi River also led to evolution of the eastern and central Chimpanzee subspecies. Divergent evolution of the apes has thus taken place due to physical separation caused by these mighty rivers, and also due to climate change that allowed ancestors of the Bonobo to cross the Congo River to its southern side.

BONOBO EVOLUTION AND RIVER CHANGE

Maps illustrating the range and evolution of the Bonobo (*Pan paniscus*) and Chimpanzee (*P. troglodytes*) in the Congo River Basin. Changing river flows resulted in evolution of these two different species separated by the Congo River, as well as evolution of the eastern and central Chimpanzee subspecies on different sides of the Ubangi River.

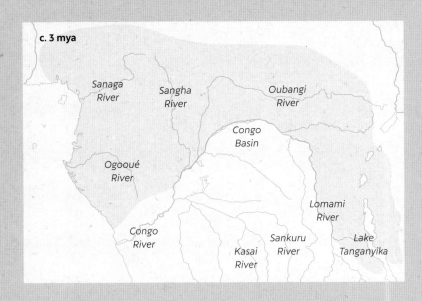

c. 3 mya

Sanaga River
Sangha River
Oubangi River
Congo Basin
Ogooué River
Lomami River
Congo River
Sankuru River
Kasai River
Lake Tanganyika

○ Range of proto-*Pan*

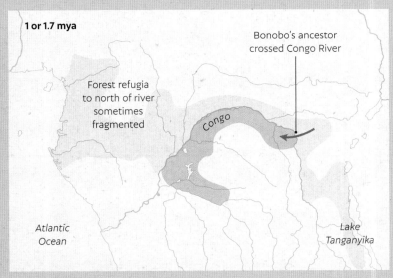

1 or 1.7 mya

Bonobo's ancestor crossed Congo River

Forest refugia to north of river sometimes fragmented

Congo

Atlantic Ocean

Lake Tanganyika

○ Range of proto-*Pan*
Forest refugia in the southern Congo Basin

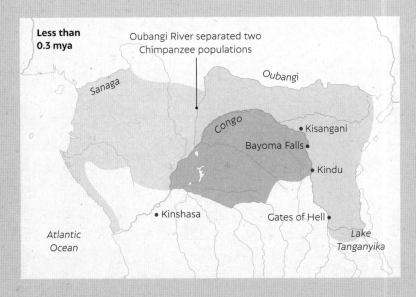

Less than 0.3 mya

Oubangi River separated two Chimpanzee populations

Sanaga
Oubangi
Congo
• Kisangani
Bayoma Falls •
• Kindu
• Kinshasa
Gates of Hell •
Atlantic Ocean
Lake Tanganyika

Pan ranges

Central African Chimpanzee
Eastern African Chimpanzee
Bonobo

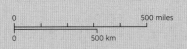

The value of rivers

Rivers not only provide humans with the most basic requirements for life, but they also support a wide range of other benefits, collectively known as ecosystem services, which underpin all aspects of human wellbeing. The economic value of these benefits is vast, but the ecosystem functions that support them are breaking down under growing environmental stresses.

▼ **Precious cargo**

A water lily harvest on the Mekong River, Vietnam. In addition to its ornamental use, the harvested lilies are used in the preparation of many foods, such as sour water lily soup.

Rivers have sustained humans for as long as we have existed, providing a wide range of benefits, including water supply, transportation, food, and access to recreational opportunities. The benefits that humans gain from river and floodplain ecosystems are collectively known as ecosystem services, and the richness of these services is astonishing. In 2014, it was estimated that the total annual economic value of the services provided by all the world's rivers, lakes, and floodplains was around US$4 trillion (2011 values). This represents 5 percent of the value of all the ecosystem services delivered across the terrestrial portion of the planet, even though the land area covered by rivers and floodplains is less than 2 percent.

The world's largest freshwater fishery

The Mekong River is the world's largest freshwater fishery, with a total annual fish catch estimated at 2.3 million tonnes (2.5 million tons, or more than the total freshwater fish catch landed in Africa) and an annual economic value of US$11 billion. The fishery is vital to millions of people for their food security and household income: some 3 million people in Cambodia alone rely on fish from the Mekong for their main source of protein. The number of people living within the Mekong River basin is expected to increase substantially in the coming decades, and dependence on the fishery will consequently also increase. However, there are already signs that climate change, fragmentation of the river network by damming (many of the Mekong's 1,148 fish species are migratory), pollution, and overfishing are having a negative impact. Specifically, while the total fish catch has increased in the last 15 years, the size and range of fish being caught are declining.

Fishers at work
Fishers on the Mekong River at Sangkhom, Thailand.

Types of ecosystem services

The Millennium Ecosystem Assessment, a major study commissioned by the United Nations in 2000, identified four broad categories of ecosystem services: provisioning, regulating, supporting, and cultural services. While the dollar figure given above conveys the vast economic value of the world's rivers and floodplains, financial estimates such as this do not convey the fundamental importance of our rivers—without them, human life would simply not be possible. Our physical and mental wellbeing depend on the critical functions provided by rivers and floodplains.

A provisioning ecosystem service is any type of benefit to people that can be directly extracted from nature. Examples from rivers and floodplains include clean water for drinking, water for washing and cleaning, water for use in industry and agriculture, and the food extracted from fisheries and grown on the world's floodplains.

Regulating services are the benefits provided by ecosystem processes that moderate natural phenomena. For rivers, examples include processes such as the natural filtration of water through aquifers or vegetation, which removes harmful contaminants and regulates water quality; the buffering of excessive erosion through the presence of riverbank vegetation; the deposition of nutrient-rich sediments on floodplains, which can sustain productive cultivation; and the storage of floodwaters on floodplains, which naturally attenuates the risk of flooding in downstream reaches. There is growing recognition that the restoration of these natural regulating services—in many places lost after decades of intensive engineering—can be an important tool in managing rivers in a more sustainable way.

▲ **Making room
for the river**

*After centuries of
human engineering,
parts of the River
Rhine floodplain in
the Netherlands have
been renaturalized,
reconnecting them to
the river and allowing
storage of floodwaters,
improving habitats,
and reducing flood
risk downstream.*

Supporting services are defined as the ecosystem services that are necessary
for the production of all other ecosystem services. Within the context of rivers and
floodplains, the basic water cycle is a key example of such a supporting service.

Cultural ecosystem services are the benefits that humans gain from their
interactions with environmental spaces. River landscapes are home to major
cultural festivals, religious sanctuaries, and rituals, and offer opportunities for
recreation through activities such as kayaking, hiking, and tourism to scenic
locations. Such cultural ecosystem services have a priceless value, underpinning
human wellbeing on many levels.

Maintaining ecosystem services

Maintaining the high value of riverine ecosystem services is essential. However,
many provisioning services are now being used by humans in a highly extractive
manner, introducing trade-offs between the different types of ecosystem services
on which we all depend in a way that can make it challenging to preserve and
sustain them. For example, overabstracting water from rivers to irrigate crops
(a provisioning service) may have serious negative impacts on the quantity of
flows in reaches downstream, reducing provisioning and regulating services there.
The overharvesting of wild populations of freshwater fish can likewise deplete
their numbers and reduce the resilience of natural populations to external shocks.
It is essential that we find more sustainable ways to manage the multifaceted
resources provided by our rivers and floodplains if we are to continue to
enjoy their material and immaterial benefits.

▲ Cultural services

The Kumbh Mela is the world's largest human gathering. It is held every four years over a 12-year cycle and attracts up to 50 million Hindus, who make the pilgrimage to bathe in the holy waters at one of four sites on the banks of the Ganges, Yamuna, Sarasvati, Godavari, and Shipra Rivers.

▶ Drought on the Murray

Intensive water abstraction to irrigate crops upstream has led to record low water levels on the downstream portions of the Murray River in southeastern Australia.

5

Rivers and Us

Population growth and urbanization

The Industrial Revolution stimulated a profound restructuring of human societies. In 1800, most people were rural dwellers and just 3 percent of the global population lived in cities. Today, more than half (55 percent) of the world's 8 billion people live in urban areas, and that will likely increase to 68 percent (6.66 billion) by 2050. Cities have always been located close to waterways, but this massive urban growth has fundamental implications for how we live alongside them.

Flooding and pollution

The rapid global growth of urban areas and populations is affecting the risk of flooding. There are two main drivers of these changes. First, the construction of roads and buildings means that urbanization substantially increases the area of impervious land surfaces. Rainfall therefore has less opportunity to soak into the ground, greatly raising the risk of pluvial flooding, in which floodwater accumulates from rainfall that cannot drain rather than from an overflowing body of water. Second, new satellite data has shown that urban growth is disproportionately focused on existing floodplains—precisely the areas that are at the greatest risk of flooding. Indeed, the total urban area located on floodplains almost doubled in the period 1985-2015. Where we choose to live is now one of the dominant drivers of increased global flood risk.

▼ **Urban growth drives flood risk**

Bamako, Mali's rapidly growing capital, sprawls over the Niger River floodplain.

RISKY CITIES

Bamako, Mali's capital, has grown in the Niger River floodplain (photo opposite). The maps below show the proportion of new urban encroachment on the world's floodplains between 1985 and 2015 as a percentage of total floodplain area (below), and a 2015 snapshot of the world's urban population exposed to river flooding, separated by income bracket (bottom).

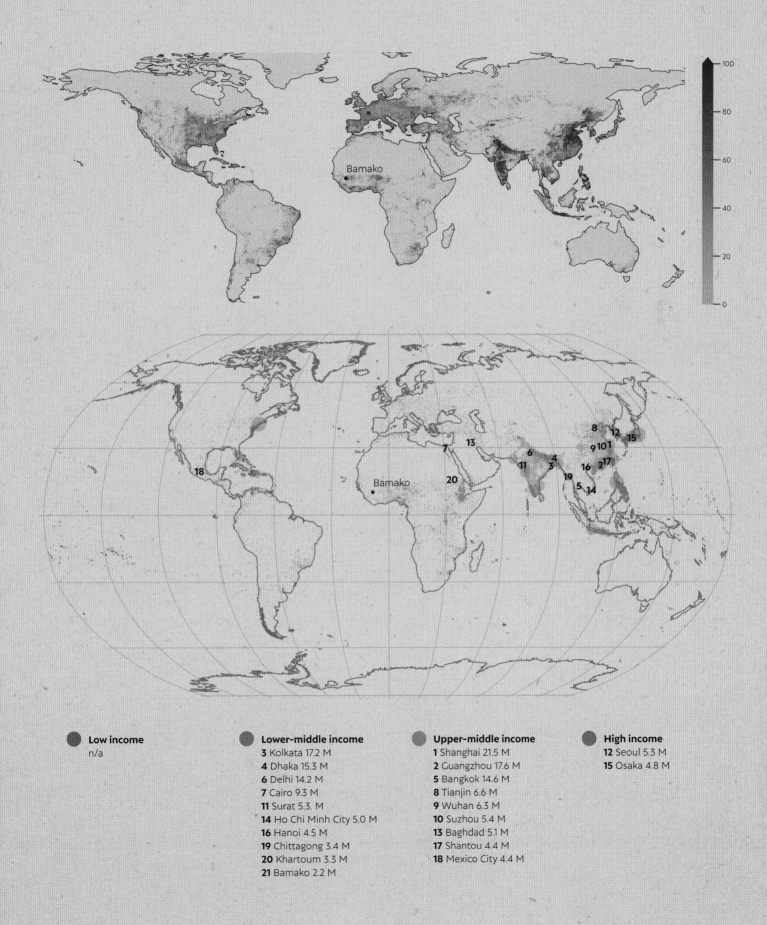

● **Low income**
n/a

● **Lower-middle income**
3 Kolkata 17.2 M
4 Dhaka 15.3 M
6 Delhi 14.2 M
7 Cairo 9.3 M
11 Surat 5.3. M
14 Ho Chi Minh City 5.0 M
16 Hanoi 4.5 M
19 Chittagong 3.4 M
20 Khartoum 3.3 M
21 Bamako 2.2 M

● **Upper-middle income**
1 Shanghai 21.5 M
2 Guangzhou 17.6 M
5 Bangkok 14.6 M
8 Tianjin 6.6 M
9 Wuhan 6.3 M
10 Suzhou 5.4 M
13 Baghdad 5.1 M
17 Shantou 4.4 M
18 Mexico City 4.4 M

● **High income**
12 Seoul 5.3 M
15 Osaka 4.8 M

Urban rivers are also highly prone to pollution. The heavy industries located in urban areas often use rivers as a convenient means to discharge waste products, while the dense human populations living here produce high volumes of excreta, which if left untreated provide a major source of contamination. A recent scourge has been the large volumes of plastic waste that we generate, which can all too easily make their way into our watercourses. The River Thames in England has some of the highest recorded levels of microplastics of any river in the world: some 94,000 microplastic particles per second are present in its water as it flows past Greenwich, London.

Urban river restoration

In the past, urban rivers have been seen as a threat to infrastructure and human wellbeing, so have often been managed by confining them within flood walls. Urban river restoration seeks to achieve a balance between the need to protect people and assets with a desire to unlock the value of the ecosystem services (see pages 125-27) that rivers bring. Restoration must provide improved access to urban open space and better habitat provision, and re-establish vegetated floodplains and floodplain ponds, which can help increase water storage capacity and thereby help protect against flooding.

◤ **Urban idyll**

The River Isar in Munich, Germany, during (left) and after (right) restoration. The restored river provides amenity value, improved habitat, and space for floodwater storage.

Climate change: Floods and droughts

Anthropogenic climate change, driven by emissions of carbon dioxide and other greenhouse gases, is intensifying Earth's water cycle and increasing the frequency and severity of floods and droughts. These changes directly affect hundreds of millions of people globally. Increasing flood flows also release and transport more sediment, leading to a changing climate that is reshaping the world's river landscapes.

Climate change and the water cycle

The emission of greenhouse gases into the atmosphere is driving worrying rises in temperature. In 2022, the World Meteorological Organization estimated that global mean temperature was already 1.15 °C (2.07 °F) above pre-industrial (1850-1900) levels. It may not sound much, but this warming is already having profound impacts on Earth's water cycle, because more heat increases evaporation over the oceans, adding to the amount of moisture in the atmosphere. In turn, this greater atmospheric moisture results in more frequent and more intense rainfall, and over land enhanced evaporation caused by the higher temperatures dries out the surface. As a result, areas with increased precipitation will also see an increase in the intensity and frequency of flooding, but areas that are far away from storm tracks will experience a greater risk of drought.

Floods and droughts

The twin hydrological extremes—floods and droughts—bring misery and hardship to millions. Flooding is the world's most serious natural hazard, impacting more than 300 million people each year, and causing loss of human life, damage to property, destruction of crops, and loss of livestock. In addition, floods bring substantial ongoing impacts to health and wellbeing due to the emotional trauma they can cause and the increased risk of illness from waterborne diseases. Drought also has major economic, societal, and environmental costs, and affects around 55 million people each year through its impacts on agriculture, water quality, and infrastructure. Mass migration can be driven by drought: by 2030, as many as 700 million people may be at risk of displacement as a result of drought.

▶ **Hydrological extremes**
Floods (top): Hurricane Harvey set rainfall records when it struck Texas and Louisiana in August 2017. The ensuing flooding killed more than 100 people and caused US$125 billion worth of damage.

Droughts (bottom): In summer 2022, the River Loire in France flowed at less than 5 percent of its average annual levels, threatening the cooling water supply to four nuclear reactors built along its banks.

One billion people already live in areas where the annual probability of flooding exceeds the one-in-100-year return period (i.e., 1 percent probability) event, the standard typically adopted as the baseline level for flood protection by water agencies around the globe. However, model predictions show that in many regions—especially in South and Southeast Asia, northeast Eurasia, eastern and low-latitude Africa, and much of South America—warming temperatures will increase the frequency and intensity of flooding. Elsewhere—notably in northern and eastern Europe, Central Asia, and southern South America—some regions will have reduced flood frequency. Overall, however, even 1.5 °C (2.7 °F) of warming by 2100 will double the number of people exposed to the one-in-100-year flood event.

DRYING UP

Model projections (below) of the impact of a 2 °C (3.6 °F) increase in global mean temperature on the return period of the present-day 50-year (2 percent probability of occurrence) drought at the end of the twenty-first century. Red and blue shades highlight areas of increasing and decreasing drought frequency, respectively. More frequent droughts make it more challenging to use river flows for basic human needs such as washing clothes.

Return period (years)

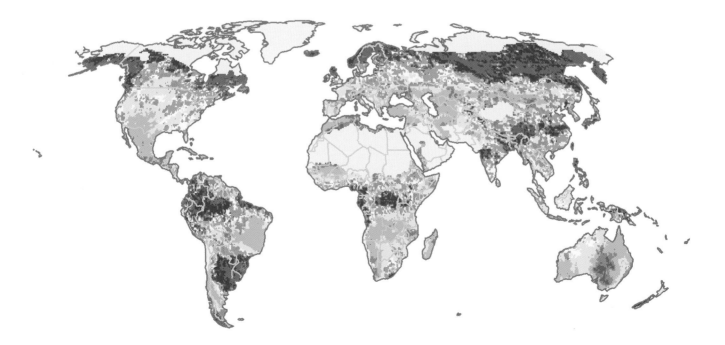

INCREASING FLOOD FREQUENCY

Model projections of the impact of a 1.5 °C (2.7 °F) increase in global mean temperature of future (to 2100) warming on the return period of the present-day 100-year flood. Blue shades show areas where this level of flooding becomes more frequent, while yellow to red shades highlight areas where flooding becomes rarer.

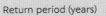

Return period (years)

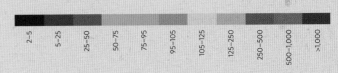

| 2–5 | 5–25 | 25–50 | 50–75 | 75–95 | 95–105 | 105–125 | 125–250 | 250–500 | 500–1,000 | >1,000 |

Gold Coast flooding
Australia's Gold Coast experienced record flooding in 2022. Here, the swollen Tweed River has overflown its banks, inundating sugarcane fields, pastures, and urban communities with sediment-laden water.

Models also show changes in the frequency and intensity of droughts: overall, two-thirds of the global population will experience increased drought conditions by 2100. The areas most affected tend to be those that are already dry, including much of Africa, Australia, the Mediterranean, southern and central United States, and parts of South America. However, around a fifth of the Earth's surface, mostly in northern latitudes and parts of eastern and southeastern Asia, will experience less drought in the future.

Muddier rivers

As the climate warms and precipitation increases in intensity, soil erosion and landslides from hillslopes become more likely, releasing more sediment into river networks. Coupled with higher flood discharges, climate change is projected to increase the rate of sediment transport, especially in high-latitude regions, the Balkans, and across much of Asia. Changes in the volumes of sediment being transported through Earth's river networks have important knock-on implications for river morphology, but they also lead to changes in water quality and threaten key infrastructure, especially by infilling water storage reservoirs.

ALTERED SEDIMENT LOADS

Percentage difference in suspended sediment transport as estimated for the last decade of the twenty-first century (2090–2099) compared with the recent past (1950–2005), for a model scenario where future mean global temperature rise is limited to 2 °C (3.6 °F) above pre-industrial levels.

Suspended sediment transport change (%)

Changes in sediment yield are especially pronounced in the Tibetan Plateau and surrounding High Mountain Asia, which feeds many of Asia's largest rivers. This region is known as the third pole because it is Earth's third-largest ice reservoir. Here, our warming climate has already driven rapid melting of the region's glaciers and permafrost, releasing higher sediment fluxes that are changing water quality and infilling hydropower reservoirs. Model projections indicate that sediment flux from the rivers draining High Mountain Asia could more than double by 2050 if we are unable to limit global warming to within the "safe" 1.5 °C (2.7 °F) target set by 196 nations in the 2015 Paris Agreement on climate change.

THE THIRD POLE IS MELTING

The river basins draining High Mountain Asia have seen substantial increases in their water and sediment loads in the last 60 years.

River discharge
Sediment flux
Glacier
Permafrost

Increase rate:
0 (% change per year)
2.6 (% change per year)

Damming

Dams provide a multitude of benefits for humans, including hydroelectric power, flood control, water supply, irrigation, fish farming, and recreation. However, dams can also change the landscape, fragment the long profiles of rivers, and alter riverine ecology. How we construct, operate, and decommission dams when they are defunct raises massive issues for the future.

A world of dams

Dams range in size from small check dams in tributary valleys to vast megadams on some of the world's largest rivers. Dams often pose major construction challenges and represent some of the most spectacular engineering accomplishments. They can also create significant political cooperation or tensions where they affect the cross-boundary transfer of water. Humans have built dams for millennia, with the earliest known dam—the Jawa Dam in Jordan—being constructed some 5,000 years ago. Perhaps more than 800,000 dams of various sizes have been built across the globe since that time. Extensive dam construction in the United States and Europe began in the late nineteenth century, with construction accelerating there in the 1950s, and in the mid- to late twentieth century in China. Worldwide, there are some 57,000 large dams over 15 m (50 ft) in height, with China and the United States having built more than 23,000 and 9,200 of these, respectively. More than 300 megadams exceed 150 m (500 ft) in height.

Dam change
Environmental change on the Xingu River, Brazil (**1**), due to construction of the Belo Monte (**2**) and Pimental (**3**) hydropower dams. A new canal (**4**) links the two new reservoirs.

May 26, 2000

July 20, 2017

N

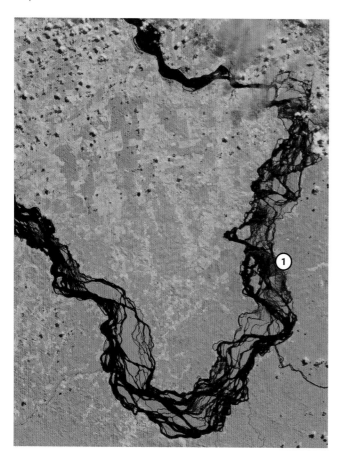

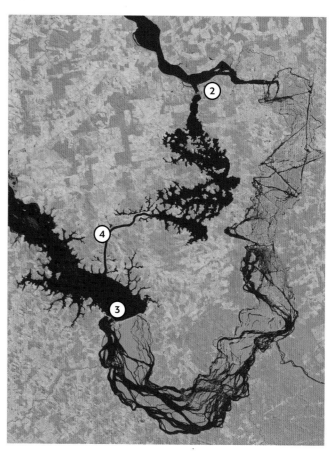

DAM LOCATIONS

The locations of 35,000 of the world's large and medium-sized dams in 2023 (upper map) and their catchment areas (lower map). Where identified, the principal uses of these dams are hydroelectricity generation (25%) and irrigation (20%), with additional main purposes including water supply, flood control, recreation, water for livestock, and navigation.

Locations of dams

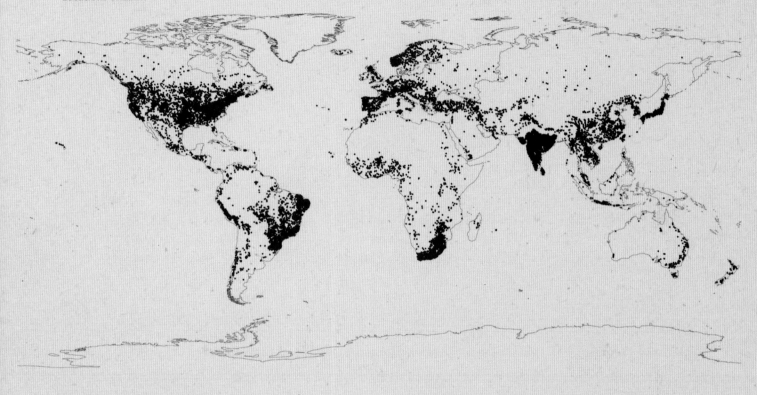

Global catchment areas of dams

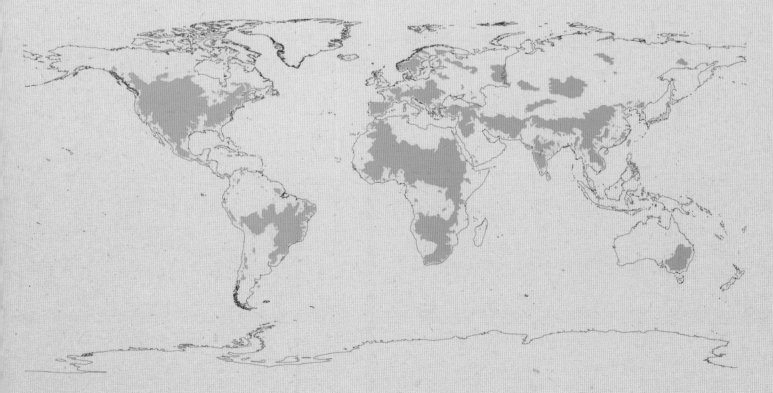

Dam the Blue Nile
Construction of the Grand Ethiopian Renaissance Dam at Bameza, Ethiopia, is creating tensions between Ethiopia, Sudan, and Egypt, potentially altering the balance of power in the region.

A muddy flush
People watch sediment flushing from the Xiaolangdi megadam, Huang He (Yellow) River, China.

It is thought that some 40–80 million people worldwide have been displaced due to dam construction, with some estimates suggesting that between 26 million and 58 million people were uprooted in India and China alone between 1950 and 1980. Hydroelectric dams have become an increasingly important source of green energy in many countries, with China, the United States, Brazil, Canada, and Russia being the largest hydropower producers. However, dams do cause substantial environmental change, and some estimates suggest that rotting vegetation following the flooding of tropical forests by damming may generate substantial greenhouse gases. The economic costs of hydropower energy from megadams have also been questioned and are thought to be commonly far worse than original estimates, with cost overruns averaging 96 percent and some exceeding 1,000 percent.

Decommissioning dams

Two major issues with dams concern their decommissioning to restore more natural flow regimes to rivers, and the trapping of sediment in the reservoir behind a dam, leading to the end of its useful life. Dam decommissioning is proceeding apace in North America (by 2022, 1,951 dams had been decommissioned there) and northwest Europe, but we have yet to grapple with the issue of what to do with very large dams when they are full. More recent dams and retrofitted dams do allow some "flushing" of sediment, both to slow reservoir sediment infill and allow the passage of sediment downstream. For most dams, however, the issue of sedimentation is one that still demands solutions.

Land-use change and agriculture

River basins have been central to the development of human civilization, agricultural practices, and the growth of urban areas. Changes in land use brought about by human occupation and population growth have impacted rivers over millennia, altering the nature of the river itself and leaving an indelible mark of humans on river valleys.

The flow of water and sediment into the rivers of the world, which drives channel change and sustains riverine ecologies, is influenced significantly by the way in which humans use and manage river drainage basins. Vegetation type and land use exert first-order controls on the flows of water and sediment, and as humans have changed land use over millennia, so these river valleys have responded. For example, deforestation can change the quantity of water and sediment fed to a river, and different agricultural practices can alter how much sediment can be eroded from the landscape. A spectacular example is the Huang He (Yellow) River, China, which has seen enormous changes in its sediment yield over the last 4,000 years. Vast increases in sediment flux came through removal of forest from the Loess Plateau across which it flows and development of agriculture, encouraging massive erosion of fine-grained loess sediments. Recent remediation schemes are dramatically returning the river's sediment load to a state similar to that of 1,200 years ago.

HUMAN-INDUCED SEDIMENT FLUX

Changes to sediment flux in the Huang He (Yellow) River over the past c. 3,000 years, as a result of human influences in the river basin catchment. Each point is the annual average for the labeled historical interval.

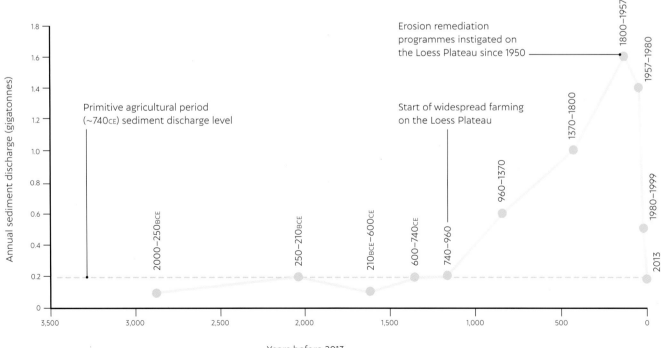

Virtual water

One aspect of water transfer from river basins that has become increasingly important for us to appreciate and quantify is the flow of virtual water—the water used and locked up in the production of crops and manufactured goods. Each of us has a water footprint—the quantity of water used to produce the goods and services we use. This may thus entail "export" of virtual water from regions that have water scarcity to those that import these goods but may be water sufficient. Such virtual water export can be between regions within a country or between countries across the globe. For example, it has been calculated that the United Kingdom uses around 185 million m^3 (3 billion ft^3) of African water each year through importing green beans from 14 African countries, which is equivalent to about 9 percent of the total annual flow of the River Thames through London. Our use of virtual water, and how this influences the rivers in the regions where the goods are produced, must be considered when planning water use and environmental change in response to water abstraction and the sustainable management of water globally.

GLOBAL FLOWS OF VIRTUAL WATER

Global virtual water flows between different regions in 2016. Flows (**bold** figures) are in 10^9 m^3 and go from left to right, showing whether regions are net exporters (*italic* figures) or importers (Roman figures) of water, with the width of the bands denoting volume of water exchange. Percentages on the left show total exported virtual water as a percentage of water used in total production, and percentages on the right display total imported virtual water as a percentage of water embodied in total consumption.

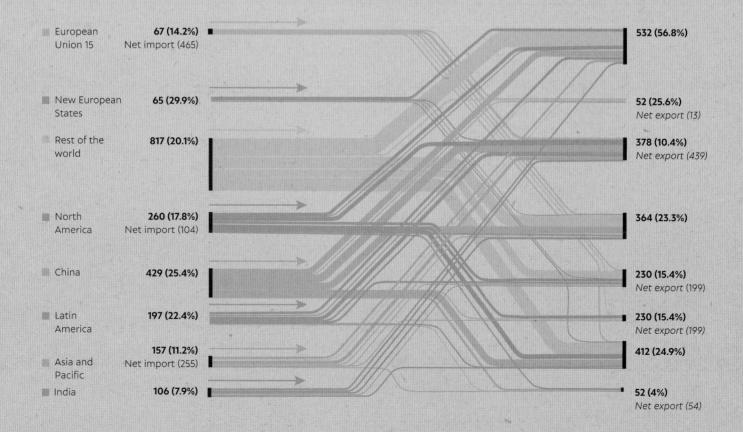

Human use of water and river engineering

Human demand has often necessitated large engineering infrastructure to control the movement of water across the landscape, prevent flooding, and provide a secure and reliable source of water for growing populations. These schemes can remake the physical geography of the landscape, substantially modifying river patterns and fluvial processes.

As population growth and urbanization proceed apace, demands for water grow—for agricultural, domestic, recreational, and industrial uses. Some regions receive less rainfall and possess fewer groundwater resources to support these growing demands, prompting diversion schemes that reroute water between areas. Such schemes are present across the globe and have supported the establishment and growth of cities such as Los Angeles in California. Los Angeles' development in the late nineteenth century led to the construction of aqueducts, which created substantial conflicts concerning water rights between the city and farmers in the source regions—the so-called Californian Water Wars. Although they provided water essential for the growth of Los Angeles, these diversions caused massive changes to the ecology of the source areas and prompted legal litigation and restoration schemes. Water diversions, therefore, do not come without both ecological and financial costs.

The world's biggest water transfer schemes

Two massively ambitious water diversion schemes are currently under way. China's South-to-North Water Transfer Project is the largest water transfer ever undertaken, with a planned completion date around 2050. Three routes, comprising 2,900 km (1,800 miles) of canals and tunnels and costing an estimated US$62 billion, will transfer water from the wetter regions of the Chang Jiang (Yangtze) River to the drier northern region and Huang He (Yellow) River valley, home to the growing region of Beijing, which has severe problems of groundwater depletion.

▼ **Diverted water**

Water diversion canal as part of China's South-to-North Water Transfer Project.

The Indian National River Linking Project aims to transfer water from regions of surplus to those of deficit through the construction of 30 canal links of total length of over 12,500 km (7,800 miles), and 300 water storages that will irrigate up to 350,000 km^2 (135,000 square miles) of land and generate 34 GW of electricity. The expected cost of this scheme is around US$120 billion with completion before 2050.

Such large-scale water transfers raise issues concerning the spread of pollution within the channel network, alterations to the flow regime, the introduction and spread of non-native species, losses in fish biodiversity, impediments to fish migration by dams and salinization, displacement of human populations, and starvation of sediment supply to downstream deltas.

Levees and embankments

River engineering schemes have included the construction of levees, embankments, dikes, and in-channel structures to prevent floodplain inundation, control the movement of water and sediment, and maintain channel depths suitable for shipping. Since a 965 km (600-mile) system of levees was constructed along the west bank of the River Nile more than 3,000 years ago, such schemes have become commonplace on many rivers worldwide. In regions where a changing climate is bringing enhanced storminess and precipitation, the question of flood magnitude and frequency is now leading to a reconsideration of the size such engineering structures need to be to achieve their original goals. However, flood embankments can disconnect floodplains from their rivers, compromising their natural ecosystem services. In some places, levees are even now being removed to restore natural floodplain function and increase resilience to climate change.

RECONNECTING RIVERS

In many areas of the world, levee removals are restoring floodplain functions, including water storage, to help protect downstream areas from flooding.

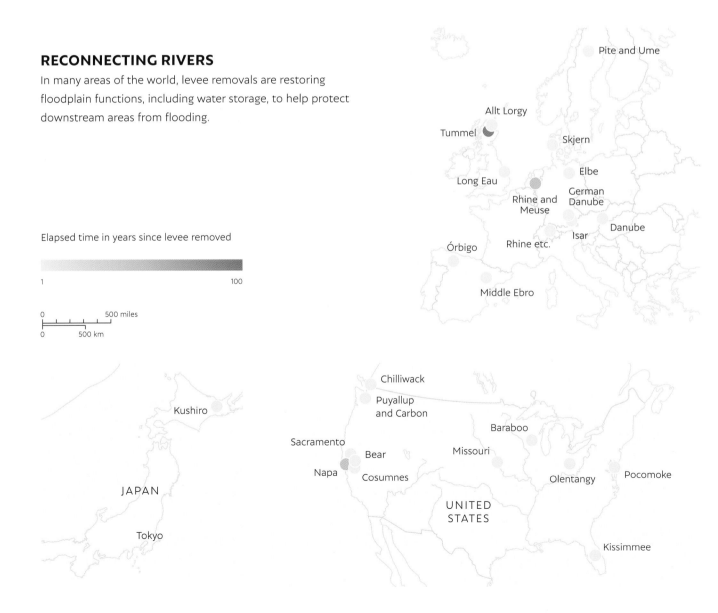

Elapsed time in years since levee removed

1   100

0   500 miles

0   500 km

Pollution

Rivers have long been used as conduits by which waste products can be removed from human settlements. As populations, urbanization, and the extent of agriculture have grown, so pollution has become a major issue within the world's rivers. In some rivers, rehabilitation efforts have shown spectacular successes, demonstrating that river clean-ups can rejuvenate aquatic ecology.

The types of pollutants that are cast into rivers are many and varied, including human and animal waste, by-products of industrial manufacturing, phosphates and nitrates washed off agricultural fields (which cause damaging low-oxygen areas in places such as the Gulf of Mexico), and heavy metals and processing waste yielded from mineral mining. Recent research has also shown that enormous quantities of plastics are being fed by rivers to the oceans, and that pharmaceuticals derived from human and animal consumption are entering rivers via sewage and slurry and giving rise to increasing bacterial resistance to antibiotics. Legacy pollutants, which are those pollutants arising from human activities and stored in river sediments, will pose a large environmental and financial risk long into the future.

▼ **Pollution prayers**
People praying amid toxic foam on the surface of the Yamuna River at New Delhi, India.

The river that caught fire

The Cuyahoga River in Cleveland, Ohio, became infamous when its water surface caught fire on June 22, 1969, even though extensive pollution had been present for decades and the river had been ablaze a total of 13 times dating back to 1868. The river had become known as the most polluted in the United States, a result of its heavy industrial use and the unregulated dumping of oils, toxins, and waste. In 1970, *Time* magazine reported that the Cuyahoga River was "chocolate-brown, oily, bubbling with subsurface gases, it oozes rather than flows" and "the lower Cuyahoga has no visible life, not even low forms such as leeches and sludge worms that usually thrive on wastes."

In the late 1960s, the Cuyahoga River, together with other waterways extensively affected by pollution, became touchstones to galvanize a growing environmental movement. Following mass public participation in the first Earth Day, on April 22, 1970, pollution such as that in the Cuyahoga was influential in establishing the United States Environmental Protection Agency that year, followed by the Clean Water Act in 1972. These have had an enormous impact on the quality of waterways in the United States. Indeed, in 2019 the Cuyahoga River once again had fish that were safe to eat—an example of what successful governance and legislation can achieve.

The fires on the Cuyahoga River have also found their way into American popular culture via artwork and music. The rock group R.E.M. wrote the song "Cuyahoga" in 1986 to highlight environmental degradation and the treatment of Indigenous Americans:

"Let's put our heads together and start a new country up
Up underneath the river bed we'll burn the river down

This is where they walked, swam
Hunted, danced and sang
Take a picture here
Take a souvenir

Cuyahoga
Cuyahoga, gone"

Burning river
The Cuyahoga River ablaze on November 1, 1952 (top) and Morgan Adler's interpretation of the fire in her 2020 painting *The Cuyahoga River is on Fire.*

Sediment mining

Rivers can provide sources of important minerals and metals for society, and have been exploited by humans for millennia for gold, diamonds, gravels, sand, and other materials. Sand provides silica, used for buildings, roads, glass, and electronics, and is hence the foundation of modern society. As demand for these resources becomes increasingly acute, so pressures on our river valleys are increasing.

▼ **Ravaged by mining**

Deforestation, gold mining, and environmental degradation along the Madre de Dios River, Peru.

Valuable minerals

Minerals yielded by erosion of the landscape can become concentrated in river deposits, especially harder grains that are more resistant to erosion, such as diamonds. Rivers may thus provide valuable sources of important minerals, termed placer deposits, including gold, diamonds, sapphires, uranium, and rare earth minerals. These minerals can be mined from both modern river channels and ancient riverine deposits, and are the source of economically vital resources. They also provide important types of employment, ranging from small-scale artisanal mining to industrial-scale extraction. However, such mining poses significant environmental concerns where the impacts of extraction are detrimental or unsustainable to the riverine ecology. This can arise from unregulated mining, lack of remediation, and pollution caused by the processing of minerals. The largest source of mercury contamination worldwide is that caused by the processing of gold.

River sand

Sands and gravels are a resource on which the economies of all countries rely, and rivers often provide a source of excellent sand for construction. River sand has angular grains, making strong concrete, and unlike marine sands it rarely contains salts, which are detrimental to concrete strength. Rivers also provide sources of aggregates that are often close to the place of use, as is the case when wetlands are infilled to create new land for buildings and urban expansion. River sands and gravels have thus become a resource of global importance in the last twenty years, but in many rivers they are now being extracted at an unsustainable rate. As an example, estimates from the lower Mekong River reveal that annual extraction is removing nine times the amount of sand that is being transported naturally by the river. River sand mining here and in many other countries—including India, China, Cambodia, the United States, and Myanmar—has been associated with ecological degradation, channel lowering, riverbank erosion, and water pollution. In addition, the mining of sand has consequences for the people living in these regions, including population displacement and migration, and crime associated with illegal mining.

MINING AND ENVIRONMENTAL STRESSES

Mining activities generate stresses that impact salmon habitats in North America's Pacific Northwest. Stressors generated by mining and associated infrastructure include changes to the hydrology, temperature, and habitats of rivers, as well as pollutants yielded to the environment. These stressors cause changes to watershed processes, habitats, and fish health.

ACTIVITIES

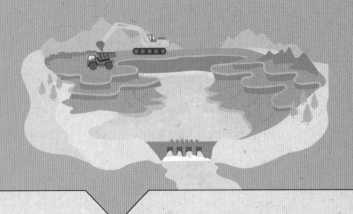

Mining
- Exploration
- Construction
- Extraction
- Processing
- Transportation
- Smelting and refinement
- Closure

Associated infrastructure
- Power plant
- Housing facilities
- Transportation corridor
- Water-control structures

STRESSORS

Altered hydrology and temperature
- Water diversion and discharge
- Ground water interception and pumping
- Altered water temperatures
- Altered natural flow regimes
- Surface water–groundwater disconnection

Habitat modification and loss
- Waste rock piles
- Tailings storage facilities
- Open pits and underground tunnels
- Filling of valleys
- Clogging by fine sediment
- Roads and stream crossings
- Removal of soils and natural habitat
- Heap leach piles

Pollutants
- Heavy metals
- Acid-generating rocks and tailings
- Chemical nutrients
- Fuels and chemical spills
- Dust
- Mine camp sewage
- Turbidity
- Noise

RESPONSES

Watershed processes

Habitat quality and quantity

Fish health and survival

Fragmentation

Dams, reservoirs, and levees, along with changes to water temperature, chemistry, and quality created by pollution, may all create barriers that impede the flow of water and sediment and the movement of species. The long profile of a river can thus become fragmented, such that the functioning of its natural ecosystem changes. In particular, species that rely on long-range transit through rivers may be severely restricted. Likewise, rivers with natural fragmentation features such as waterfalls may undergo ecosystem change when these barriers are removed, as with the creation of extensive reservoirs.

River fragmentation can influence the passage of migratory species, thereby restricting their range and abundance, and can change the distribution of natural habitats within the river basin, thereby modifying their ecological functioning. This may affect anadromous migratory fish (adults that live in the sea and migrate upriver to spawn in freshwater) such as sea trout and salmon, and catadromous species (which migrate to the sea to spawn but grow in rivers) such as some eels. Fragmentation may involve both creating barriers to the upstream and downstream migration of species, as well as separating the channel from its floodplain, which can also lead to large changes in ecosystem behavior. For example, many species of fish rely upon migration to floodplain ponds, which act as spawning grounds and provide a place for young fish to grow. Restricting this movement by constructing levees may therefore impede this vital part of their life cycle.

How many rivers are fragmented?

A 2019 study found that globally only 37 percent of rivers longer than 1,000 km (620 miles) remain free-flowing over their entire length, with very long free-flowing rivers restricted to the Arctic and the Congo and Amazon River basins. Very few free-flowing rivers remain in densely populated areas, with dams and their reservoirs being the leading cause of losses in connectivity between different sections of a river's long profile.

Other research in 2020 estimated that there are 1.2 million instream barriers (comprising dams, ramps, weirs, sluices, culverts, and fords) in the rivers of 36 European countries—a mean density of one barrier every 1.35 km (0.84 miles). The current European Union Biodiversity Strategy aims to reconnect 25,000 km (15,500 miles) of Europe's rivers by 2030, demanding that the role of these barriers be fully understood. Dam decommissioning has begun in some countries, including the United States, and is seeking to reconnect rivers in longitudinal profile as well as with their floodplains. The Room for the River program in the Netherlands (see page 381), begun in 2007 and with a target completion date of 2022, aims to lower the level of floodplains, creating water buffers for high water flows, relocating levees, increasing the depth of side channels, and constructing flood bypasses. Not only will this facilitate improved resilience to flooding, but it will also decrease river fragmentation.

FREE-FLOWING AND FRAGMENTED RIVERS

A map of the world's free-flowing rivers and those fragmented rivers impacted by reduced connectivity. The connectivity status index (CSI) quantifies river connectivity ranging from 0% to 100%.

River status

Free-flowing rivers (CSI ≥ 95% over entire length of river)

VL   L   M   S

Good connectivity status (CSI ≥ 95% over parts of river)

VL   L   M   S

Impacted (CSI < 95%)

VL   L   M   S

No flow

VL Very long river (> 1,000 km)
L Long river (500–1,000 km)
M Medium river (100–500 km)
S Short river (10–100 km)

Blocking dolphins
River fragmentation can impede the migration of species such as the Amazon River Dolphin (*Inia geoffrensis*).

Non-native and invasive species

As humans have spread across the globe, we have taken with us flora and fauna that may become established in new areas and outcompete and displace indigenous inhabitants. Such non-native species—termed invasive where they cause ecological harm—may spread especially rapidly through river valleys and generate swift and pronounced ecosystem change. Ecosystem alterations may also be amplified by global climate change, which is affecting the temperature of river waters and the geographical ranges in which species can live and adapt.

Rivers have provided some of the most useful corridors for humans to access the interior of continents, and have been vital routes for exploration, transport, and trade over millennia. River basins have also offered pathways through which species that are not native to a particular region may spread. These species may be introduced by humans either on purpose or accidentally. This phenomenon has been termed biological pollution, and has occurred through human exploration, colonial expansionism, and world trade, including aquaculture and stocking of fish.

ZEBRA CROSSINGS

The rapid spread of the Zebra Mussel (*Dreissena polymorpha*) in the United States between 1989 and 1994.

0   200 miles
0   200 km

Distribution of Zebra Mussel

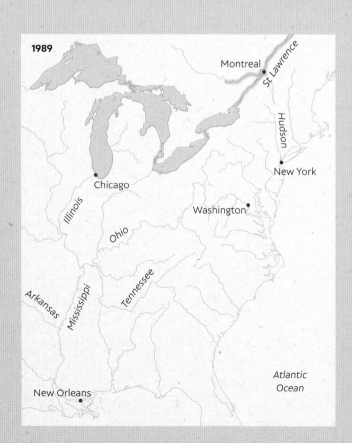

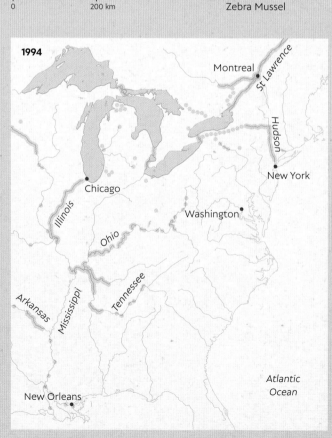

Harmful invaders

The rapid spread of non-native and invasive species can be exacerbated by their introduction at seaports and transport along river channels on boats. For example, the freshwater Zebra Mussel (*Dreissena polymorpha*), native to southern Russia and Ukraine, likely arrived in the North American Great Lakes in the 1980s as a stowaway in ballast water discharged by ships. It then spread rapidly across the large rivers of the eastern Mississippi drainage basin, and has been responsible for outcompeting native species by filtering out the algae these indigenous mollusks need for growth. After its arrival, the mussel's first passage beyond the Great Lakes was in 1991, when it crossed New York State through the Erie Canal and Mohawk River into the Hudson River. Although the mussel can migrate naturally by swimming and transport in water flows, its extremely rapid spread was enabled by transit in ballast water and by adult mussels attaching to the hulls of commercial ships, barges, and recreational boats.

Other examples of invasive species that have spread through waterways outside their native range include the aquatic plant Waterthyme (*Hydrilla verticillata*) in the United States; the Signal Crayfish (*Pacifastacus leniusculus*) in Europe, Japan, and California; Eucalyptus and Tamarix trees in North America; the alga Didymo (*Didymosphenia geminata*) in rivers of Patagonia in South America and in Australia and New Zealand; the South American Water Hyacinth (*Eichhornia crassipes*) across large areas of North America, Europe, Asia, Australia, and Africa; the American Mink (*Neogale vison*) in Argentina, Chile, and northwest Europe; and the Asian Clam (*Corbicula fluminea*) in rivers of northwest Europe, South America, northwest United States, and Canada.

Invasive species can cause major environmental change, and measures to remove them, or prevent their spread, are expensive. A 2022 study estimated that since 1960 the global costs of dealing with all invasive species (not just those affecting river basins) totaled US$95 billion in management expenditure and a staggering US$1,131 billion in mitigating the damage they cause. In the United States, some 70 percent of extinctions of native species in the past century can be linked to the impact of non-native organisms. Managing the ingress and spread of non-native species in the world's rivers is therefore clearly a matter of prime ecological and economic importance.

▲ **Leaping invaders**
Non-native Bighead Carp (Hypophthalmichthys nobilis) *and Silver Carp* (H. molitrix)*, both native to Asia, in the Illinois River, United States.*

The politics of rivers

Rivers bring both benefits and hazards, and have long been managed in attempts to maximize the former and minimize the latter. However, the connected nature of river networks, coupled with the inexorable downstream flow of water, sediment, and nutrients, means that gains and losses are often traded off between different communities and countries.

Rivers as borders

Rivers have long been used as convenient natural territorial delimiters: at least 23 percent (58,588 km/36,404 miles) of the world's interior (non-coastal) national borders are set by rivers. However, river borders may be a source of tension. Rivers that form borders between nations can become hazardous crossing points for migrants, while the fact that rivers naturally change their course has long been a flash point. In Act 3, Scene 1 of William Shakespeare's *Henry IV, Part 1*, the conspirators Edmund Mortimer, Harry Percy (Hotspur), and Owen Glendower plot against the king. Discussing how to split England after their planned victory, Mortimer suggests using the courses of the River Trent and River Severn to define their territories. Hotspur notes that a loop of the River Trent meanders far into his territory and suggests straightening it out to get more land. Glendower is ready to fight over this, his reply suggesting that in 1597 Shakespeare understood the fundamental nature of a river to meander and migrate: "Not wind? It shall, it must. You see it doth."

Conflict over dams

Hundreds of millions of people in 11 countries depend on the Nile, the world's longest river. Since 1999, the Nile Basin Initiative has governed shared use of the river's precious water, but tensions rose in 2011 when Ethiopia announced construction of a 6,000 MW hydroelectric dam on the Blue Nile. This will significantly affect flow and make it challenging for Egypt, the most downstream and historically powerful Nile Basin nation, to meet its water use requirements.

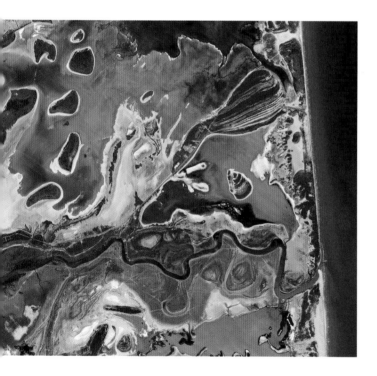

▲ **Border line**

The Rio Grande drains into the Gulf of Mexico, dividing the southernmost tip of Texas in the United States (north) from the state of Tamaulipas in Mexico (south). In 2022, more than 150 migrants died in their attempt to cross the river from Mexico to the United States.

Social inequalities of flooding

Who floods? Answering this question is vital to direct the specific assistance needed to help communities recover after flooding. The burden is greatest where the twin drivers of risk—population growth in flooded areas (which increases exposure to flooding) and climate change (which drives increased hazard)—are largest. In the United States, the annual average economic losses due to flooding (US$32.1 billion under current climate conditions) are borne mainly by poorer white communities. However, demographic shifts and climate change mean that future risk will disproportionately impact Black communities. Such inequities are well illustrated in the city of Los Angeles. Here, ultra-high-resolution flood models reveal that non-Hispanic Black and Hispanic residents are, respectively, 79 percent and 17 percent more likely than non-Hispanic white residents to be exposed to deep (>1 m/3 ft) flooding.

INEQUITABLE FLOOD RISKS

In Los Angeles, California, the 100-year return period (1 percent annual chance) flood hazard zone intersects with different communities over the city. This means that the flood exposure representativeness (FER—the fraction of a population by race or ethnicity living in a flood zone divided by the fraction of the same group within a region) highlights large inequities in exposure to flooding. For example, the FER for Non-Hispanic Black communities (left) is much larger than for Non-Hispanic White communities (right).

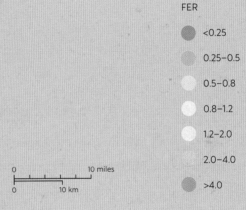

FER

- <0.25
- 0.25–0.5
- 0.5–0.8
- 0.8–1.2
- 1.2–2.0
- 2.0–4.0
- >4.0

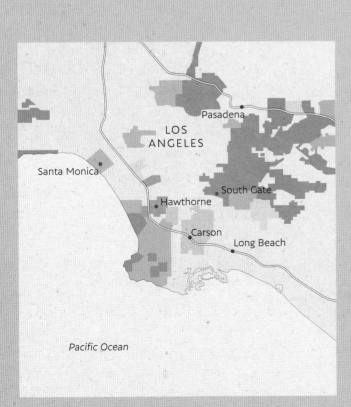

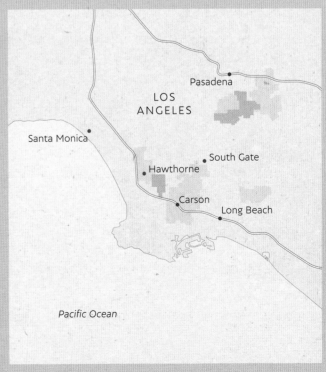

Anthropogenic change in the world's rivers

How are we to manage the world's rivers to ensure their sustainability? Not only do we need to identify the types of anthropogenic stresses that a river may experience, but we must also assess their scale of effect, whether different stresses may interact, and the pace of likely resultant change. Like first responders at the site of an accident, we need to triage the river to ensure we are treating the symptoms that need the most urgent care and action first, and prioritize our plan of treatment.

Timescales of change

As we have seen, the world's rivers are being influenced by a wide range of human-caused stresses that can change their geomorphology and ecology, potentially leading to ecosystem collapse. Some of these stresses act over timescales of hours to days, including pollutant spills from industrial complexes; others operate over years to decades, such as the spread of non-native species; and yet others take decades or centuries—for example, the changing magnitude of floods and droughts due to climate warming, and sea-level rise at the downstream ends of rivers.

Resilience thresholds

We can view the action of a particular anthropogenic stress in relation to a resilience threshold condition, beyond which the river will change. The long-term background stress may be increasing due to climate warming, with changes in the frequency and magnitude of flow resulting from extreme events potentially crossing this threshold. In addition, reductions to the resilience threshold may be caused by other anthropogenic stresses, such as dam construction, which may thus yield ecosystem change at an earlier time. It is also likely that rivers will be subject to multiple, rather than singular, stresses. We therefore also need to assess whether these stresses are additive, offset one another, or even produce a reaction opposite to that of the individual stresses. River triage must include an assessment of all these complex mechanisms and feedbacks.

The context of anthropogenic stresses

How we respond to changes in riverine habitats and functioning is also determined by the overarching background that sets the societal, economic, and political context for each river. For example, global health epidemics, conflicts, and wars, as well as political ideologies, have all provided shocks and constraints to how rivers are viewed, used, and managed, thereby exerting major influences on many river basins. How we manage and govern rivers that flow through several countries, reach across climatic zones, and affect very different populations, together with their natural resources, is a matter of critical global importance. Yet achieving such successful governance is essential if the amazing biodiversity of riverine corridors is to be sustained.

▼ **Toxic kill**

Fish killed due to toxic pollution in the Oder River, Germany, 2022.

ANTHROPOGENIC STRESSES ON RIVERS

The overarching societal, economic, and political context of anthropogenic stresses on the world's rivers, along with their different types and timescales of effect.

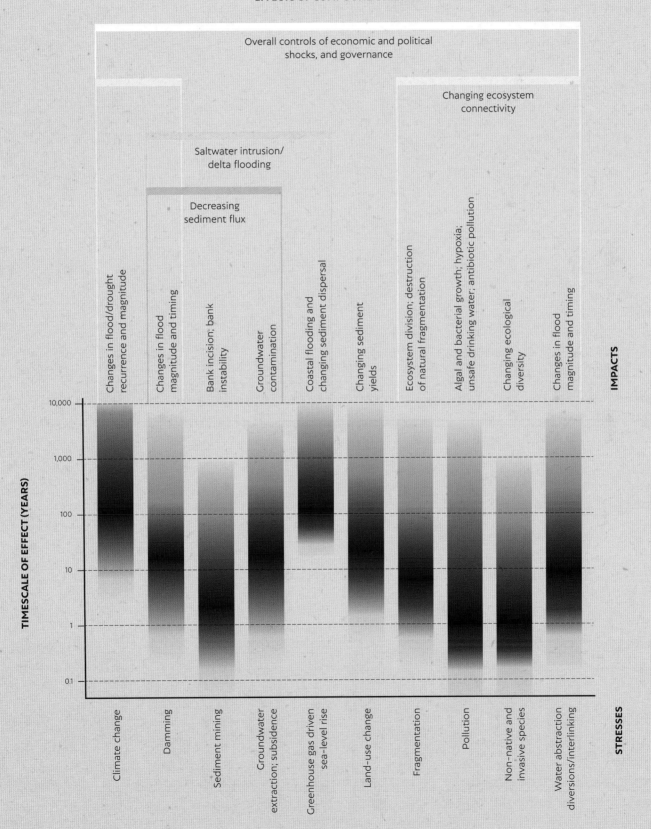

EFFECTS OF COMPOUND STRESSES

Overall controls of economic and political shocks, and governance

Changing ecosystem connectivity

Saltwater intrusion/ delta flooding

Decreasing sediment flux

IMPACTS

- Changes in flood/drought recurrence and magnitude
- Changes in flood magnitude and timing
- Bank incision; bank instability
- Groundwater contamination
- Coastal flooding and changing sediment dispersal
- Changing sediment yields
- Ecosystem division; destruction of natural fragmentation
- Algal and bacterial growth; hypoxia; unsafe drinking water; antibiotic pollution
- Changing ecological diversity
- Changes in flood magnitude and timing

TIMESCALE OF EFFECT (YEARS)

10,000 — 1,000 — 100 — 10 — 1 — 0.1

STRESSES

- Climate change
- Damming
- Sediment mining
- Groundwater extraction; subsidence
- Greenhouse gas driven sea-level rise
- Land-use change
- Fragmentation
- Pollution
- Non-native and invasive species
- Water abstraction diversions/interlinking

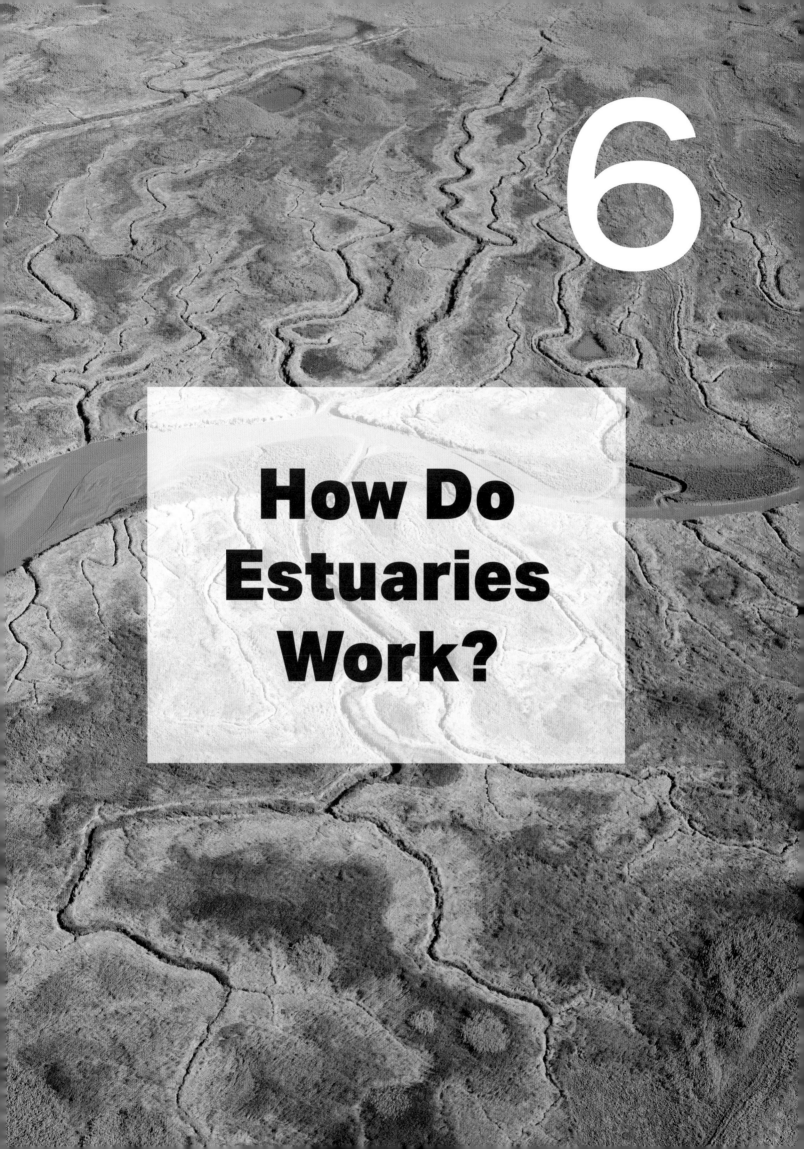

6

How Do Estuaries Work?

The importance of tides

Tides move water levels up and down, creating space for intertidal habitats to form. Tidal currents play a key role in estuarine circulation, siltation, exchange of nutrients, and the distribution of marine life. The greater the tide, the larger its influence on an estuary.

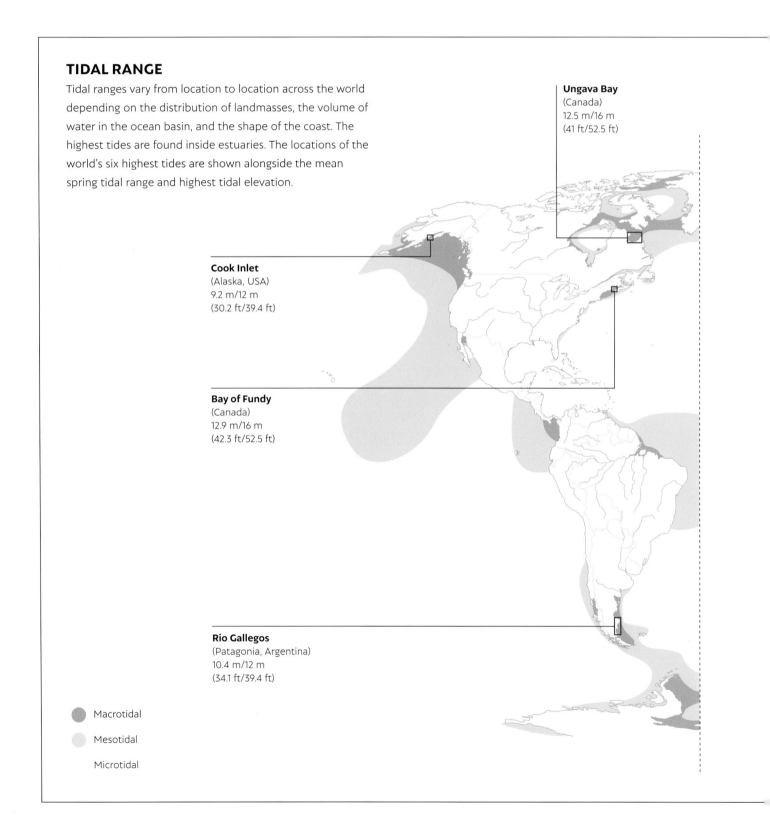

TIDAL RANGE

Tidal ranges vary from location to location across the world depending on the distribution of landmasses, the volume of water in the ocean basin, and the shape of the coast. The highest tides are found inside estuaries. The locations of the world's six highest tides are shown alongside the mean spring tidal range and highest tidal elevation.

Ungava Bay
(Canada)
12.5 m/16 m
(41 ft/52.5 ft)

Cook Inlet
(Alaska, USA)
9.2 m/12 m
(30.2 ft/39.4 ft)

Bay of Fundy
(Canada)
12.9 m/16 m
(42.3 ft/52.5 ft)

Rio Gallegos
(Patagonia, Argentina)
10.4 m/12 m
(34.1 ft/39.4 ft)

● Macrotidal

○ Mesotidal

Microtidal

In most places, the tide will rise and fall twice a day. The vertical change in water level between low tide and high tide is called the tidal range, which can be categorized as either microtidal (when it is less than 2 m/6.5 ft), mesotidal (2-4 m/6.5-13 ft), or macrotidal (more than 4 m/13 ft). The term hypertidal is used for places where tidal variation is very large (more than 6 m/20 ft). Intertidal habitats are a key characteristic of estuaries, with mudflats developing in areas exposed during low tides and saltmarshes or mangroves developing in areas flooded during high tides.

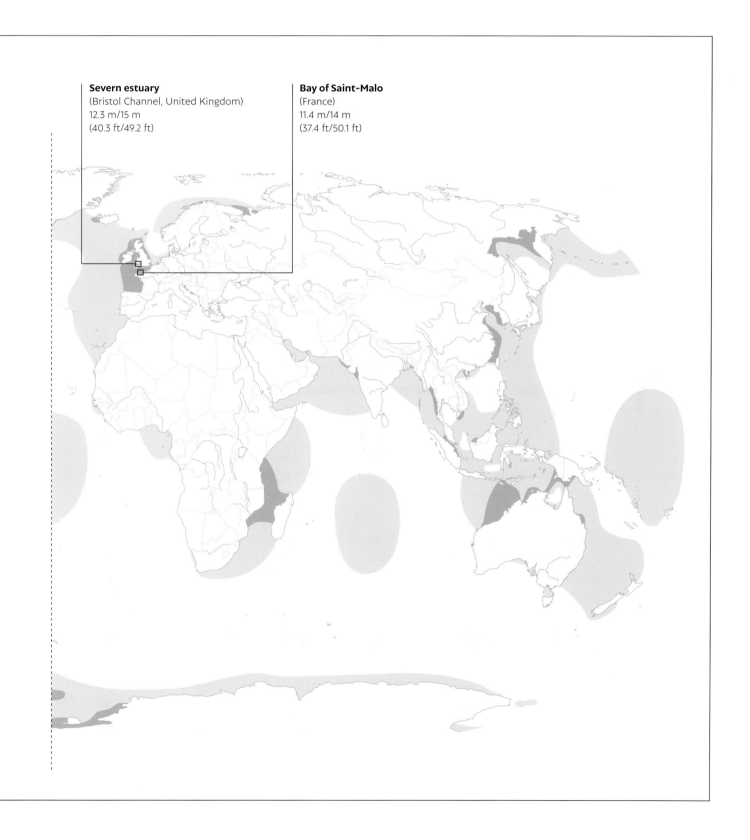

Severn estuary
(Bristol Channel, United Kingdom)
12.3 m/15 m
(40.3 ft/49.2 ft)

Bay of Saint-Malo
(France)
11.4 m/14 m
(37.4 ft/50.1 ft)

Extraterrestrial influence

Astronomic tides are caused by the gravitational pull exerted by the moon and sun on the Earth's surface. Gravitational forces increase with increasing body mass and decrease by a factor of three with distance. Although the moon has much less mass than the sun, it is much closer to Earth, and thus has double the sun's influence on tides. The tidal elevation varies through time as the distances between the moon, the sun, and the Earth change. Although astronomic tides can be predicted accurately, water levels at the coast are also affected by weather conditions, which are more difficult to predict.

Tides are smaller when the sun and moon are at 90 degrees in relation to the Earth (during half-moon phases), as their gravitational pulls are in opposition to each other; these are called neap tides. Twice a month, when the sun and moon are aligned in relation to Earth (during full and new moons), tidal variations are larger; these are called spring tides (see page 27). The highest spring tides occur twice a year, around the spring and autumn equinoxes (March or September), when the sun and moon are over the Earth's equator. These tides are even higher every four-and-a-half years or so, when the moon passes closest to Earth—these events are called super tides.

▼ **Renewable energy**
This 16-m (52.5-ft) diameter turbine was installed on the Bay of Fundy, Canada, seabed in 2016 to harness energy from the powerful tidal currents. The project failed in 2018 when the turbine stopped working and the manufacturer entered liquidation, an indication of what an engineering challenge such projects are.

The influence of location

Tides are perceptible only in very large waterbodies and have little effect in those that are not connected to large ocean basins. The Mediterranean Sea, Baltic Sea, and North American Great Lakes, for example, have very small tides. On the other hand, tides are amplified in shallow and narrow areas connected to the oceans, such as estuaries. Hypertidal ranges usually occur in the narrower areas of funnel-shaped estuaries.

As tidal levels vary across locations, water tends to flow from higher to lower water elevations, creating tidal currents. The currents are stronger during spring tides, as there is a greater volume of water moving between the low and high tides. Tides move large volumes of seawater and create so much energy that the friction with the seabed in shallow waters causes the Earth's rotation to slow down a thousandth of a second every century.

HYPERTIDAL RANGE

The Bay of Fundy in Nova Scotia, Canada, has the world's greatest tidal range. Here, tidal amplification is evident as the tidal range varies from 6 m (20 ft) at the entrance to 16 m (52 ft) at Minas Basin. For comparison, the range is only slightly over 2 m (6.5 ft) on the open coast at Halifax.

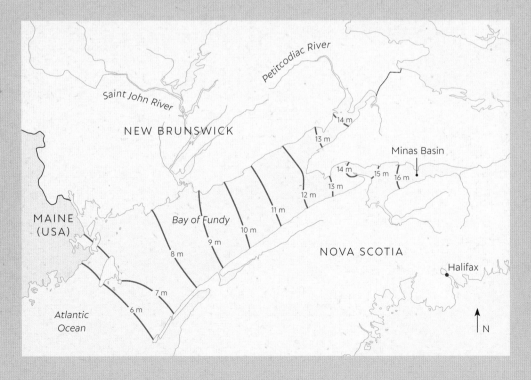

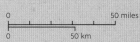

Tidal bores

In some hypertidal estuaries, the incursion of the highest tides causes friction with the estuary bed and margins in ways that favor the development of tidal bores. Worldwide, these bores are known to occur in around a hundred estuaries, which are usually long and funnel-shaped, and have an average channel depth similar to the local spring tidal range. Only the highest tides (around the equinox) form bores, but river discharge and wind conditions can prevent their formation. A tidal bore looks like a turbulent, fast-moving wave (or series of waves) that suddenly causes the water level to rise as the flood tide moves up the estuary. The strong current stirs up sediment from the bottom of the estuary and can cause rapid erosion of its margins, sweeping away trees, houses, and people on the way. Tidal bores can carry large quantities of estuarine sediment up the river.

As an early warning, tidal bores produce a roaring sound that may be heard for over an hour before the wave is seen—the tidal bore in the Amazon is called Pororoca, meaning "roaring sound" in the local Tupi language. Despite the dangerous conditions, tidal bores attract surfers seeking a very long ride. One surfer was able to ride the Bono tidal bore on the Kampar River in Sumatra, Indonesia, for 17.2 km (10.7 miles) in 2016. The bore was more than 2.4 m (8 ft) high at its peak and traveled at 20 km/h (12 mph).

The tidal bore on the Qiantang River in eastern China, known as the Silver Dragon, is the world's largest. It is celebrated in a festival that attracts thousands of viewers every September. The bore reaches up to 4 m (13 ft) high and 3 km (2 miles) wide, and travels faster than 20 km/h (12 mph).

Tidal bores attract surfers even in Alaska (water temperature around 4 °C/39 °F) as shown here. At the Turnagain Arm of the Cook Inlet near Anchorage, the bore can reach 3 m (10 ft) in height and travel at speeds over 24 km/h (15 mph).

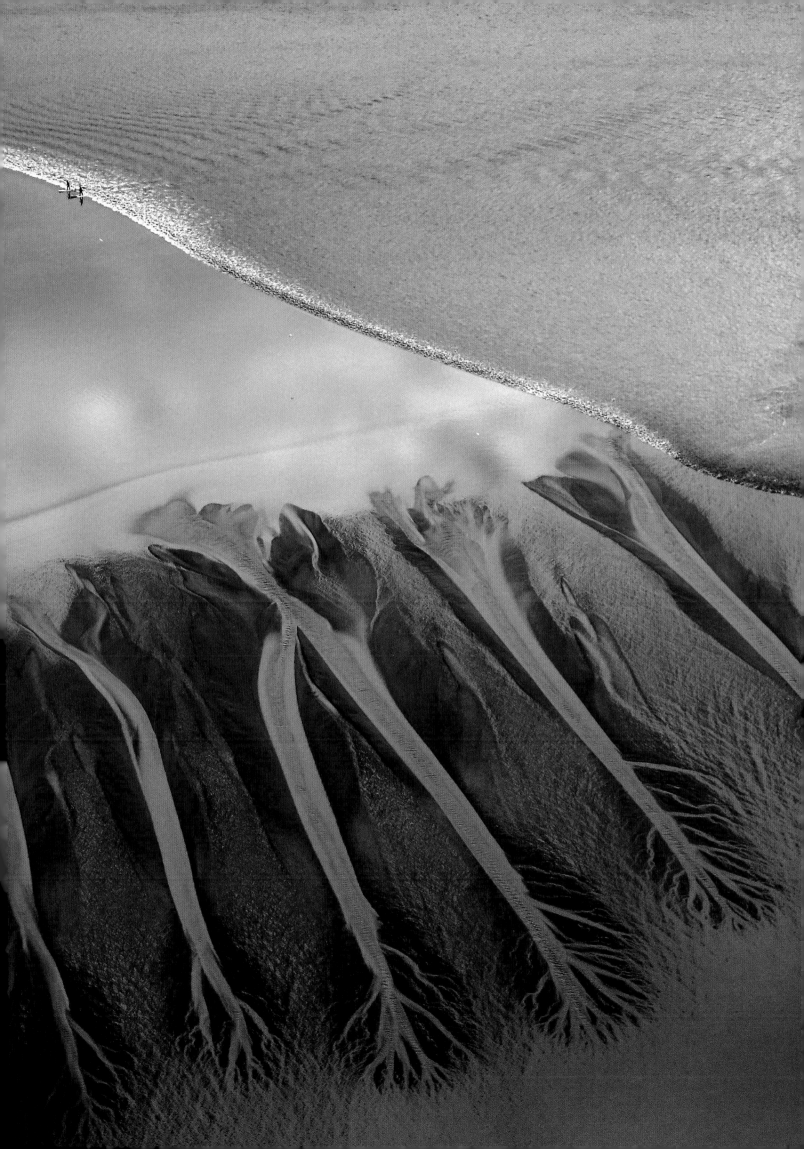

The balance of power

Estuaries could be described as the battleground of two opposing powers: the river moving fresh water seawards; and the tides moving seawater landwards. What happens in an estuary depends mostly on whether the river discharge or the tidal currents are winning this battle.

Typically, an estuary can be divided into three parts based on the relative dominance of the marine or fluvial influence. The lower estuary (nearest the sea) is dominated by marine conditions, which have a decreasing influence further inland. Here, waves and tidal currents can form deposits of sand, and seagrass beds develop if conditions are favorable. Although it is still subjected to tidal variations, the upper estuary (at the landward end) is dominated by freshwater conditions, with the river's influence reducing seawards. The middle estuary is characterized by a blend of river and tidal energy and the mixing of fresh water and seawater.

Both the river and seawater carry organisms, sediments, nutrients, oxygen, and pollutants into the estuary. Particularly in the middle estuary, physical, chemical, and biological processes act like filters, determining what stays in the estuary (and for how long) and what goes out. Fine sediments, nutrients, and some pollutants tend to settle here, with this process assisted by the presence of salts. The boundaries between the upper, middle, and lower estuarine zones are not well defined and can shift when conditions favor either the river discharge or the incursion of tides.

River Thames, London
In this false-color satellite image, urban areas are shown in pink, agricultural areas are red, and green shows areas of vegetation.

1 Tidal limit
2 Upper estuary
3 London
4 Middle estuary
5 Lower estuary
6 Thames estuary

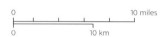

0   10 miles
0   10 km

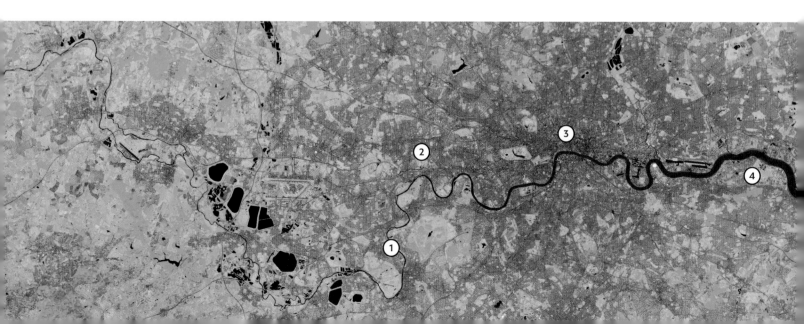

FLUID INFLUENCE

In urbanized estuaries, such as the Thames in England, the river and tidal flows are often altered by land reclamation, channelization, and flood control measures, which constrain the types of habitats that develop in and around the estuary. While tidal currents and river flow are important across the length of estuaries and their influence is balanced in the middle estuary, fluvial processes dominate in the upper estuary and marine processes prevail in the lower estuary as described in the graph below.

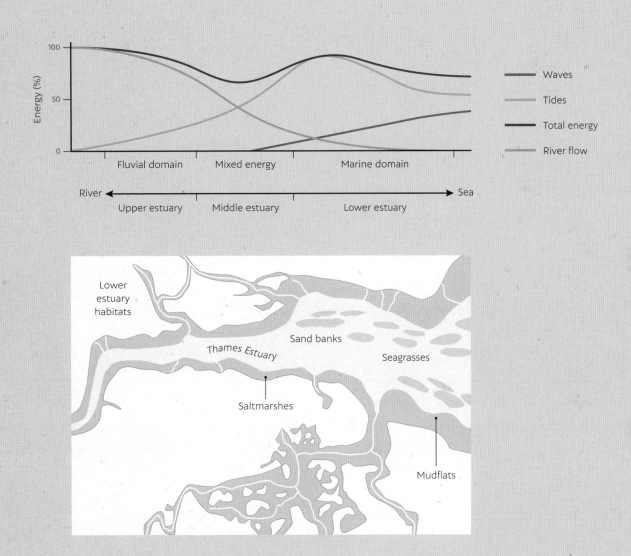

Waves
Tides
Total energy
River flow

Energy (%)
100
50
0

Fluvial domain   Mixed energy   Marine domain

River ← → Sea

Upper estuary   Middle estuary   Lower estuary

Lower estuary habitats

Thames Estuary

Sand banks

Seagrasses

Saltmarshes

Mudflats

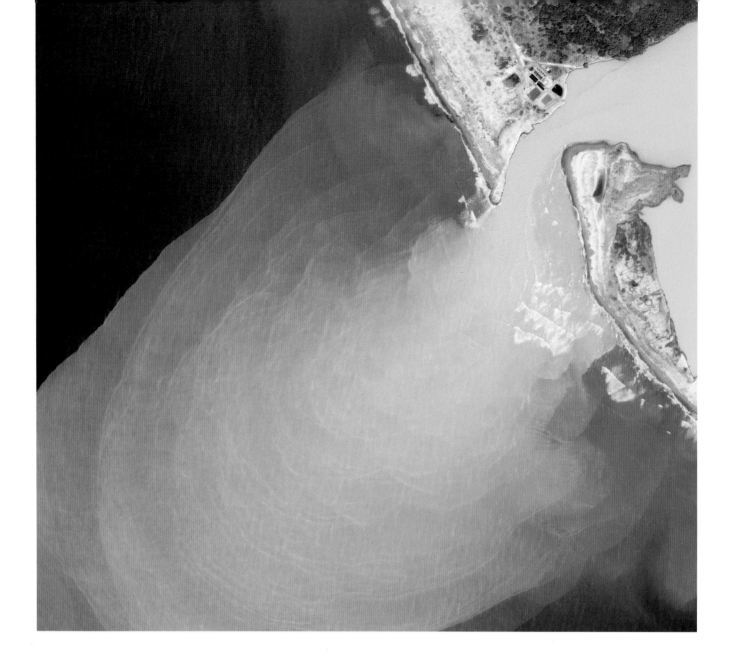

▲ **Underneath**

High river discharges can extend the freshwater influence further down the estuary. While less dense water can flow out to sea at the surface (as seen in this estuary of the Rio Baluarte, Mexico, a few days after the passage of Hurricane Jimena, September 9, 2009), tides can still push seawater into the estuary underneath this surface flow.

Water density

The difference in density between fresh water and saltwater is only about 2.5 percent, but it is a key factor influencing flow in estuaries. Water density increases with increasing concentration of salts and suspended sediments and decreasing water temperature. The less salty and less dense fresh water will flow on top of the saltier and denser seawater, with a limited amount of mixing resulting from friction between the two. Therefore, the water column in most estuaries has saltier water at the bottom and less salty water at the top.

Whether saltier water or fresh water will dominate the water column (or along the estuary) depends on the speed of the tidal current, the discharge of fresh water, and how much water mixing occurs. For the waters to mix, turbulence is needed to break the boundary between fresh water and seawater. This mixing is mostly driven by strong tidal currents, although effects caused by the shape of the estuary and persistent winds can also be important. In the same estuary, conditions can vary through time, changing which parts of the estuary will be dominated by fresh water, brackish water, or seawater.

SALT

Changes in salinity at four moments illustrate how the presence of fresh water, brackish water, and seawater varies in the water column along the lower 10 km (6 miles) of a narrow estuary. When river discharge is low, there is a greater seawater presence and more water mixing (brackish water), even at halfway to low water. A high river discharge will reduce the amount of brackish water (less mixing) and will flush out the seawater faster during the ebbing tide.

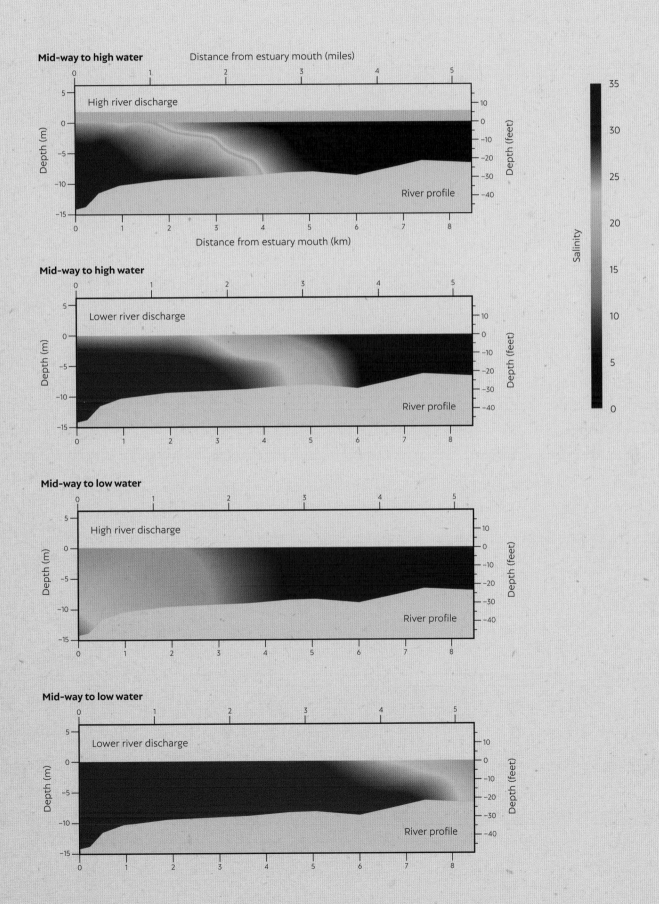

Sand

Waves and tides bring sand into the lower estuary, giving clear evidence of marine influence. This is commonly seen in bar-built estuaries, such as the Noosa estuary on Australia's Gold Coast.

Tidal currents

Rising tides moving into an estuary are called flood tides, while receding tides moving out are ebb tides. For a brief period at each side of high and low tide, the tidal current almost stops before changing direction; this period is known as slack water. At this time, the extremely low flow velocities allow even very fine suspended sediments to settle through the water column, potentially forming mud deposits. Depending on the depth and shape of the estuary, tidal currents are fastest halfway between, or just before, high and low tides. The speed of tidal currents is usually less than 4 km/h or around 1 m/s (2.5 mph or around 3 ft/s), but they are faster when the estuary is narrower, as the same volume of water is squeezed through a smaller area. For this reason, tidal currents can be very strong when passing through narrow inlets. Beyond the inlet, the current slows down as it spreads over a larger area, and this causes the larger sediments carried by the current (usually sand) to settle. As a result, sandy bars or a flood tidal delta form just landward of the inlet, or an ebb tidal delta forms just seaward of it. The finer sediments can move in suspension further up the estuary or out to sea.

Extremes

Tropical estuaries can look very different during the wet and dry seasons. As freshwater input reduces through the dry season, the estuary becomes increasingly dominated by marine conditions. If the river flow becomes very weak and the tides are not strong, the estuary's connection with the sea might be interrupted. In this situation, small amounts of seawater may intermittently enter the estuary through underground infiltration or by overtopping the sand barrier during high tide. The lack of circulation can cause water-quality issues, as salinity increases due to evaporation and oxygen is consumed by living organisms and the decay of organic matter. Such conditions persist until the connection with the ocean is re-established, either artificially, or naturally by wave erosion, increased river flow, a very high tide, or storm surge.

In contrast, periods of extreme rainfall or rapid snowmelt in the catchment area of an estuary can greatly increase river discharge, limiting the marine influence to the lower estuary. During these times, the capacity of the estuary to retain sediment and nutrients is reduced due to lower salinity (less flocculation), and a faster outflow moving material out to sea before it can settle.

The passage of tropical cyclones (known as hurricanes in the Atlantic and typhoons in the Pacific) can cause a double effect on estuaries. The strong winds and lower atmospheric pressure can cause storm surges, which increase water levels at the coast—and even more so in the confines of estuaries. Storm surges can cause severe erosion and coastal flooding, particularly when they coincide with high tides. During these events, tidal currents are enhanced and the sea can reach further inland, where the sediment eroded from the estuary bed and margins will be deposited. Tropical cyclones also cause intense rainfall over large inland areas, often leading to flooding. After a few days, these large volumes of water flow into the estuary, carrying debris, sediment, and nutrients, and large amounts of this material exit to the sea. Although cyclones may cause human fatalities and considerable financial damage, they also help to enrich coastal and marine waters with nutrients that are needed to stimulate biological productivity.

Flood
This Landsat 8 satellite image combines visible light and infrared data to highlight the amount of organic matter carried by rivers to estuaries along the North Carolina coast of the United States and into the sea on September 19, 2018, five days after the passage of Hurricane Florence, which caused intense rainfall.

1 New River
2 White Oak River
3 Adam's Creek
4 Cape Lookout
5 Atlantic Ocean

Colored dissolved organic matter

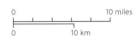

Less   More

0   10 miles
0   10 km

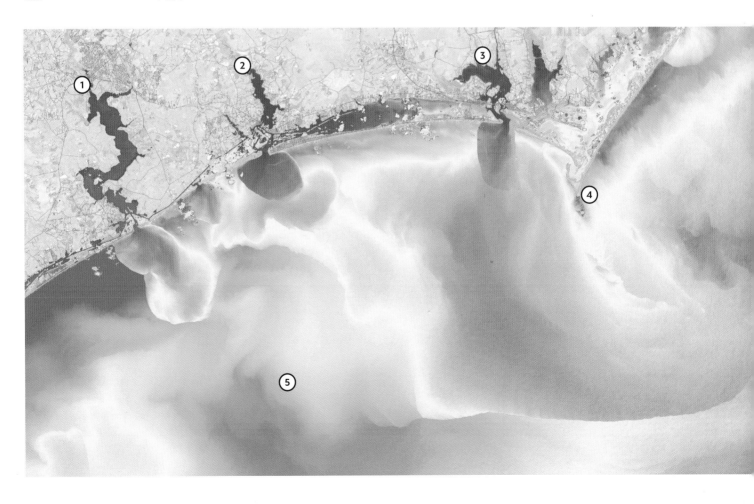

Estuarine waters

Estuarine waters are considered to be transitional waters between fresh water and seawater. This transition is often described in terms of changes in salinity, which play an important role in estuarine circulation and the concentration of sediments and nutrients.

Whether fresh water and seawater mix readily, or do not mix at all, is a key factor influencing what happens in an estuary. The variation in salinity is a simple indicator of the different types of estuarine circulation. Most estuaries have positive circulation, where saltier water flows into the estuary along the bottom and less salty water flows out along the surface. There are four patterns of positive circulation according to the level of water mixing: well-mixed, partially mixed, stratified, and highly stratified estuaries. Some estuaries show negative or inverse circulation, with saltier water flowing out of the estuary. As the river discharge and the speed of the tidal currents vary at different time scales, the circulation pattern in an estuary can vary when conditions change.

Inverse circulation

In arid areas or during prolonged droughts in tropical areas, evaporation can exceed the total input of fresh water in the upper estuary. Evaporation increases the concentration of salts in the remaining water, which will eventually become denser, sinking to the bottom of the estuary. This denser, saltier water will flow out to sea along the bottom, while seawater will flow into the estuary along the surface. In contrast with positive estuaries, the loss rather than the gain of fresh water in inverse estuaries is a driving force of water circulation.

Well-mixed estuaries

Also known as tidally dominated estuaries, well-mixed estuaries have tidal currents that are strong enough to mix with fresh water throughout the water column as they enter the estuary, the result being that there is no vertical gradient in salinity. However, the salinity does vary along the estuary, decreasing upriver. Strong vertical mixing usually occurs in broad, shallow, and macrotidal estuaries.

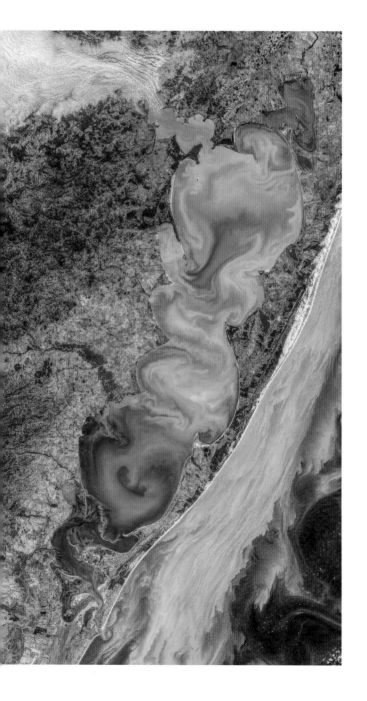

◀ **Trapped**

Strong southerly winds combined with relatively lower freshwater input push seawater 120 km (74.5 miles) into the

Patos Lagoon (south Brazil), reducing the seaward flow of sediment-rich freshwater (lighter colours).

MIXING WATERS

Five patterns of estuarine circulation occur depending on the amount of water mixing taking place, which in turn depends largely on the speed of tidal currents. These patterns are named according to the changes in salinity through the water column. Salinity is shown in parts per thousand.

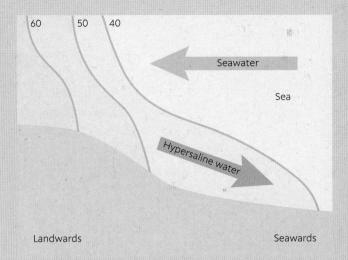

INVERSE CIRCULATION

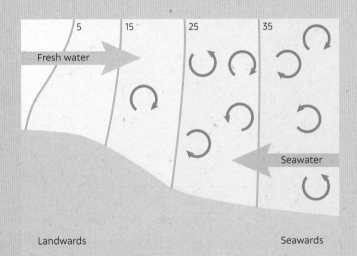

WELL MIXED

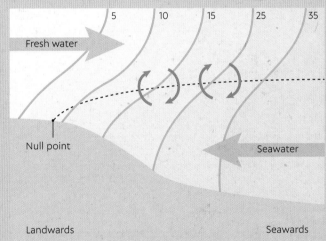

PARTIALLY MIXED

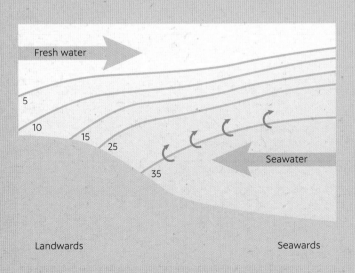

STRATIFIED

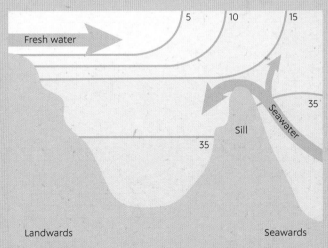

HIGHLY STRATIFIED

Partially mixed estuaries

Stronger tides will generate turbulence due to friction at the estuary bed and also between fresh water and seawater, causing more effective mixing of seawater upward and fresh water downward. As the brackish water moving out of the estuary carries a greater proportion of seawater, this loss is compensated by more seawater entering the estuary along the bottom, thereby strengthening the landward flow. This creates a vertical gradient in salinity that is less pronounced than in stratified estuaries. At a point in the upper estuary, the volume of water moving out along the surface is compensated by water moving in along the bottom, and there is no net flow of water seaward or landward. At this point, a turbidity maximum forms where the concentration of sediment is highest, and this moves up and down the estuary with the tides. The depth at which the turbidity maximum occurs increases further up the estuary until it is at the estuary bed. This is called the null point and defines the landward limit for the deposition of river and marine sediments carried by the denser estuarine water.

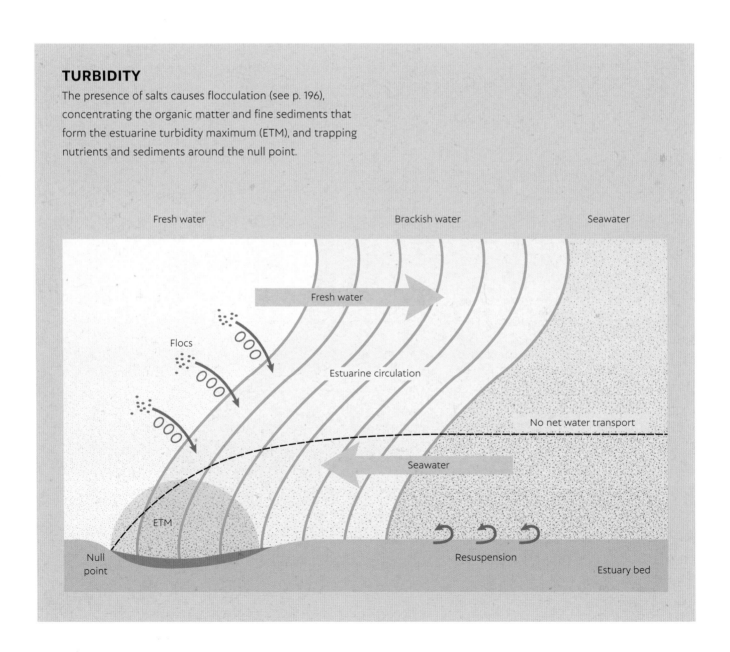

TURBIDITY

The presence of salts causes flocculation (see p. 196), concentrating the organic matter and fine sediments that form the estuarine turbidity maximum (ETM), and trapping nutrients and sediments around the null point.

Fresh water

Brackish water

Seawater

Fresh water

Flocs

Estuarine circulation

No net water transport

Seawater

ETM

Resuspension

Null point

Estuary bed

Stratified estuaries

These estuaries are also known as salt-wedge or river-dominated estuaries. Facing resistance from the river discharge, seawater enters the estuary wedged between the seabed and the fresh water that is moving seawards. Freshwater discharge here is considerably greater than the tidal current, and the flow forms distinct layers, with limited upward mixing of seawater into the fresh water between the two layers. This creates a halocline, which is a sharp change in salinity between the surface layer (lower salinity) and the bottom layer (higher salinity). The salt wedge can move further inland during spring tides or periods of lower river discharge. River sediments will dominate, with the larger particles settling through the halocline and the finer sediments being transported out to sea (see page 175).

Highly stratified estuaries

Highly stratified estuaries are also known as fjord-type estuaries because many fjords display this kind of water circulation. However, not all fjords show highly stratified circulation, and it can also occur in other types of estuaries. Like stratified estuaries, these estuaries have a pronounced vertical change in salinity. However, a salt wedge does not form here, as the incursion of seawater into the estuary is limited by the presence of a physical underwater barrier near the estuary mouth. The circulation is mainly of fresh water flowing out of the estuary over this barrier, which makes it even more difficult for saltier waters to enter. Fjords are shallower at their mouth due to glacial deposits left behind by the receding glacier, and this restricts seawater flow along the bottom. If they stay stagnant for too long, the deeper waters inside a fjord can become hypoxic (with low oxygen levels) or even anoxic (depleted of oxygen).

▲ **Horizontal variations**

Changes in salinity may occur across the width of the estuary, as seen in this summer image of the Affall River in Landeyjarsandur, Iceland, with fresh water (brownish color) and seawater exiting and entering the estuary side by side.

Estuary measurements

Many estuaries benefit from regular monitoring to assess the effects of human activities and natural events on water quality and the health of their ecosystems. Instruments can be installed on site or deployed from boats for precise measurements at specific locations, while satellite data are used to assess conditions over large areas.

Measuring water quality

Water quality is usually assessed through measurements of salinity, temperature, dissolved oxygen, turbidity, pH (how acidic or alkaline the water is), and nutrients such as nitrogen and phosphorus. These are good indicators of environmental quality because they affect primary productivity and the presence and behavior of aquatic life. Salinity, temperature, and pH can be measured with instruments either permanently installed or deployed from a boat. For other indicators, water samples must be taken and then analyzed in a laboratory. Water-quality monitoring is very important in estuaries where oysters and other commercial shellfish are farmed, as these food sources accumulate pollutants that can affect human health.

The presence of excess nutrients is a critical issue in estuaries. Nutrients enter rivers and estuaries naturally through the weathering of rocks and decomposition of organic matter, and artificially through domestic and industrial wastewater and water runoff from agricultural and urban areas. Although nutrients are needed for the development of plants, an excess of nitrogen and phosphorus can cause eutrophication, which is the increased growth of phytoplankton (leading to an algal bloom) and the zooplankton that eat them, critically reducing oxygen levels. Phytoplankton are microscopic plants that form the base of the aquatic food web. Like other plants, they contain chlorophyll, a green pigment essential for the process of photosynthesis. The amount of chlorophyll in water is therefore an indicator of the concentration of phytoplankton and the amount of primary productivity taking place.

Satellite data

▶ **Satellite images**
Landsat 8 visible light data contrast seawater (blue) and the sediment and nutrient concentrations (yellow to red) in the Ord River estuary in Western Australia. Shortwave and near-infrared data show mangroves (dark green) and other vegetation on land (light green).

Some satellites carry sensors specifically designed to capture conditions in the atmosphere, on land, or in the ocean. They register radiation at the Earth's surface and in the atmosphere, including in the ultraviolet, visible light, infrared, and microwave bands. Landsat 8 sensors, for example, capture radiation across 11 bands of infrared and visible light, detailing every 15-100 m (50-330 ft) of the Earth's surface in the satellite's orbit and returning to the same location every 16 days. The Terra and Aqua satellites pass over the same location every 1-2 days and capture 36 bands of radiation at a spatial resolution of 250 m (820 ft).

Satellites can capture changes in water quality because plankton and sediment absorb and reflect light differently than clear water. Sensors can capture chlorophyll concentrations, reflecting the health of vegetation on land and the concentration of phytoplankton in water. The difference between what is captured by the satellite sensors and what is expected from clear water is an indication of turbidity.

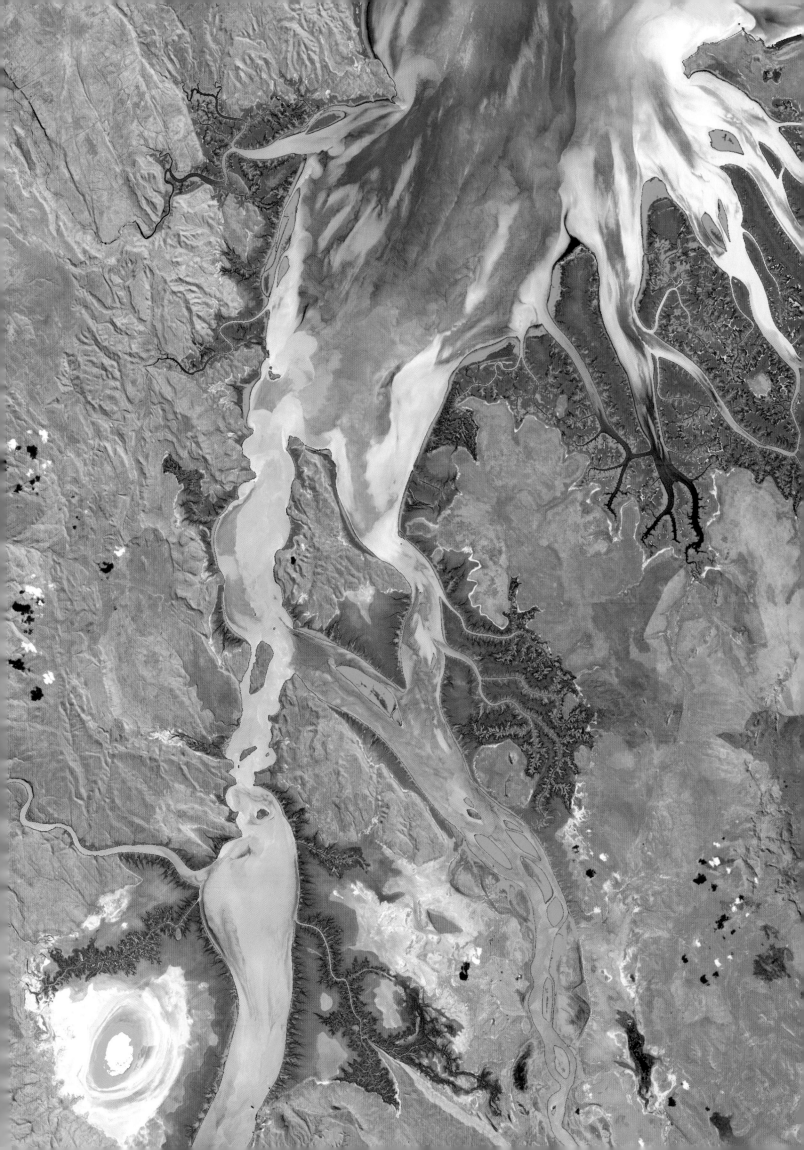

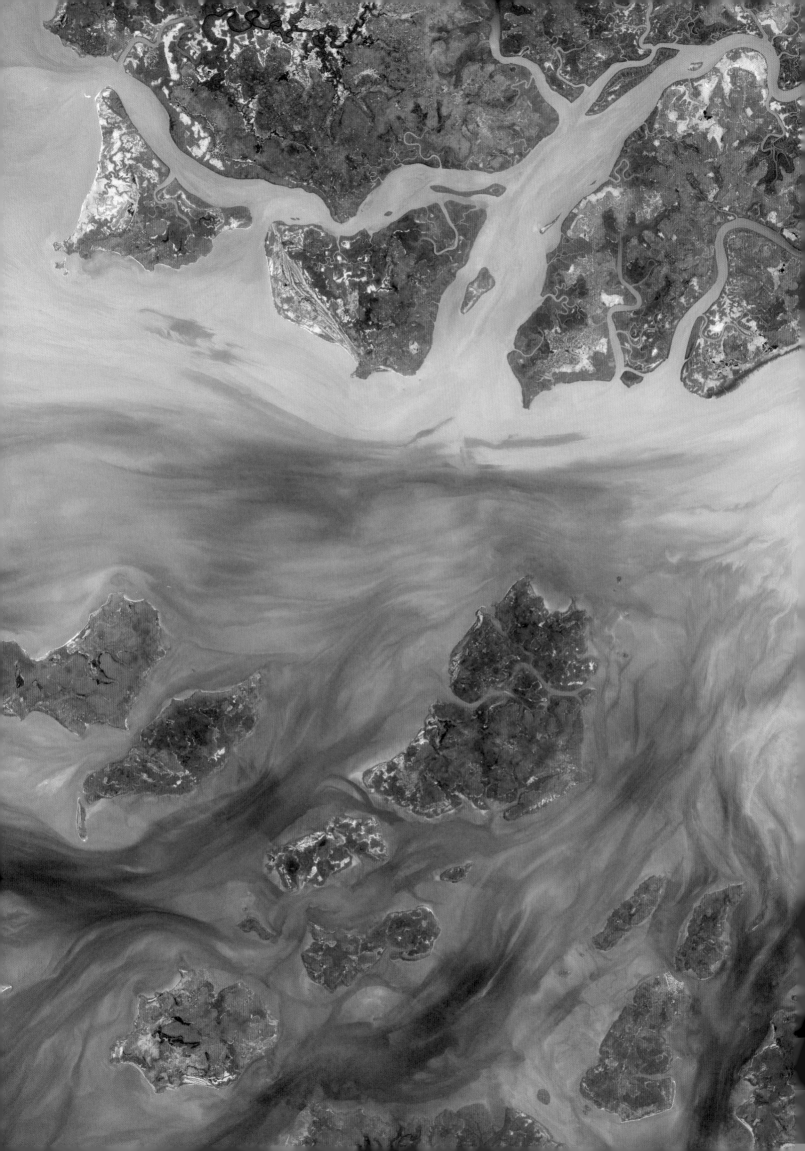

7

The Anatomy of Estuaries

Important estuaries and estuarine cities

The privileged location of estuaries at the interface between rivers and the sea has attracted human settlements since the early ages. In the past, many settlements were shaped and then eventually abandoned due to changes in the estuaries on which they were located. Today, human occupation reshapes estuaries.

MEGACITY GROWTH

Estimates of population growth indicate that the number of megacities around estuaries will increase from 17 to 20 by 2030 (estimated 2030 populations given in parentheses after 2020 population in millions), with more than 355 million people living in these conurbations alone.

5–9.9 million

10–14.9 million

15–19.9 million

20–24.9 million

25–29.9 million

30 million and above

New York-Newark
18.8 (20.0)

Rio de Janeiro
13.5 (14.4)

Buenos Aires
15.2 (16.5)

In 2017, the United Nations estimated that about 40 percent of the global population lives within 100 km (60 miles) of the coast, and 10 percent lives in low-lying coastal areas below 10 m (30 ft) altitude. The global coastal population in 2022 is thus almost 3.2 billion, with many living in urban areas around estuaries.

The world currently has 35 megacities (cities with more than 10 million inhabitants), 17 of which developed around estuaries. Their population alone exceeds 250 million people, and this number does not include other large estuarine conurbations, such as London (9.3 million people), Nanjing (8.9 million), Hangzhou (7.6 million), and Houston (6.3 million), to name just a few. Most of these cities are located on coastal plain estuaries.

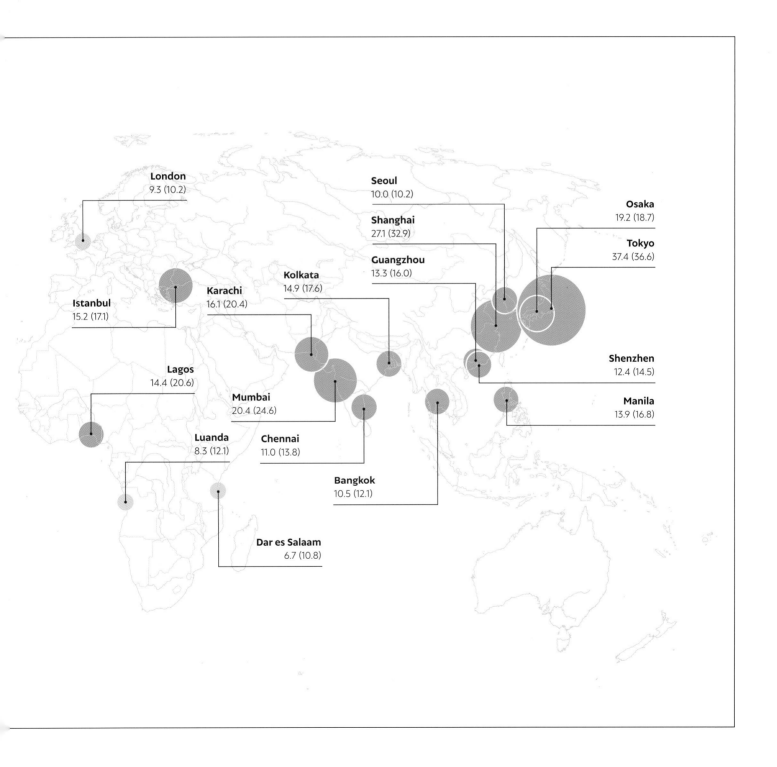

London
9.3 (10.2)

Seoul
10.0 (10.2)

Osaka
19.2 (18.7)

Shanghai
27.1 (32.9)

Tokyo
37.4 (36.6)

Guangzhou
13.3 (16.0)

Istanbul
15.2 (17.1)

Kolkata
14.9 (17.6)

Karachi
16.1 (20.4)

Shenzhen
12.4 (14.5)

Lagos
14.4 (20.6)

Mumbai
20.4 (24.6)

Manila
13.9 (16.8)

Luanda
8.3 (12.1)

Chennai
11.0 (13.8)

Bangkok
10.5 (12.1)

Dar es Salaam
6.7 (10.8)

▲ Growth

Land reclamation has led to a 90 percent loss of the tidal flats in Tokyo Bay over the last 100 years, contributing to ecological imbalances that lead to the development of harmful algal blooms (HABs). Tidal flats host bacteria that naturally inhibit algal growth.

Urbanization pressures

In the past, the largest cities were in the global north, but today most are in the global south. Here, the rates of urbanization and population growth continue to accelerate, while in many developed countries they have decelerated. Thirteen out of the 17 estuarine megacities are located in the global south, nine of them in Asia. By 2030, there will be three new megacities around estuaries, two of them in Africa: Luanda in Angola and Dar es Salaam in Tanzania.

Urbanization, especially at the scale of megacities, causes tremendous changes in the physical, chemical, and ecological characteristics of estuaries, to the point that many lose the amenities that attracted human occupation in the first place. Once ecosystem services are lost, human ingenuity is needed to replace the benefits that once were provided by nature. In megacities, solutions to water pollution, flooding, biodiversity loss, and the implications these have on human health can be difficult to come by.

Tokyo's harmful algal blooms (HABs)

Tokyo was once called Edo, which means "estuary", reflecting its location in the northern part of Tokyo Bay, where the rivers Edo, Arakawa, and Sumida provide freshwater input. Human occupation here started 5,000 years ago, and the city was already prominent in medieval times—as indicated by the construction of Edo Castle, built in 1457 and still used as the Tokyo Imperial Palace. In the 1880s, Tokyo was the first city in the world to pass the mark of 1 million inhabitants, and today it is the world's most populous city, with more than 37 million people living in the metropolitan area.

Through the course of its history, Tokyo has been affected by natural hazards, including earthquakes and tsunamis. In common with many other estuaries around Japan, Tokyo Bay also routinely needs to cope with harmful algal blooms (HABs). These "red tides" are toxic to aquatic organisms and humans, and sometimes can be blue or green depending on the color of the dominant algae.

HABs occur when large amounts of nutrients enter a waterbody, triggering overgrowth of toxic plankton species. In some places they are caused naturally, when upwellings of ocean waters rich in nutrients are pushed toward the coast by winds and ocean currents.

However, they occur more frequently where both natural processes and human activities combine to cause a proliferation of algal blooms. The plankton can use all the oxygen in the water, causing anoxic conditions that lead to mass mortality of aquatic life. The toxic plankton can also enter the food chain, affecting people who eat contaminated fish and shellfish.

HABs causing mass fish mortality and human deaths have been reported in Japan since the thirteenth century. Today, toxin levels in commercial fish species are carefully monitored, but outbreaks of seafood poisoning still occur. In Tokyo Bay, degradation of estuarine habitats and pollution from intensive aquaculture and industrial and domestic effluents cause algal blooms 50 times per year on average, about ten of which are HABs.

Algal blooms occur most often in warm waters, as higher temperatures stimulate plankton growth. Climate change may therefore increase the occurrence of HABs and expand the reach of toxic algal species to new areas. In October 2021, HABs caused large mortalities of salmon and sea urchins along the coast of Hokkaido, an area not usually affected by algal blooms, causing damage to the local seafood industry at a cost of an estimated US$55 million.

Bloom from space
This satellite image taken on June 14, 2019 captures about 100 km (60 miles) of an algal bloom off the coast of Hokkaido, Japan, that in total covered an area more than 500 km (300 miles) long and 200 km (120 miles) wide.

How are estuaries formed?

Estuaries are often described as the area where the river meets the sea. More precisely, they form where fresh water and seawater mix, creating brackish conditions. Many estuaries are found in coastal lagoons, bays, or the lower reaches of rivers where there is tidal influence.

▶ **Hypersaline**
Low freshwater input has favoured sedimentation that has cut off the connection of the Kuyalnik estuary, Ukraine, from the Black Sea. The estuary is drying out and its salinity reaches 300 ppt.

Coastal areas are shaped by the interactions between geological controls, sea-level changes, and hydrodynamic forces (waves, tides, and river flows). Geology controls the topography, types of rocks present and, combined with climatic conditions, the weathering of rock into sediment and the role of gravity in removing them. Sea level and geology together determine the "accommodation" space available for sedimentation at the coast. The energy of hydrodynamic forces and the amount and type of sediment reaching the coast also shape the features that will form.

If more sediment reaches the coast than can be removed by hydrodynamic forces, the accommodation space will be filled, and deltas or coastal plains develop. Where hydrodynamic forces are able to remove sediment at a faster pace than it can accumulate, or sediment supply to the coast is low, and there is mixing of fresh water and seawater, an estuary forms instead.

Types of estuaries

Estuaries can be categorized as one of four types depending on their origin or geomorphology: bar-built estuaries, coastal plain estuaries, fjords, and tectonic estuaries. The last two types are generally deeper and steeper than the first two, which tend to be relatively shallow and low-lying. These four types describe estuaries as coastal waterbodies that are partially enclosed by land and have a permanent or temporary connection with the sea, as indeed most estuaries do. However, this classification is based on the shape and origin of estuaries, disregarding the mixing of waters. The discharge of some rivers can be so powerful that fresh water can flow into the sea for many kilometers. Fresh water and seawater will meet and mix mostly outside the confines of estuarine margins, as happens in the Amazon and Congo, where brackish water can be traced for hundreds of kilometers beyond the river mouths.

SALINITY

The illustration shows the types of water according to the
concentration of salts (or salinity), which is measured as parts
per thousand (ppt). Estuaries are usually brackish, but salinity
varies through time as water inputs change.

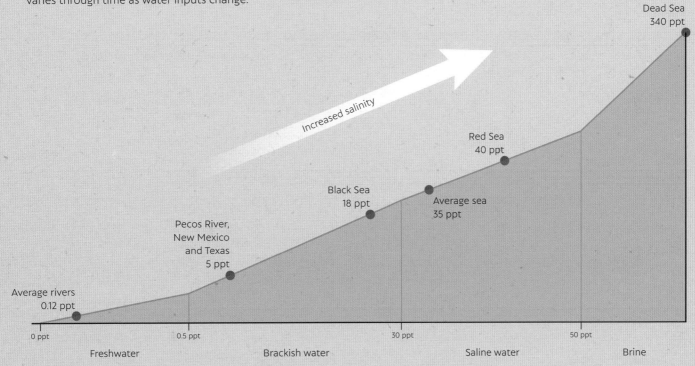

Increased salinity

Dead Sea
340 ppt

Red Sea
40 ppt

Average sea
35 ppt

Black Sea
18 ppt

Pecos River,
New Mexico
and Texas
5 ppt

Average rivers
0.12 ppt

0 ppt   0.5 ppt   30 ppt   50 ppt

Freshwater   Brackish water   Saline water   Brine

▲ **Wave-dominated estuaries**

Bar-built estuaries are predominantly sandy near their inlet, where waves and longshore currents are the main hydrodynamic forces at play and freshwater flow and tidal currents tend to be limited. This is the estuary of the Little River, Maine.

Bar-built estuaries

Bar-built estuaries occur where currents generated by waves create an elongated deposit of sand or gravel at the mouth of a river (or lagoon or bay receiving freshwater input), restricting the connection with the sea. These currents are known as longshore currents or littoral drifts, and they form when waves reach the shore at an angle, forcing water to move parallel to it. The elongated deposit (known generically as a sand bar) is generally aligned with the direction of the current. Bar-built estuaries are found on coasts with relatively high wave energy, common in eastern South Africa, Brazil, Australia, southern and eastern United States, and parts of Europe. In locations where wave energy and freshwater input vary seasonally, the sand bar is usually connected to land (and is then called a spit); examples are found in estuaries on the east coasts of South Africa and Australia. In other locations, the estuary will be formed by one or more sand bars not connected to land (known as barrier islands), as seen along the coast of Texas and the eastern United States seaboard. Barrier islands form along coasts where relative sea level has been rising for thousands of years.

At times, bars might grow across the estuary's entrance, creating a barrier beach that cuts off the connection with the sea. Such features form when waves and the longshore current bring much more sediment to the entrance of the estuary than the estuarine flow or tides can remove. The inlet may reopen naturally after heavy rainfall or major storm events. Sometimes, barrier beaches can last for extended periods, particularly during dry months in tropical and subtropical areas, when the freshwater flow into an estuary can be very weak. This combination of factors is common along the east coast of South Africa. Water quality issues can arise when the connection with the sea is cut off for prolonged periods. In some cases, an artificial inlet is created to restore the tidal flow into the estuary and improve water quality. On the other hand, barriers can be removed completely during extreme rainfall or storm surges, and this can create different types of problems, such as increasing the risk of flooding inland.

Coastal plain estuaries

Also called rias, coastal plain estuaries are drowned river valleys that form when rising sea levels flood existing river valleys with seawater. Rias are currently widespread, as sea levels have been rising along many coasts worldwide since the Last Glacial Maximum. At this time (about 18,000 years ago), ice cover formed in the last glacial period was at its greatest extent and sea levels were more than 120 m (400 ft) lower than today, as large volumes of water were trapped in glaciers. Shorelines shifted hundreds of kilometers seaward, transforming large areas of previously shallow seas (the continental shelf) into coastal plains. River valleys expanded on these coastal plains, adjusting to the changes in baselevel. (For more on this, see page 58.)

Glaciers retreated when temperatures started to increase, releasing the water previously locked up as ice back to the sea. Sea levels rose rapidly for thousands of years but reduced to the current rate from around 2,000–5,000 years ago. As this process continues, river valleys fill with seawater, forming rias.

▼ **Rias**
Coastal plain estuaries vary in shape but typically are similar to a funnel, being narrow at the river end and widening toward the sea, as shown here at the mouth of the Avon River in Devon, England.

*The steep walls and calm
waters of fjords offer
breathtaking views.
Geiranger Fjord and
Nærøyfjord in Norway
are together listed as a
UNESCO World Heritage
Site for their stunning
scenic beauty.*

Fjords

Fjords are drowned glacial valleys. Glaciers are slow-moving bodies of ice and
sediment, and cut a valley in the ground where they pass. The retreat of glaciers can
leave behind deep, steep valleys, with an accumulation of sediment at the retreating
end that forms a barrier or sill. Fjords form when these depressions are filled with
seawater. The presence of fjords is limited to higher-latitude areas where glaciers
once dominated, such as in Scandinavia, Iceland, Scotland, New Zealand, northern
North America, and southern South America.

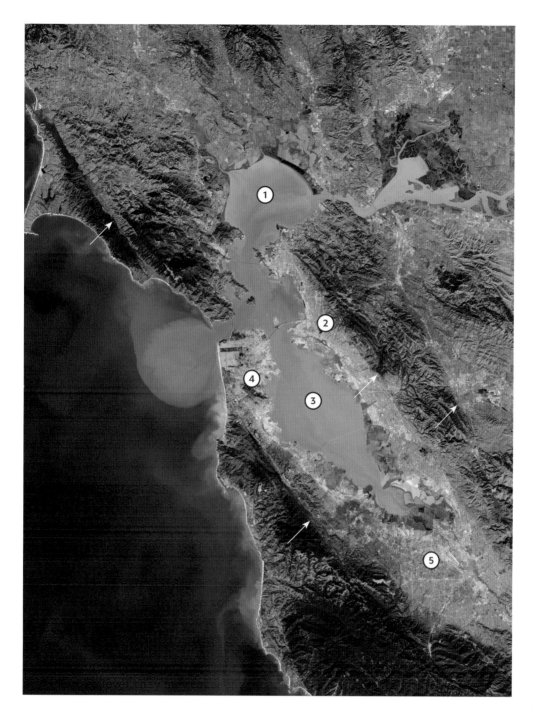

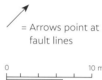

= Arrows point at fault lines

0   10 miles
0   10km

Tectonic estuaries

This type of estuary forms when tectonic movements create a depression along fault lines or grabens, which are valleys of tectonic origin. If these depressions are below sea level, they may cause the incursion of seawater and force convergence of river drainage into the valley, thus forming an estuary. San Francisco Bay (West Coast of the United States) is one of the best-known tectonic estuaries. Along the coast of California, tectonic activity has caused a series of faults to accommodate the movement between the Pacific and North American tectonic plates (see page 28). These plates slide against each other, creating tension that is released in the form of fractures (faults) and earthquakes. San Francisco Bay is a tectonic estuary formed due to the relative movement along seven major faults that run almost parallel to one another in this area.

The diversity of estuarine shapes

Estuaries come in many sizes, shapes, and depths. They occur in areas of small and large tides and small and large river discharges, and in a range of climates. Estuaries create a variety of habitats that attract freshwater, estuarine, and marine species.

DIVERSITY OF FORM

These four examples, taken from the United States (Galveston Bay), France (Arcachon Bay), South Africa (Tugela estuary), and Australia (Hastings estuary), illustrate the diversity of shapes and sizes of bar-built estuaries. Other estuary types are also diverse.

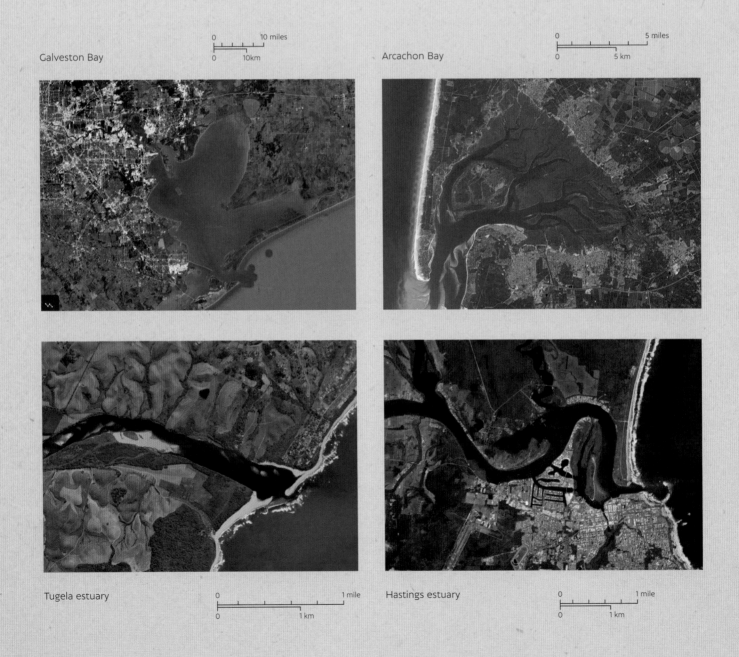

Galveston Bay

0   10 miles
0   10km

Arcachon Bay

0   5 miles
0   5 km

Tugela estuary

0   1 mile
0   1 km

Hastings estuary

0   1 mile
0   1 km

Estuaries are particularly diverse because they are formed and shaped by many processes interacting across a range of timescales. Meteorological conditions greatly influence what happens over days and weeks, as rainfall affects river flows, and winds form waves and, together with changes in atmospheric pressure, can create storms. Climate determines what happens over seasons. On a longer timescale, changes in relative sea level are important across many decades, centuries, and millennia. Geological processes occurring over thousands and millions of years set the foundation that determines which coastal types and landscapes will form in each location. Astronomical processes are also important, as they create tides, but the tidal range in each location is determined by the shape and size of oceanic basins and seas. Each place on Earth has a particular combination of these settings, which determines whether estuaries will form and, if they do, what shape they will have and the types of ecosystems that can develop there.

The largest estuaries

Estuaries are so varied that it is difficult to find consensus about which is the largest, smallest, or deepest. Their boundaries fluctuate and their size can be described using area, volume, or length. The largest in area tend to be coastal plain estuaries, while the deepest are undoubtedly the fjords. The St. Lawrence estuary in Canada is often said to be the longest in the world. It extends 400 km (250 miles) from Île d'Orléans, near Quebec City, to the river mouth at the Gulf of St. Lawrence, with brackish waters traced more than 1,200 km (750 miles) to the Strait of Belle Isle. It has the second-largest freshwater discharge in North America—an average of 12,000 m³/s (420,000 ft³/s), equivalent to five Olympic-sized swimming pools per second at Quebec City—and the tidal range is around 3 m (10 ft). The St. Lawrence estuary has a typical funnel shape but the geological settings here are complex. The river valley developed in a depression that was formed tectonically millions of years ago and subsequently filled with ice during glaciations over the last 2.5 million years.

▼ **Marine life**
The St. Lawrence estuary supports a huge range of living organisms, from migratory birds to colorful underwater creatures. It is a major feeding ground for many marine mammals, such as Fin Whales (Balaenoptera physalus), while another 12 whale species also visit the estuary.

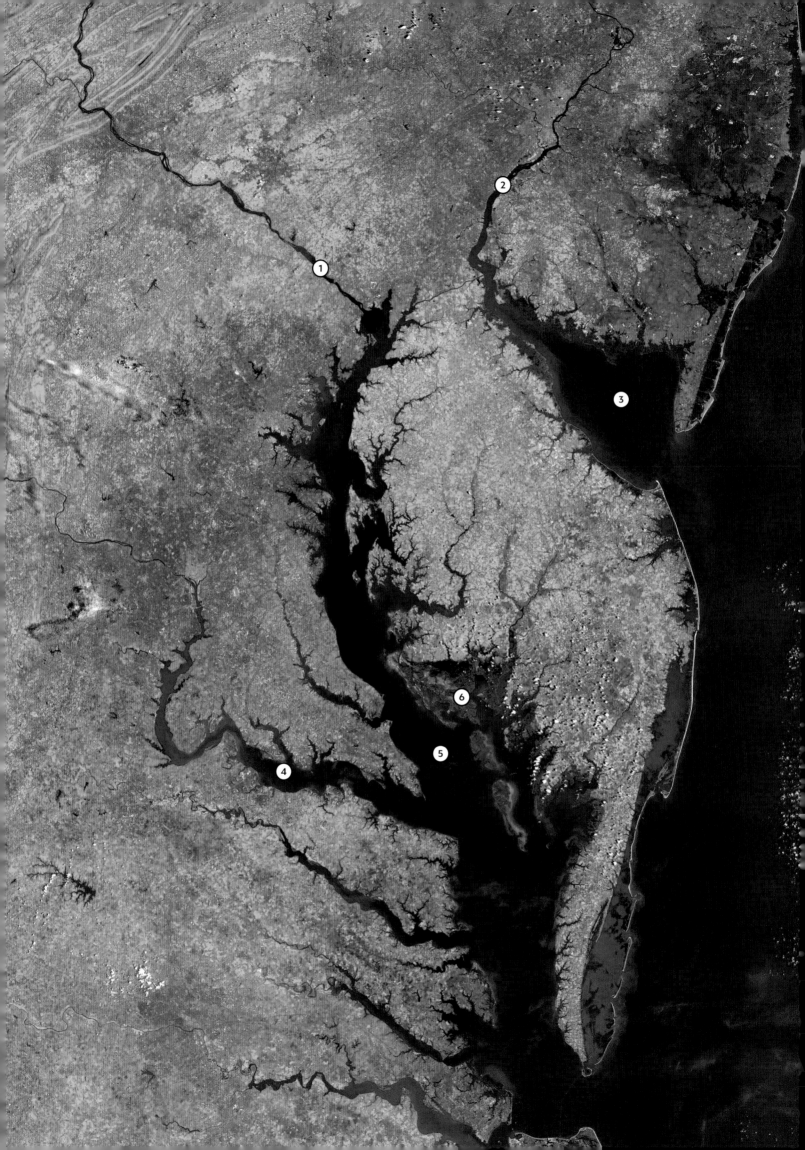

Others consider the Gulf of Ob in Siberia to be the world's largest estuary. It measures 850 km (530 miles) long, and has an area of 41,000 km² (16,000 square miles) and a catchment of 3.3 million km² (1.3 million square miles). The average freshwater discharge into the Gulf of Ob is 16,800 m³/s (5.9 million ft³/s), comprising about 15 percent of the total runoff to the Arctic Ocean. The Gironde estuary in France is the largest in western Europe at 75 km (47 miles) long and with an area of 635 km² (45 square miles), while the Chang Jiang (Yangtze) estuary in China is the largest in Asia, with a length of 150 km (93 miles), an area of around 2,200 km² (850 square miles), and an average discharge of more than 28,600 m³/s (940,000 ft³/s). Both the Gironde and Chang Jiang estuaries have a typical funnel shape and a tidal range of around 5 m (16 ft).

Chesapeake Bay

Chesapeake Bay is the largest estuary in the United States and considered to rank third in the world. This ria formed when rising sea levels flooded the Susquehanna River valley, and due to the drainage system of the area it does not have a typical funnel shape. A human population of more than 18 million live in the catchment area, including the cities of Washington, DC, and Baltimore. Due to sediment input, the bay is very shallow, with an average depth of 6.5 m (21 ft). Chesapeake Bay is a biodiversity hotspot, receiving about a third of the Atlantic Coast migratory bird population. It was the first estuary in the United States to have an integrated catchment bay restoration program.

The deepest estuaries

Norway has about 1,200 documented fjords, including some of the most widely known. However, the world's deepest known fjord is Skelton Inlet on Antarctica's Ross Ice Shelf, with a maximum depth exceeding 1,900 m (6,200 ft). This is nine times the depth of the deepest part of the Congo River, which is the deepest river in the world (see page 86). The longest and deepest fjord in Norway is the Sognefjord, often called the "king of fjords." It extends more than 205 km (127 miles), its maximum depth exceeds 1,300 m (4,200 ft), and it is surrounded by mountains that reach 1,000 m (3,300 ft) in altitude. Considering the range of elevations from the top of the mountains that flank the estuary to its maximum depth, then the Sognefjord might indeed be the king.

Chesapeake Bay
This Landsat 5 satellite image highlights the drainage of the many rivers discharging into both margins of Chesapeake Bay on the northeast coast of the United States. The dark-reddish areas on the eastern margin of the bay (6) are marshes that support thousands of birds.

1 Susquehanna River
2 Delaware River
3 Delaware Bay
4 Potomac River
5 Chesapeake Bay
6 Marshes

A fate to fill

Estuaries are sediment traps. In geological terms, most estuaries are short-lived, gradually transforming into marshes. Their lifetime is determined by the rate at which they infill with sediments. Many estuaries have formed in the last 10,000 years, yet many have also disappeared over that time frame.

Siltation in estuaries

Most estuaries are muddy. Mud is a term used to describe deposits formed from a mixture of silt and clay, which are sediments of very small size (called fine-grained sediments). Deposits of mud are often indicative of low-energy environments, where fine suspended sediments can sink slowly through the water column. This accumulation of sediment is called siltation, in a reference to silts, the size of the sediments that dominate in many estuaries.

In estuaries, siltation is intensified by flocculation. This is a complex biogeochemical process involving clay minerals, salts, and organic matter, in which clay particles bind together to form flocs. The flocs are larger and heavier than individual clay particles, and so will settle faster, thus increasing the rate of sedimentation. Flocculation also plays a role in the concentration of nutrients, heavy metals, and microplastics in estuarine sediment, as these tend to bond with flocs.

Due to the geochemical bonding that takes place in the presence of salts, only very fast tidal or river currents can remove the mud deposits that settle on the estuary bed. This bonding causes mud particles to stick together and, for this reason, silts and clays are frequently cohesive sediments. As fast flows are uncommon in estuaries, mud deposits tend to accumulate through time. Without this geochemical bonding, estuaries would not be as muddy as they are, and the estuarine sediments would not be as rich in organic matter and, in some cases, pollutants.

▶ **Mudflats**

The receding tide exposes mudflats and the network of creeks that mark this intertidal environment. Mudflats are very important feeding grounds for wading birds. In sediment-rich estuaries, they can soon gain elevation and be colonized by vegetation.

▶ **Support**

Estuarine habitats provide feeding and roosting grounds for waders, such as these flocking Dunlin (Calidris alpina) at the Hayle estuary in Cornwall, England.

Estuaries offer sheltered waters, connection to the sea and, sometimes, to inland areas through navigable rivers, and access to fresh water. These favorable conditions have attracted human settlement and the installation of ports and harbors through millennia. However, estuaries tend to become shallower through time due to siltation, and this has sealed the fate of many ancient harbors and is a common problem for modern ones.

Human activities can accelerate siltation. Deforestation and poor soil management in river catchments can increase the amount of sediment reaching estuaries, while land reclamation may reduce the area of estuaries at rates that equate to hundreds of years of natural siltation. Although dredging can actively remove sediment buildups, it is a palliative measure that is unlikely to change the fate of estuaries in the longer term.

The fate of ancient ports

Sheltered waters are needed to ensure the safety of seafarers, vessels, and their cargo, but they invite sedimentation. In modern times, dredging operations are often needed to keep channels deep enough for safe navigation. In the past, early forms of engineering were used to create shelter for boats, but there was no easy way to mitigate siltation. As a result, many historical ports were abandoned because they became too shallow. In some places, harbor activities were moved to different parts of the estuary as siltation progressed, but this is only a temporary delay of an inevitable fate.

About half of the known ancient harbors in the Mediterranean were abandoned or moved to a different location, mostly due to siltation, in common with many other places across the Old World. The port of Chester, on the margins of the tidal River Dee in northwest England, was the busiest port in the region from Roman times until it fell victim to siltation. Boats used to moor just outside the city walls, where Chester Racecourse is located today, but by the fourteenth century siltation had moved the river channel some distance away. Siltation defeated multiple attempts to keep the port alive through dredging and relocation of the moorings to different sites in the estuary. The River Dee was canalized and new quays were constructed in an attempt to renew activities, but the port of Liverpool started to dominate local trade from the seventeenth century. Today, the River Dee is still tidal up to the weir in the center of Chester, but the port no longer exists.

▲ **Silted up**

The Dee estuary in northwest England is heavily silted up and reaching the late stages of the life cycle of an estuary. This aerial view shows that part of the lower estuary has transitioned into marshland.

The fall of Ephesus

Some 6,000 years ago, the area where the ancient Greek city of Ephesus developed was at the margins of a large gulf formed by sea-level rise. Through time, sediments infilled the gulf, pushing the shoreline further west and forming estuaries and floodplains. Human settlements in the area can be traced back more than 7,000 years, successively shifting locations to follow the movement of the shoreline.

Ephesus was an important city in ancient Greece, as shown by the construction of the impressive Temple of Artemis around 2,800 years ago. This was double the size of the Parthenon in Athens and considered one of the Seven Wonders of the Ancient World. The temple was twice destroyed and rebuilt, only to succumb in 401 CE during the Christian occupation of the city. Only one of its columns still stands on site.

The Temple of Artemis is thought to have been built near the city's port, on the estuary of the River Kaystros (also known as the Cayster or Küçük Menderes). The port was central to the prosperity of the city, which was one of the most important trading centers between the West and East. However, by the Roman period the shoreline had shifted due to siltation. The city surrounding the temple was abandoned due to a combination of this loss of the harbor and the spread of malaria—the development of marshes in the shallowing waters may have facilitated the proliferation of mosquitoes.

Dried up

Harbour Street, extending from the Roman theater in Ephesus toward the estuary, was built in the fourth century CE to connect the city with its port. Right: The brownish area at the end of the road is the wetland that developed where the estuary once was. Below: This artist's representation shows Ephesus port and its connection with the sea in Roman times.

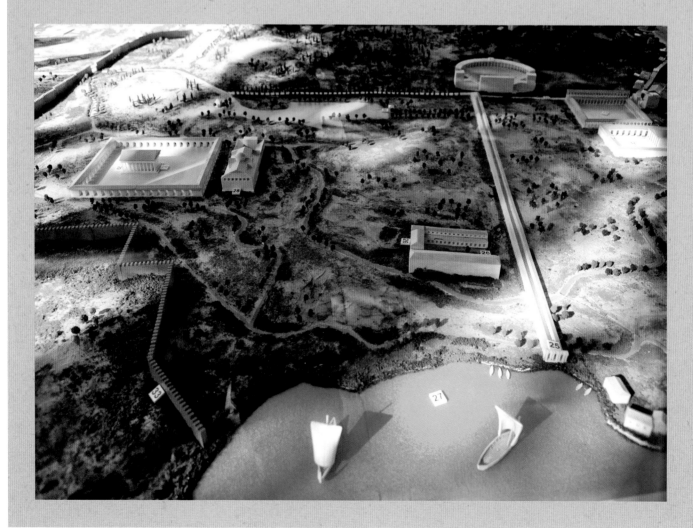

Under the Romans, a new settlement was built at the site and later prospered—the remains of the theater, aqueduct, and Harbour Street still visible today are examples of their impressive architecture. Although siltation of estuaries is a natural process, deforestation and land reclamation for the construction of Ephesus likely accelerated this cycle. In 262 CE, the Goths attacked and destroyed the city, and although it was later rebuilt again, siltation had diminished its cultural and economic importance.

Today, the ruins of the ancient city are about 6 km (4 miles) inland from the Aegean Sea coast, south of Izmir in modern-day Türkiye (Turkey). Ephesus is recognized by UNESCO as a World Heritage Site, not only due to its well-preserved Greek, Roman, and early Christian architecture, but also as an outstanding example of a cultural landscape shaped by environmental change.

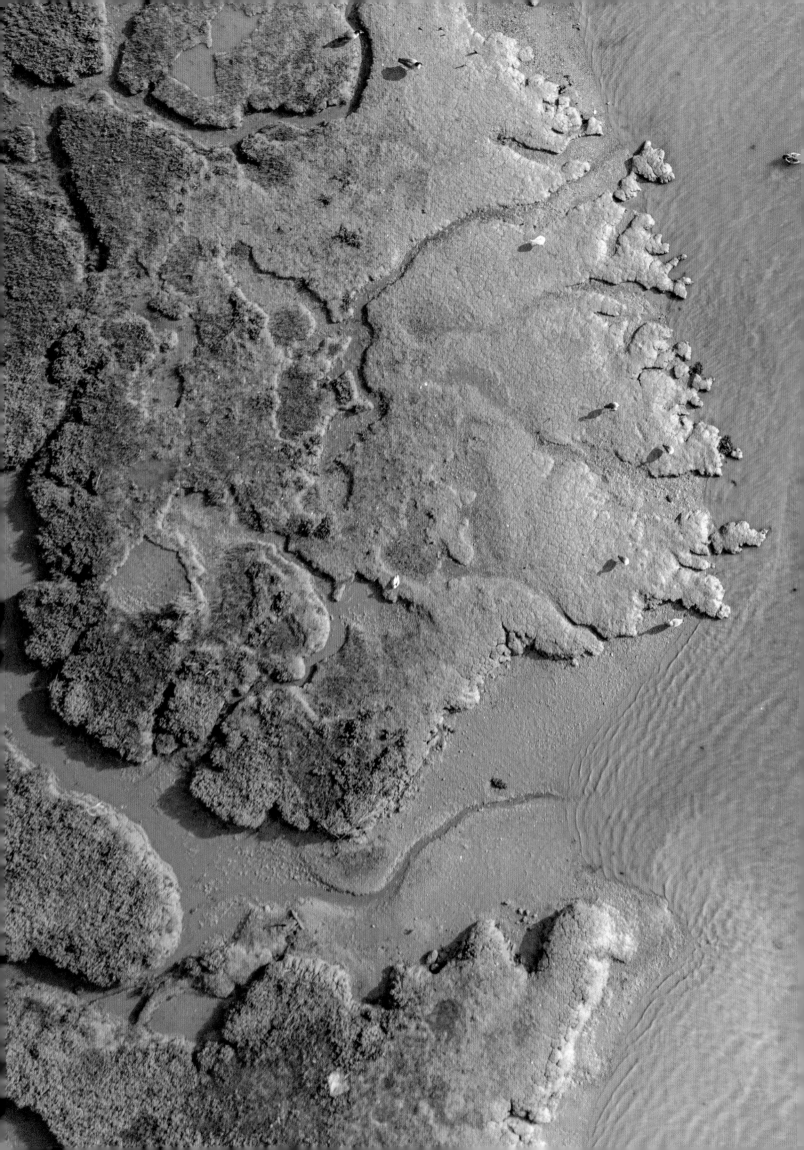

8

Estuarine Ecology and Biodiversity

Productive waters

Estuaries are considered the nurseries of the sea. Many aquatic animals, including commercially important species, rely on these sheltered and nutrient-rich waters to reproduce, spawn, grow, and hide from predators during their early stages of life. The intertidal margins are also vital breeding, roosting, feeding, and overwintering areas for migratory waterfowl and shorebirds.

The combination of muddy and sandy grounds, shallow and deeper waters, varying gradients of salinity, and relatively low wave energy found in estuaries provides ideal conditions for the development of some of the most productive ecosystems in the world. Estuaries are home to saltmarshes, mangroves, oyster reefs, seagrass beds, and mudflats, and the many species of plants and animals that depend on them. Estuarine biodiversity varies from microscopic bacteria to large marine mammals and includes species that have developed strategies for living with large variations in salinity.

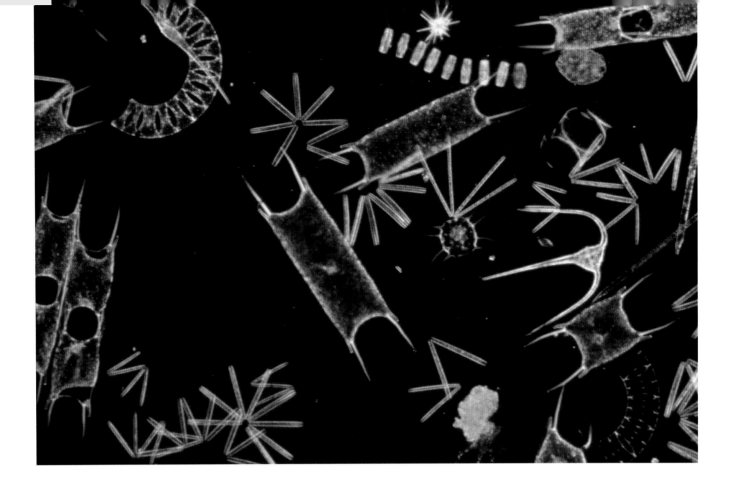

The base of the food chain

Life thrives in estuaries, and this is mostly due to the presence of abundant microscopic life. The concentration of nutrients brought by rivers and the decomposition of organic matter produced by estuarine vegetation stimulate the growth of phytoplankton, microscopic algae that are at the base of the food chain. Zooplankton (microscopic animals, which include larvae of fish and crustaceans) feed on the phytoplankton and, in turn, serve as food for larger organisms in the estuarine food web. On the margins, birds and terrestrial predators feed in shallow waters or the estuary bed during ebbing tides. Plankton are organisms that live in the water column and move with the flow, as they are unable to swim actively. Organisms living near or at the bottom, on or in the sediment, are called benthos.

Microbial life is extremely important for the health of estuaries and is usually dominated by a group of microscopic algae called diatoms, which have many beneficial roles. Diatoms tolerate large variations in salinity and are ubiquitous across the lower and upper estuary, with some species living in the water and others in the sediment. They are a key primary food source and support the development of estuarine ecosystems, consuming less oxygen than they produce and thereby increasing the concentration of oxygen in the water. Oxygen is needed for bacteria to recycle dead organic matter and release nutrients back in the water. Diatoms can photosynthesize at relatively low levels of light and thus start to consume nutrients before other algae, reducing the concentration of nitrogen and phosphorus in the water, and the chances that harmful algal blooms (HABs) will develop.

▲ **Microscopic**
Examples of diatoms and dinoflagellates that give a glimpse of the variety of phytoplanktonic species that make up the important microbial life of estuaries. This image was taken under a microscope with dark-field illumination, where the background appears dark.

◄ **Fisheries support**
Estuarine intertidal and shallow-water vegetation provides food and refuge for fish, shellfish, and other fauna. These mangroves, growing on a sandy bed and in clear water, are characteristic of tropical lower estuaries found along the Caribbean coasts.

Diversity of species

Estuaries have fewer species than adjacent marine environments, as many organisms cannot survive the large variations in salinity and turbidity experienced here. However, the species that are adapted to the variable estuarine conditions can be found in large numbers due to abundant food and less competition with other species. Generally, the number of species increases with the size of the estuary, water temperature, primary productivity, and a more constant connection with the sea. The combination of these factors contributes to the variations in biodiversity found in estuaries around the world. At the local scale, salinity and the type of sediment are the most important characteristics influencing species richness within estuaries.

Areas with higher concentrations of fine sediments in suspension and on the estuary bed have a lower diversity of species. Fine sediments take much longer to settle than sand, causing turbidity in the water. Turbidity reduces the penetration of light and, consequently, the number of primary producers and the organisms that feed on them, with cascading effects on the predators higher in the food chain. Murky waters also affect the ability of predators to see their prey, forcing them to move away to hunt in clearer waters. Although some species are adapted to live in muddy environments, high concentrations of fine sediment can clog the feeding and respiratory systems of filter-feeder organisms such as oysters, clams, and some species of crustaceans and fish. Some large animals are also filter feeders, including Whale Sharks (*Rhincodon typus*), manta rays, Humpback Whales (*Megaptera novaeangliae*), and flamingos.

COMPLEX LIFE

The European Eel (*Anguilla anguilla*) is a Critically Endangered fish with a tough feat to achieve. The eels start life in the Sargasso Sea and drift eastward more than 5,000 km (3,000 miles) to reach western Europe and northern Africa (yellow arrows). They live here in brackish and fresh water for 20–30 years, and pass through three stages before reaching maturity. They then start their migration westward back across the North Atlantic to the Sargasso Sea, where they spawn and die. Efforts are being made to restore their populations in European rivers.

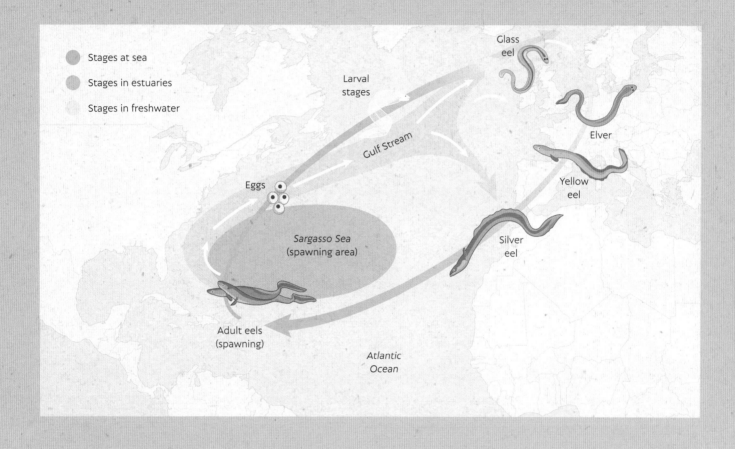

Marine species predominate in estuaries, except in very low salinities. The number of species decreases with the decrease in salinity from the seaward edge of the lower estuary to the upper estuary. Biodiversity is lower at salinities of 5-10 parts per thousand (ppt), as fewer species are adapted to these intermediate conditions. In salinities below 5 ppt, species richness increases again due to the presence of freshwater species. Although estuaries receive many marine and freshwater visitors, only a few species are truly estuarine and spend their life in brackish waters. These species can live in a wide range of salinities but tend to concentrate in areas of intermediate salinity, where they benefit from less competition for habitat and food. Some marine species visit estuaries to feed, while others pass through on their migration to spawn. Salmon and sturgeon are marine fish that breed in fresh water, while some species of freshwater eel spawn in seawater. Shrimps, herring, whiting, plaice, and cod are marine species that spend their juvenile stages in estuaries, where the availability of shelter and food helps them to reach adulthood.

▲ Nursery

The juvenile Lemon Shark (Negaprion brevirostris) finds shelter in the root of mangroves in Eleuthera in the Bahamas. The sharks remain in brackish waters until maturity, moving to deeper open waters as adults. Unlike Bull Sharks (Carcharhinus leucas), they cannot tolerate fresh water.

A pinch of salt

Many organisms have little or narrow tolerance to changes in salinity because they are unable to regulate the concentration of salts in their body. If they move to an environment with lower salinity, their body cells will gain water to dilute the internal salt concentration and consequently their cells will swell. Some may survive a limited time in such conditions but, ultimately, will perish if their cells start to burst. If they move into an environment with higher salinity, their cells will lose water and can dehydrate. Special adaptations are needed to survive varying salinity.

Many marine invertebrates, including sharks and rays, are able to produce organic compounds to balance the internal and external concentrations of salts. Other species of fish can gain or expel the required amount of salts or water through their gills to achieve an equilibrium with the external salinity. Saltmarsh plants use a similar strategy, depending on their needs, and can retain salts in their tissues or excrete them through their leaves.

Some marine predators have evolved specialized adaptations for living in low-salinity waters for long periods, where competition from other predators is reduced. One example is the Bull Shark (*Carcharhinus leucas*), which spends time in brackish waters and gives birth here, using estuaries as nursery grounds. These sharks are known to survive for months in fresh water and have been seen in rivers far inland, including the Amazon and Mississippi. When they are living in brackish and fresh water, Bull Sharks eliminate water from their body by producing urine much more frequently and at a very diluted concentration compared to when they are in seawater. And to retain the concentration of salts in their body, they recycle urea (a nitrogen-rich waste product) in their kidneys rather than eliminating it immediately in their urine.

Sea cow heaven

Maintaining a healthy estuarine system is crucial for the conservation of manatees (genus *Trichecus*) and Dugongs (*Dugong dugon*). These are docile, slow-moving herbivores and, although more closely related to elephants than to bovines, they are nicknamed sea cows. Manatees and the Dugong belong to two different families within the order of aquatic mammals called Sirenia. 'Siren' is a root word that means mermaid in many languages (as in the Spanish *sirena* and French *sirène*). You might not see the similarity between these mammals and mermaids, but the Italian explorer Christopher Columbus did. In January 1493, a few months after he first landed in the New World, he described seeing three mermaids, "not half as beautiful as they are painted," around the shores of today's Dominican Republic. This journal account was the first written record of manatees in the Americas.

There are four species of sea cow, but not all are marine creatures—the Amazonian Manatee (*Trichecus inunguis*) lives only in fresh water. The two other species of manatee, the West African Manatee (*T. senegalensis*) and West Indian Manatee (*T. manatus*), are found in marine, brackish, and fresh water. In contrast, Dugongs live only in marine and brackish waters.

SEA COWS

Global distribution of sea cows.

- Florida Manatee (*Trichechus manatus latirostris*)
  Status: endangered

- Antillean Manatee (*Trichechus manatus manatus*)
  Status: endangered

- Amazonian Manatee (*Trichechus inunguis*)
  The only manatee restricted to fresh water
  Status: vulnerable

- West African Manatee (*Trichechus senegalensis*)
  Status: vulnerable

- Dugong (*Dugong dugon*)
  Status: vulnerable

- Dugong (Nansei subpopulation)
  Status: critically endangered <10 mature adults (Okinawa)

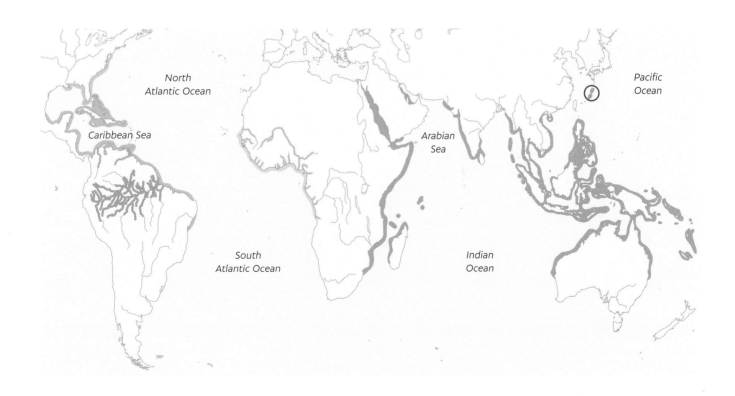

▲ **Underwater**

Manatees stay underwater most of the time, and show only their nostrils when they come to the surface to breathe every 15 minutes or so. Small fish tend to stay around manatees to eat the algae that grow on them.

Except for the Amazonian Manatee, sea cows depend greatly on estuaries for shelter, food, and access to fresh water. They feed mostly on seagrasses and less so on algae and mangroves. From time to time, manatees need to ingest fresh water to regulate the salt concentration in their body, and do so by drinking or eating vegetation with a high water content. During local dry seasons, when freshwater input to estuaries is reduced and salinities are higher, manatees roam upriver feeding on freshwater plants. They stay in the lower estuary and in coastal waters during wet seasons, when salinities are diluted by increased river flow and rainfall.

Endangered

The manatee and Dugong populations have been decreasing due to a range of human pressures, from hunting and injuries caused by boats, to environmental degradation and loss of seagrass habitats. Dugongs and Amazonian Manatees are listed as Vulnerable on the International Union for Conservation of Nature (IUCN) Red List of Threatened Species. In 2019, the distinct group of Dugongs (subpopulation Nansei) that live along the coast of Okinawa in Japan was estimated to comprise fewer than ten animals, leading to their categorization as Critically Endangered. Both subspecies of the West Indian Manatee were assessed as Endangered in 2008, as their populations were estimated to include fewer than 2,500 mature individuals and likely to reduce by 20 percent over the next 40 years. Conservation efforts are in place, and the population of the Florida Manatee (a subspecies of the West Indian Manatee, *Trichecus manatus latirostris*; see box) is recovering, although with large fluctuations.

Saving the Florida Manatee

In the summer, Florida Manatees disperse along the United States coasts between Texas and Massachusetts. As they are sensitive to temperatures below 20 °C (68 °F), they seek the warmer waters of Florida in the winter, congregating around outflows from power plants and thermal springs. Prolonged colder spells can lead to starvation and be fatal, as manatees can exhaust local pastures when they are unable to forage further afield. Cold spells, loss of seagrass beds, and poisoning by harmful algal blooms have led to an increase in the mortality of manatees in Florida, with a record of 1,100 deaths in 2021 and a concentrated effort to feed starved animals with lettuce in the first months of 2022. Such events show the precarity of these populations.

The Crystal River, flowing into the Gulf of Mexico in northwest Florida, has a coastal plain estuary with unique natural and cultural features. Here, springs constantly supply fresh water at 22 °C (72 °F), creating a safe winter refuge for hundreds of Florida Manatees. In the United States, manatees are protected by law and refuges are established for the conservation of protected species. The Crystal River National Wildlife Refuge was created in 1983 to safeguard the areas around the springs within Kings Bay, and is the only such refuge in the country focusing on the protection of the Florida Manatee. From November to March, when the number of manatees is at its greatest, areas designated as sanctuaries are closed to the public.

The Florida Park Service estimates that about 75 percent of the state's recreational and commercial fisheries are found in the Crystal River estuary, where both mangroves and saltmarshes act as nursery grounds and provide shelter. Although saltmarshes tend to occur in temperate areas and mangroves in warmer temperatures, here the subtropical climate and the mixing of fresh waters from springs with seawater from the Gulf of Mexico favor the growth of both. In addition, part of this area is considered a national landmark, and an archaeological state park was created to protect the archaeological evidence that people have lived here for almost 2,000 years—one of the longest continuous human occupations in Florida. The start of human settlements coincided with the development of the coastal wetlands, which offered plenty of food.

Visitors
The Crystal River National Wildlife Refuge receives more than 400,000 visitors each year. From April to October, visitors can snorkel, swim, or kayak to get close to manatees.

Life in the mud

Mud accumulates along the margins of estuaries, forming mudflats that are exposed during the ebbing tide. These expanses of thick, sticky, smelly mud might not be visually pleasing to the human eye, but they are bursting with life and provide vital food for shorebirds and other fauna.

Mudflats are unvegetated intertidal habitats that are covered and uncovered regularly by tides. Living here can be more stressful than in other parts of the estuary. Organisms must be able to adjust not only to variations in salinity, but also to wetting and drying, and to associated fluctuations in temperature and food supply. To minimize the effect of these changes, most organisms spend their life buried in the mud and have adaptations to survive in low-oxygen conditions. Bivalve mollusks such as clams and cockles cope with these large variations by closing their shells and relying on the water and oxygen they are able to retain inside.

Under the surface, hypoxic conditions develop rapidly as living organisms and the decay of dead organic matter take up the available oxygen. The burrows and tunnels made by mollusks, crustaceans, and worms help bring water, oxygen, and nutrients deeper into the mud. Without this increased flow of oxygen, many organisms would perish and the decomposition of organic matter would halt, greatly affecting the nutrient cycling that supports the estuarine food web. Many estuarine mudflats have been designated as areas of conservation for their role in supporting internationally important bird populations.

Biofilms

Mudflats contain a large concentration of microorganisms, predominantly diatoms and bacteria. They contribute to carbon fixation and recycle nutrients through the decomposition of organic matter. At low tide, diatoms move to the surface of the sediment, where light enables them to photosynthesize—their primary productivity can be higher than that of planktonic diatoms. These microscopic organisms have adaptations to survive the conditions on top of mudflats. They produce a type of carbohydrate that aggregates bacteria and organic matter, forming temporary communities that maximize the exchange of nutrients and genetic resources. Migrating shorebirds often rely on mudflats as stopovers, where they can rest and restore their energy.

◥ **Recharging**

Worms and other fauna found in mudflats are rich in fatty oils and are a key source of sustenance for fish, crabs, and migratory shorebirds, such as this juvenile Bar-tailed Godwit (Limosa lapponica).

Individuals of this species are known to fly more than 13,000 km (8,000 miles) nonstop on their migration from Alaska to Australasia, but on their return north they do make a stop, on mudflats in Southeast Asia.

▶ **Biofilms**

Biofilms are visible during low tide as a greenish-gold coating on some mudflats, as seen here in San Francisco Bay in California.

They make up more than half of the diet of some migratory birds and are an important food for crabs, fish, and other aquatic fauna.

▼ **Yancheng National Nature Reserve**

In Jiangsu province, this reserve is part of China's largest coastal wetland. It is internationally important for migratory birds and one of the largest habitats for the rare Père David's Deer or "Milu" (Elaphurus davidianus).

Balancing interactions

Life on the mudflats involves complex interactions between species and the environment. The biofilms produced by diatoms have a binding effect that stabilizes sediment and reduces mudflat erosion. This binding creates a more substantial habitat for other species to colonize, but conditions depend on the balance between primary producers and predators farther up the food chain. Animals that eat diatoms or the biofilm can reduce the stability of the mudflats; they include amphipods, a group of crustaceans that are commonly found in estuaries scavenging for diatoms. The concentration of diatoms can reduce greatly where amphipods are abundant, which in turn can affect the production of biofilms. Some migratory shorebirds prey on amphipods, helping stabilize the number of diatoms and, therefore, the production of biofilms, which they also eat.

Protecting mudflats to save migratory birds

The loss or degradation of mudflats can have detrimental consequences on the populations of migratory birds. Pollution, land reclamation, and port construction are some of the factors leading to the decline of mudflats worldwide. Mudflats along the Yellow Sea, between China and South Korea, are fast disappearing, threatening the bird species that migrate between the Arctic and Australasia along the East Asian-Australasian Flyway. This trend might reverse with the creation of migratory bird sanctuaries along the Yellow Sea coast to protect what are considered to be the world's largest mudflats.

In 2019, two sites (around 185,000 ha/700 square miles) in China's Jiangsu province were listed as UNESCO World Heritage Sites, due to their outstanding universal value as a vital and irreplaceable habitat for more than 400 species of birds and a critical stopover for more than 50 million individual migratory birds. Sites along the South Korea coast were similarly listed in 2021. The migratory birds that use these mudflats include many Endangered species and the Critically Endangered Spoon-billed Sandpiper (*Calidris pygmaea*), whose population has declined rapidly in past decades and in 2021 was estimated to include only 500 birds.

▲ **Monitoring**

*This Spoon-billed Sandpiper (*Calidris pygmaea*) is carrying a transmitter, which records data that help researchers gain an understanding of the timing and patterns of migration of this Critically Endangered species. Researchers can then identify sites that are important to the birds and therefore need to be protected. More than half of the global population of the Spoon-billed Sandpiper uses the mudflats along China's Yellow Sea as a stopover and molting site.*

The value of estuaries

Estuaries provide a range of benefits—or ecosystem services—to people who visit or live around these shorelines, and to society more widely. Important industries that support local and national economies, such as fisheries and ports, depend on estuaries. In addition, estuarine plants remove carbon from the atmosphere and reduce the risk of flooding, and estuaries provide opportunities for recreation.

▼ **Subsistence**

It is estimated that 210 million people living in coastal areas depend on mangroves as a source of food and income. This woman harvests oysters near mangroves (seen in the background) in Senegal.

Ecosystem services is a term used to describe all the benefits that we humans obtain from nature. In general terms, they include all the "products" we extract from nature, such as air, water, soil, food, and wood; and the "services" we receive, including regulation of temperature, pollination, cycling of nutrients, removal of carbon from the atmosphere, and recreation opportunities (see also Chapter 4). As we have seen (Chapter 6), estuaries are important nursery grounds for a range of aquatic species, are critical stopovers for migratory birds, and have been culturally significant as sites of human settlement for thousands of years, including the majority of the world's megacities today. In the United States, for instance, estuarine regions are home to 40 percent of the human population and account for 47 percent of the economy and 39 percent of employment.

Estuaries have undeniable natural and cultural value, offering a range of ecosystem services, some with direct economic value (such as fisheries) and others that are more difficult to quantify. Some are arguably priceless—for example, the archaeological remains of an ancient estuarine port, or the benefit to our wellbeing of visiting an estuarine shore where saltmarshes develop and many species of birds and other fauna can be seen. Ecological economists try to estimate the worth of ecosystems based on the contribution their ecosystem services provide to the economy. Using this approach, estuarine ecosystems are among the highest valued.

Natural capital

Estuarine ecosystem services are estimated to be worth around US$29,000 (in 2007 values) per hectare per year (US$11,740 per acre per year), which is the same value as seagrass beds. Their services include recreation, storm protection, food, habitat, and nutrient cycling. Mangroves and saltmarshes, which are often found along the intertidal estuarine margins, are worth US$194,000 per hectare per year (US$78,540 per acre per year). Their role in protecting coastlines against the impact of waves and flooding, together with their capacity to store carbon and filter pollution, greatly contributes to their value. Only coral reefs have an estimated value higher than estuaries, seagrasses and mangroves, reaching US$352,000 per hectare per year (US$142,000 per acre per year). These estimates are averages of values obtained from local assessments undertaken around the world, where variations between places can be large. Not all ecosystem services can be translated into a monetary price. Thus, these average values give only an idea of the order of magnitude of the contributions these ecosystems make to the economy, and they can change as new assessments become available.

▲ **Artisanal**

Although the farming of fish and algae is expanding, small-scale artisanal fishing is a key source of sustenance and income for more than 110 million people in developing countries around the world. The fish traps used by the Tsonga people in the Kosi Bay estuary in South Africa are designed to let small fish escape and direct the larger fish to the trap, to be collected at low tide.

The importance of nature to our wellbeing and survival has been greatly overlooked and undervalued. Economic development has relied on the use of natural resources but failed to account for the true cost of using these resources, which has led to the widespread degradation of valuable ecosystems, such as those found in estuaries. Translating their value into monetary terms can draw attention to how much they contribute to the economy, helping lead to policies that can halt or reduce human pressures.

Fisheries

Every book or article about estuaries includes a reference to their importance for fisheries. Estuaries support fisheries by serving as breeding or nursery grounds for non-resident commercial species, such as prawns. They are also home and feeding grounds for resident species, some having direct commercial value (including oysters and cockles) and others that are part of the food chain of commercial species. Intertidal habitats provide organic matter and nutrients that support the food web within the estuary and beyond. The survival of fisheries depends largely on the presence and connectivity of the services they need throughout their life cycle across the brackish and marine habitats they inhabit.

In 2020, 112 million tonnes (123 million tons) of animals were harvested in marine waters, with about 70 percent taken through fishing and 30 percent through aquaculture. It is estimated that fisheries and aquaculture support about 600 million livelihoods worldwide, the majority in developing nations.

Marine tigers

Asian Tiger Prawns (*Penaeus monodon*) are consumed worldwide but are a native of the Indo-West Pacific coasts of Australia, Southeast Asia, and southeast Africa. They spend their juvenile phase in mangrove-lined estuaries and move to open marine waters when reaching maturity. The prawns burrow into the seabed during the day and feed at night on detritus, polychaete worms, mollusks, and other benthic fauna. Degradation of estuarine systems and overfishing have reduced wild populations, yet at the same time Asian Tiger Prawns have become one of the most farmed species of prawns. Aquaculture reduces fishing pressures on wild populations but causes other impacts, such as degradation of mangroves, eutrophication, and the risk of spreading diseases to wild animals. There has been a decline in the importation of Asian Tiger Prawns by the European Union, from more than 55,000 tonnes (60,600 tons) in 2016 to 35,000 tonnes (38,600 tons) in 2021, partly because consumers are increasingly seeking products from certified sustainable sources, which rules out many producers.

Collecting cockles

Humans have harvested clams, cockles, and other bivalve mollusks from mudflats for thousands of years. Some species are collected to serve as bait, while others are intended for human consumption. During low tide, these bivalves can be found through the holes they make on intertidal flats when burrowing. In many countries, their collection is permitted for personal consumption and following regulations regarding shell size, quantity taken, and timing of harvest, but a license is usually needed for commercial harvesting.

As the burrows of these bivalves are shallow, small numbers can be collected easily using a trowel, while a garden rake can be used to expose a larger number of shells. After collection, the shells are placed on a tray or in a sieve and the sediment is brushed off. Commercial operations employ many workers, who rake large areas of the intertidal zone and use tractors or quadbikes to transport the harvest.

However, collecting mollusks is not without its dangers. Walking on mudflats at low tide can be treacherous due to the risk of getting stuck in the mud and being caught by rapidly rising waters. In 2004, tragedy struck in Morecambe Bay in northwest England, when 23 illegal immigrants from China died after being caught by the incoming tide while picking cockles at night. They had been brought to Europe to work for a gangmaster and had been sent out onto the mudflats for hours to pick cockles. They were given incorrect information about the tides and had no familiarity with the area and poor knowledge of English. The gangmaster was found guilty of manslaughter and sentenced to 14 years in prison.

▼ **Harvest**

Licensed commercial picking of Common Cockles (Cerastoderma edule) in Southport, northwest England. In the United Kingdom, cockles have been a source of food and income for almost 2,000 years, with yields reaching hundreds of tonnes per month in some locations. The shellfish are sold fresh or cooked in boiling water, or pickled in jars.

9

Estuaries and Us

Food security

Estuaries support most of the world's marine wild fisheries and aquaculture production. With the decline in fisheries due to overfishing and the effects of habitat loss and pollution, aquaculture has spread worldwide as a solution to secure food supply. In your local supermarket and restaurants, you are certain to find fish, prawns, and shellfish that were farmed, some in distant countries.

MARINE AND BRACKISH AQUACULTURE

Thirty-five countries had marine and brackish aquaculture production exceeding 50,000 tonnes (55,100 tons) in 2021, and together were responsible for more than 99 percent of the world's marine aquaculture. China and Indonesia are the largest aquaculture producers worldwide, and the top producers of marine species.

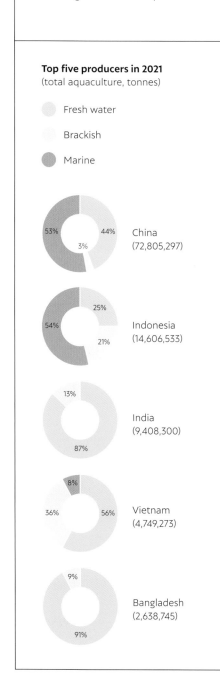

Top five producers in 2021
(total aquaculture, tonnes)

Fresh water

Brackish

Marine

China
(72,805,297)
44% / 3% / 53%

Indonesia
(14,606,533)
25% / 21% / 54%

India
(9,408,300)
13% / 87%

Vietnam
(4,749,273)
8% / 56% / 36%

Bangladesh
(2,638,745)
9% / 91%

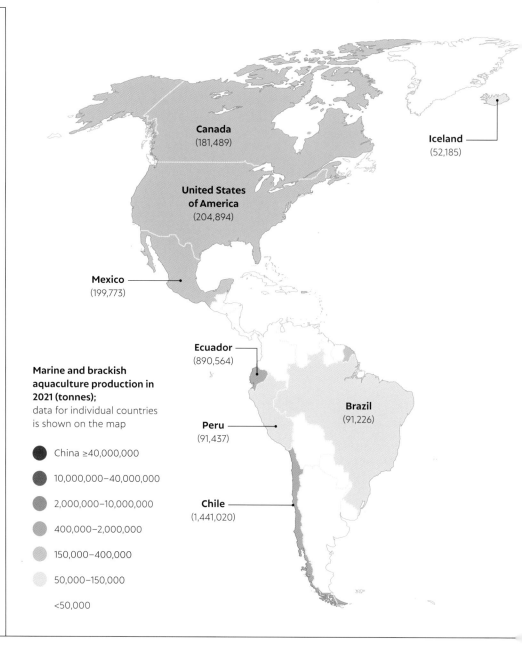

Marine and brackish aquaculture production in 2021 (tonnes);
data for individual countries is shown on the map

- China ≥40,000,000
- 10,000,000–40,000,000
- 2,000,000–10,000,000
- 400,000–2,000,000
- 150,000–400,000
- 50,000–150,000
- <50,000

Canada
(181,489)

Iceland
(52,185)

United States of America
(204,894)

Mexico
(199,773)

Ecuador
(890,564)

Brazil
(91,226)

Peru
(91,437)

Chile
(1,441,020)

More than 215 million tonnes (237 million tons) of freshwater and marine fish, shellfish, algae, and other aquatic organisms were produced worldwide in 2020, with marine species accounting for 56 percent of this. While wild fisheries account for 70 percent of the total production of marine species, 83 percent of freshwater species production comes from aquaculture. In the 1970s, aquaculture represented only 5 percent of the total global production, but by 2020 the figure stood at 122 million tonnes (134 million tons) of aquatic organisms, or 57 percent of global production.

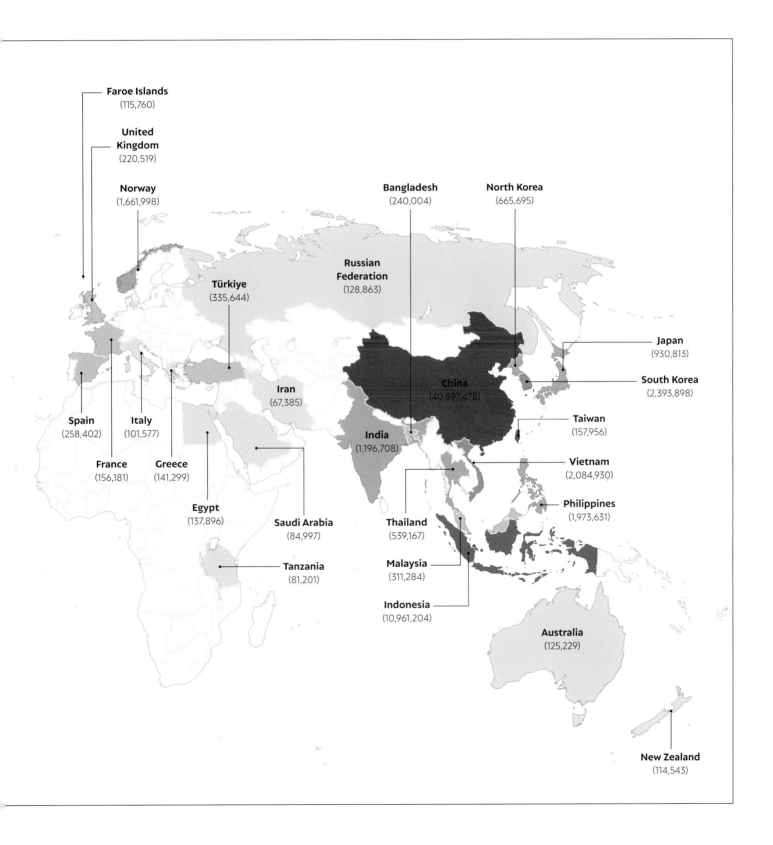

Faroe Islands
(115,760)

United Kingdom
(220,519)

Norway
(1,661,998)

Bangladesh
(240,004)

North Korea
(665,695)

Russian Federation
(128,863)

Türkiye
(335,644)

Japan
(930,813)

South Korea
(2,393,898)

Iran
(67,385)

China
(40,897,478)

Spain
(258,402)

Italy
(101,577)

India
(1,196,708)

Taiwan
(157,956)

France
(156,181)

Greece
(141,299)

Vietnam
(2,084,930)

Egypt
(137,896)

Philippines
(1,973,631)

Saudi Arabia
(84,997)

Thailand
(539,167)

Tanzania
(81,201)

Malaysia
(311,284)

Indonesia
(10,961,204)

Australia
(125,229)

New Zealand
(114,543)

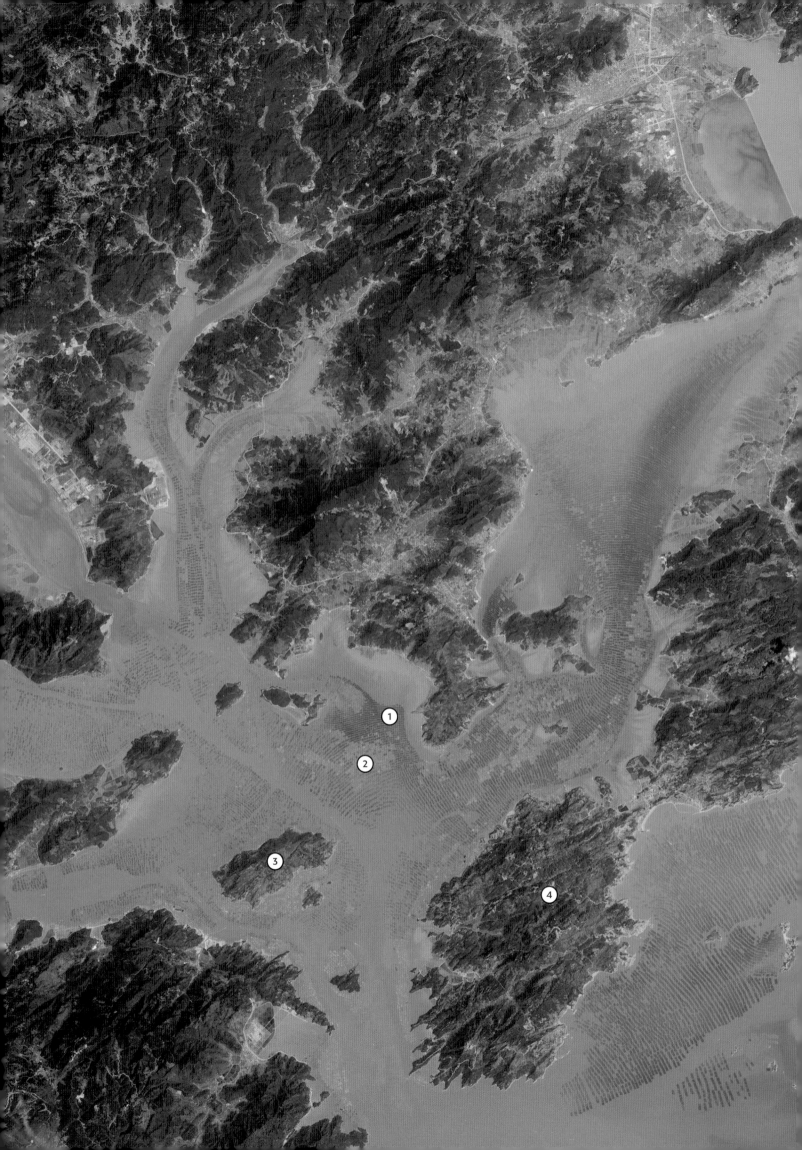

Environmental impacts

Although aquaculture is vital for food security, its expansion often causes environmental problems. Some species, such as prawns, are often farmed in ponds dug close to the sea, and this has been a major cause of estuary degradation and mangrove deforestation. Other species are farmed in estuarine or sheltered coastal areas—in baskets or enclosed nets in the case of salmon, or grown on ropes placed underwater, as with oysters and mussels. This type of aquaculture does not lead to losses of intertidal habitats, but it can affect water quality and local biodiversity and restrict other water-based activities. The detritus produced by fish and their uneaten food creates an excess of nutrients that can lead to eutrophication and algal blooms, particularly where the concentration of fish is high.

Extensive aquaculture

Aquaculture is ubiquitous in Sansha Bay, southern China, where seaweed farms, arrays of fish cages, and floating villages form a mosaic of human interventions. Cage farming started in this area as an effort to produce Large Yellow Croaker (*Larimichthys crocea*), a fish that was once abundant in the bay but whose population was reduced by more than 90 percent in the 1980s through overfishing. Aquaculture efforts in the area started to intensify around Qingshan Island as part of a plan to produce the popular croaker.

In the early 2020s, it was estimated that more than 300,000 cages growing croaker and 40,000 baskets containing abalone occupy these waters, which are among some of the most densely farmed in the world. The floating cages are anchored to the seabed. Aquaculture expansion around the bay also includes seaweed farming (mostly kelp and laver growing along underwater ropes), oyster beds, and a diverse production of octopus, bass, and other species grown in cages. A large extent of the bay's shallow intertidal zone is covered by bamboo poles used to dry seaweeds after harvest.

Sea farming
Aquaculture is extensive within Sansha Bay in China's Fujian province. This Landsat 8 satellite image of April 8, 2017, shows seaweed farming as darker areas and the cages containing fish, abalone, and other marine species in lighter shades.

1 Seaweed farming
2 Cages
3 Qingshan Island
4 Dongchong Peninsula

◀ **Export**

Indonesia is the country with the largest mangrove coverage in the world and one of the highest rates of mangrove loss due to the development of aquaculture ponds.

The floating villages on Sansha Bay are mostly occupied by so-called boat people, who have a long history of living on the water. They are thought to be descended from the Yue ethnic groups of south China, and around 1,300 years ago were driven to live on their fishing boats to avoid conflicts with the Cantonese people. For a long time, they were segregated and not allowed to go ashore. Nowadays, some have moved to the mainland and the younger generations are less interested in keeping up the old traditions.

The mudflats of Xiapu county, in the northern part of Sansha Bay, are marketed as one of the best coastal landscapes for photography in China, particularly at sunrise and sunset, when the varied aquaculture sites are highlighted. It is estimated that photography-related tourism brings more than 400,000 visitors to this area each year. Many tours are on offer, and include vantage points and spots where farmers or fishermen can help stage picture-perfect idyllic rural scenes.

The cages and bamboo poles in the waters of Sansha Bay are now so densely packed that current speed has been halved in the intertidal zone and reduced by a factor of three in deeper waters. Slower currents take longer to flush out nutrients and pollutants from the bay, increasing the risk of eutrophication. The floating structures also limit the amount of light in the water column, and this reduces photosynthesis and slows the decomposition of organic matter. All of these impacts have altered the local ecosystem, reducing biodiversity in the bay.

◀ **Boat people**

Floating villages found in the estuaries and bays of China's Fujian province are home to a minority ethnic group who subsist through fishing and aquaculture. About 7,000 people live on boats or wooden houses built on pontoons and plastic barrels in Sansha Bay.

▶ **Seaweed**

Algae farming in Nusa Lembongan, Indonesia. Although China produced 58 percent of the 36 million tonnes of seaweed harvested globally in 2020, algae farming has spread worldwide, offering income to coastal communities. Seaweeds are increasingly being used in cosmetics, biofuel, and pharmaceutical products.

▶ **Variety**

Colorful circular nets add to the diversity of photographic opportunities in Sansha Bay, China.

Invasive estuarine species

Through human activities, animals and plants have been introduced to areas outside their natural range—many of them unintentionally. Such species may travel long distances unnoticed in ships' ballast water, attached to boats or floating plastic, or in natural products transported by people.

Plants and animals that are found in areas outside their natural range are called non-native or introduced species. Some of them become naturalized and are able to reproduce and develop populations in their new surroundings without causing major changes to the environment and local ecosystem. In some cases, non-native species thrive and spread quickly, causing harm to the new environment or the native biota due to predation, spread of new diseases, or competition for food, oxygen, or space; these are often termed invasive species. Non-native species usually become invasive where conditions are favorable for their growth, and there are no natural predators or diseases that can control their spread. About 14 percent of the species listed in the World Register of Introduced Marine Species are known to be invasive in some areas.

THREAT

Invasive species are one of the major threats to ecosystems, affecting more than 85 percent of the world's ecoregions. In coastal areas, hotspots for invasives are found around the North Sea, the English Channel, the Mediterranean, and the northeastern Pacific—all areas of intense maritime traffic.

Number of invasive species

No data   1–2   3–7   8–15   16–30   31–56

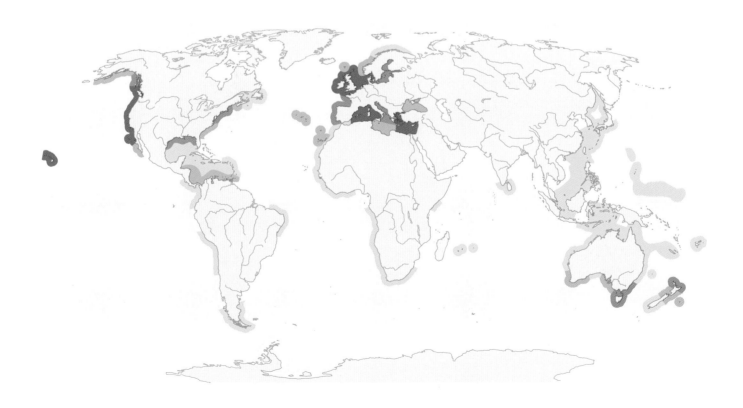

WORLD TRAVELERS

The pathways of invasive marine species closely match the major global shipping routes, indicating the importance of ships as carriers of these species.

Invasive marine species pathways

● From NW Atlantic

● From NE Atlantic

● From Asia

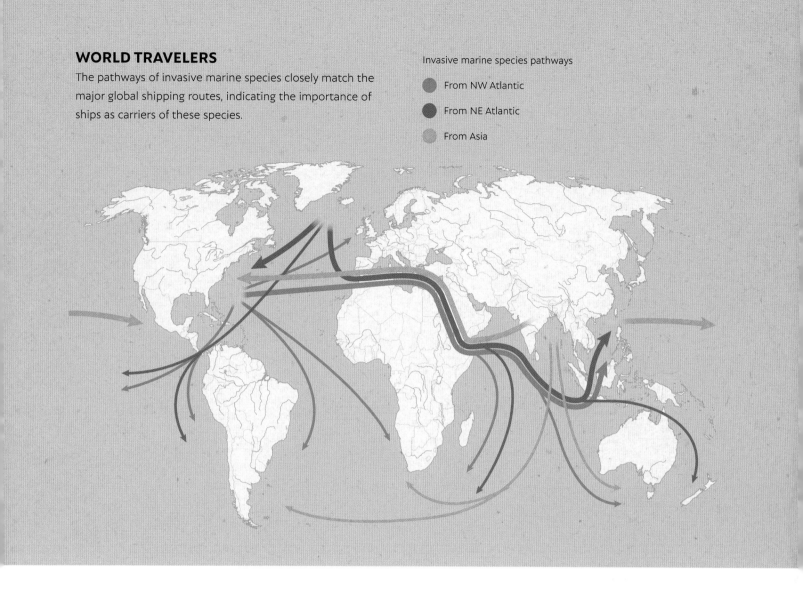

From port to port

Fishing gear, aquaculture, and ballast water used to balance the weight of ships are major vectors for the spread of non-native species to estuaries globally. When ships are carrying less or no cargo, they take up water from around the port to increase their stability during navigation. They then release the ballast water in the next port where cargo is loaded, which can introduce non-native organisms and pathogens that have survived the journey into the surrounding waters. These may then spread throughout river catchments. It is estimated that as many as 7,000 species are currently being transported in ballast waters worldwide in any given one-hour window. Some bays and estuaries around busy ports are now dominated by introduced species, such as San Francisco Bay in California, where more than 230 non-native species are found. The International Convention for the Control and Management of Ships' Ballast Water and Sediments of 2004 established international standards that came into force in 2017 with the aim of halting the introduction of potentially harmful species. Some of these measures include restricting the discharge of ballast water to open water, and ballast water treatment to kill or sterilize organisms before they are released into non-native areas.

▲ **Invasive release**

Many aquatic species reach non-native areas via the release of ship's ballast water in estuaries and coastal waters.

Unwanted arrivals

Invasive aquatic species include a wide range of organisms, from microscopic dinoflagellates that can cause harmful algal blooms, to seaweeds, bivalves, crustaceans, and fish. Most invasive species can live in a wide range of conditions, facilitating their spread and adaptation to new locations.

Despite the harmless appearance of comb jellies, one species, the American Comb Jelly (*Mnemiopsis leidyi*), which is native to temperate and subtropical estuaries of eastern South and North America, has greatly impacted the fisheries of the Caspian and Black Seas. This carnivorous species eats zooplankton, including fish eggs and larvae, outcompeting native predators and decreasing the biodiversity and biomass across the food chain. Catches of the Anchovy Kilka (*Clupeonella engrauliformis*) reduced fourfold after the arrival of the American Comb Jelly in the Caspian Sea. The decline in anchovy numbers also affected the population of their predators, which included other commercial fish and seals. The population of the American Comb Jelly itself decreased sharply in the Black Sea after the arrival in 1997 of another predator, the Brown Comb Jelly (*Beroe ovata*), also through ballast water. This allowed recovery in the abundance of zooplankton and the rest of the food chain.

From Asia to the West

Some of the top invasive species affecting freshwater ecosystems in North America and northern Europe—such as the Zebra Mussel (*Dreissena polymorpha*), native to Russia (see Chapter 5), and the Chinese Mitten Crab (*Eriocheir sinensis*)—were also introduced via ballast water. The Chinese Mitten Crab is native to eastern Asia and is farmed and commercialized for human consumption in China. These crabs can live in a wide range of salinities, which facilitated their spread through estuaries and rivers across Europe, Canada, and the east and west coasts of the United States. The crabs spend their early life at sea as planktonic larvae and enter estuaries as juveniles, while adults live in rivers or estuaries, migrating to mate in brackish waters. They affect native species through competition, predation, and changes to the

▼ Chinese Mitten Crabs

*Adult Chinese Mitten Crabs (*Eriocheir sinensis*) are easily identified by their hairy legs and furry claws, and in some non-native areas, such as Canada and the United Kingdom, they are the only crab found in rivers. Their burrowing can destabilize river and estuarine banks, with fencing needed to strenghen banks as seen here along the River Thames in England.*

environment. The invaders are also a vector of diseases, transmitting a pathogen that is fatal to European crayfish species and lung fluke parasites to humans who consume undercooked contaminated crabs. Attempts to eradicate Chinese Mitten Crabs by capturing them during their migration have failed.

Containing invasion

Eradication of invasive aquatic species can be very difficult and costly, and is practically impossible if they become naturalized, which means they can reproduce and spread in their new environment. Eradication usually involves mechanical removal or the application of chemical products. However, this is attempted only when the invasive species occupies a relatively enclosed small area and the chances of success outweigh the potential impacts that the removal method may have on the native biota. The deliberate introduction of non-native species to control invasive species is usually avoided due to the wider consequences should they themselves also become invasive. Containing the invasive species by preventing its spread, or by suppressing the population in sensitive areas to reduce impact, are the most common strategies employed when resources are available. Once an invasive species arrives in a new area, it can spread to other local watercourses via boats and equipment. Therefore, public campaigns offer a cost-effective strategy to reduce the spread of invasive species.

The tropical green alga *Caulerpa cylindracea*, native to Australia, has spread along the coast of 12 countries across the Mediterranean Sea since 1990, when it was first reported in the area. It outcompetes other local algae and seagrasses, rapidly colonizing areas that are degraded and reducing local biodiversity. In France and Spain, attempts were made to eradicate the invasive alga from natural reserves where colonization was incipient. Although a high rate of success was reported, the technique is applied only in small areas at a time, making the cost and effort required unviable where colonization is widespread.

◀ **Hitchhiking**
Warning campaigns that identify invasive species and alert recreational and commercial users to inspect and clean their boats and equipment on site are used worldwide to reduce the chances of people unintentionally spreading these species to new areas.

Impacted from afar

Their connections through water make estuaries prone to receiving pollutants from local and distant sources, both from activities that take place at sea and on land around the catchment. Pollutants concentrate in the sheltered waters and muddy sediments of estuaries. Most estuaries worldwide are affected by human activities at some level.

Oil spills

Oil spills usually occur due to accidents involving ships, pipelines, and oil rigs caused by human error or natural hazards. Crude oil derives from fossilized marine plankton that was buried many millions of years ago. Although this organic matter eventually degrades in the sea in sunlight and through the action of marine bacteria, until it does so it can have severe impacts on marine and estuarine biota. Oil is lighter than water and floats, forming a thick sticky layer on the sea surface. Over time and through the action of currents, any oil spills at sea will reach coasts and estuaries. Primary productivity is reduced as the oil blocks light from entering the water column and can cover the roots and leaves of intertidal plants, such as mangroves and saltmarsh species. Many animals die from ingesting oil or due to suffocation.

◀ **Float**
The Deepwater Horizon oil rig was located in the Gulf of Mexico 80 km (50 miles) off the Louisiana coast. For 87 days between April and July 2010, it leaked more than 554,000 tonnes (610,000 tons) of crude oil, impacting 11,000 km² (4,250 square miles) of the sea surface and 2,000 km (1,200 miles) of coast between Florida and Texas. The reddish color seen here is an indication that dispersants were used to break up the oil.

▲ **Ashore**

About 1,000 tonnes (1,100 tons) of oil reached the coast around Grand Port Bay, Mauritius, in 2020 causing the country's worst ecological disaster due to the impacts on the sensitive ecosystem of the Pointe d'Esny wetland, recognized as internationally important by the Ramsar Convention.

▼ **Clean-up**

Floating barriers, known as booms, are often used to contain the spread of oil spills. The oil can then be mechanically scooped from the water's surface, as here in Mauritius.

Accidents with oil tankers can cause spills of many thousands of tonnes and cost billions in damage and clean-up, as happened with the Exxon Valdez. In 1989, it crashed into Bligh Reef in Alaska's Prince William Sound, spilling 37,000 tonnes (40,800 tons) of crude oil along 2,000 km (1,200 miles) of coastline and killing more than 250,000 seabirds, 3,000 otters, 300 seals, and 22 Orcas (*Orcinus orca*). According to Exxon Mobil, it paid more than US$4.3 billion in compensation, fines, and other costs. It was the largest oil spill to affect United States coasts until the Deepwater Horizon disaster in 2010.

Pollution

Estuaries are receptacles of water and sediment, and any pollutants they may carry. Some pollutants are transported by runoff from the land in the river's catchments, while others are intentionally dumped into watercourses or result from accidental spills. Some plants and filter-feeder animals, such as oysters, mussels, and clams, act as natural filters and help to clean the water, but in the process they themselves can accumulate pollutants in their tissues.

Toxic mud disaster

On November 5, 2015, failure of a tailings dam released 63 million m³ (2.2 billion ft³) of toxic mud into the Doce River catchment, causing the worst ever environmental calamity in Brazil, known as the Mariana disaster. The tailings from iron ore mining contained high concentrations of mercury, lead, arsenic, copper, zinc, cadmium, and manganese. These heavy metal pollutants bind through ion exchange to clay minerals, the finest sediment in mud, and can travel long distances suspended in water and then become concentrated where the mud settles. The metals can then enter the diet of animals feeding on contaminated sediment, and the burrowing of these animals and the decay of organic matter can release metals into the water column and be consumed by other organisms.

In its path, the toxic mud carried down the Doce River buried villages, caused 19 fatalities, affected the water supply to more than 700,000 people, and killed 14 tonnes (15.5 tons) of fish and other macrofauna. Floating barriers were unsuitable to contain the contaminated sludge, which reached the Doce River estuary in Espírito Santo state and the Atlantic Ocean, from where it spread north up to the coast of Bahia, the third state affected by this disaster. The concentration of metals in the tissues of fish rendered them unsafe for human consumption two years after the disaster, and after four years contamination levels in estuarine sediment still posed a high risk of causing adverse effects on the biota.

The Mariana dam was co-owned by the Brazilian company Vale and the Anglo-Australian BHP Billiton. It had shown signs of failure twice before the disaster, yet no plan was put in place to prevent or respond to a dam collapse. Four years later, another dam owned by Vale collapsed 120 km (75 miles) from Mariana, releasing 12 million m³ (425 million ft³) of contaminated sediment and killing 270 people. Poor management and failed regulatory measures contribute to making tailings dams a disaster waiting to happen. Unfortunately, this issue is not exclusive to Brazil, and 75 percent of environmental disasters related to mining worldwide are caused by the failure of tailing dams.

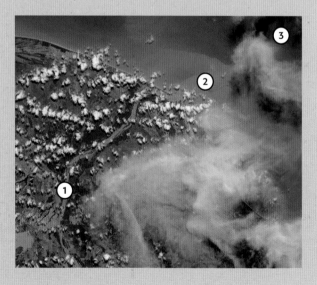

Toxic
Land-based activities far from the coast can contaminate estuaries and coastal waters. Toxic mud released by the collapse of the mining tailings dam on the Doce River in 2015 in Minas Gerais, Brazil, flowed downstream and crossed the state border into Espírito Santo. Seventeen days later, it reached the estuary and coastal waters, having contaminated soil and water along 650 km (400 miles) of the river's catchment.

1 Doce River
2 Contaminated sediment plume
3 Atlantic Ocean

0   5 miles
0   5 km

The contaminants in the bodies of filter feeders will be passed on to the predators that feed on them and increase in concentration through the food chain, a process called bioaccumulation. Some toxins that bioaccumulate, such as mercury and DDT, a type of pesticide, can cause serious illnesses and mortality in animals at the top of the food chain, such as fish, birds, otters, and people. Mercury poisoning causes serious permanent neurological problems that can lead to paralysis and death, which became known as Minamata disease following the environmental disaster in the eponymous Japanese city. In 1956, thousands of humans, cats, and dogs in Minamata were affected by severe neurological symptoms caused by methylmercury poisoning. The mercury was flushed into the waters of Minamata Bay through wastewater from a chemical factory and accumulated in the muddy sediment, from where it entered the food chain through the consumption of contaminated fish and shellfish.

Eutrophication

Eutrophication (see Chapter 1) is one of the most widespread issues affecting urban and rural estuaries worldwide. Nitrogen and phosphorus are common nutrients used in fertilizers and are found abundantly in runoff from overfertilized agricultural land, golf courses, and discharges of sewage treatment plants. While nutrients are not pollutants per se, in excess they lead to algal blooms, which can cause severe environmental problems. The overgrowth of algae can deplete the water of oxygen, due to increased uptake by organisms that feed on the algae and by bacteria as they break down dead algae. Low concentrations of oxygen cause stress on aquatic organisms, and large mortality can occur if the water becomes anoxic, which can impact recreational and commercial fisheries. Besides unpleasant odors and water discoloration, some blooms of dinoflagellates and cyanobacteria produce harmful toxins. Paralytic shellfish poisoning (PSP), for example, can cause symptoms ranging from nausea and numbness of the face to muscular paralysis and death in people who eat shellfish contaminated with toxins produced by certain species of dinoflagellate.

▲ **Sea sparkle**

A "red tide" caused by the dinoflagellate Noctiluca scintillans *caused the closure of beaches around Sydney, Australia, in November 2017. Blooms of this species do not cause serious human health issues, but they do produce ammonia and can cause hypoxia, affecting local ecosystems. This species is also known as "sea sparkle," due to its bioluminescence, as seen here in Hong Kong waters in 2015.*

Restoring habitats

Estuarine habitats are being lost at faster rates than forests due to multiple human pressures. Habitat loss, degradation, and fragmentation have caused the decline of fisheries and biodiversity that depend on the health of estuaries and their connection with freshwater and marine habitats.

Critically Endangered sturgeons

Sturgeons spend most of their lives at sea and migrate to spawn on gravel riverbeds, which have clean water that is rich in oxygen. After hatching, they slowly move downstream and spend two to three years in brackish waters, using estuaries as nursery grounds before moving out to sea. Sturgeons have existed for more than 200 million years and the 27 species alive today are considered the world's most endangered group of animals. Seven of the eight species of sturgeon native to Europe are listed as Critically Endangered on the International Union for Conservation of Nature Red List, which triggered a pan-European plan to save them from extinction.

▶ **Exhibit**

In the 1970s, large sturgeon were still being caught in the United Kingdom, usually as bycatch in trawlers, as demonstrated by this 40 kg (88 lb) sturgeon caught in Lyme Bay, Dorset in 1975, and a 3-m (10-ft) long one weighing 230 kg (507 lb) caught off Dogger Bank in the North Sea in 1977, and later exhibited in Harrods, the famous department store in London.

For thousands of years, humans have caught sturgeons for their meat and eggs, the latter known as caviar. Pollution, habitat degradation, barriers that obstruct migration, overfishing, and bycatch in commercial trawlers and gillnets are the key threats facing these extraordinary fish. Ironically, having caused their decimation, humans are now their only hope of survival.

Saved from extinction

Once widespread across Europe, the Common Sturgeon (*Acipenser sturio*) is now one of the rarest of the sturgeon species. It can reach 6 m (20 ft) in length and live more than 100 years. The Garonne and Dordogne Rivers, which discharge into the Gironde estuary in France, host the only remaining population of Common Sturgeon, but the fish here have not reproduced in the wild since 1994. The species was saved from extinction by captive breeding to restock the population, with the release of thousands of fish in the Gironde catchment having taken place since 2008.

As sturgeons take 15–20 years to mature, population recovery can involve many decades of continuous and appropriate support. France and Germany have made great efforts to reintroduce native sturgeon species into the wild, but this is only part of the solution. Some populations of sturgeons migrate across national boundaries, and so their survival depends on international cooperation and commitment to restore habitats from the rivers to the sea. Measures to halt illegal trade and bycatch, which continue to be a major threat to their survival, are also required.

▲ **Revival**
On May 21, 2016, the ongoing initiative to re-establish the population of Baltic Sturgeon (Acipenser oxyrinchus) reached the mark of one million juvenile fish released in the River Oder, Germany.

Where are the oysters?

Today, oysters are associated with upmarket seafood restaurants, but in the past they were a common food for many coastal communities. Archaeological evidence from Australia indicates that these shellfish have been part of the human diet for more than 12,000 years. Oysters have been cultivated for human consumption since Roman times, but in the wild they form reefs that provide habitat for many organisms, attracting fish and birds and creating a biodiverse ecosystem. They also improve water quality, as each adult oyster can filter 150 liters (40 US gal) of water per day, and they reduce coastal erosion by retaining sediment and dampening waves. However, more than 85 percent of shellfish reefs have disappeared globally due to overharvesting, siltation, poor water quality, and overall degradation of estuaries. These reefs are one of the most threatened marine ecosystems.

In the 1800s, oysters were sold everywhere in New York City, known then as the oyster capital of the world. Natural oyster beds covered 900 km^2 (350 square miles) from Jamaica Bay to the lower reaches of the Hudson River, and due to overharvesting, they were replaced by cultivated beds in the early 1800s. Water pollution started to affect production, and the last commercial oyster bed closed in 1927.

Many countries are making efforts to reinstate shellfish reefs, following the example of the United States, where oyster reef restoration started in the late 1990s. Poor water quality can prevent the development of oyster reefs, and restoration can succeed only after pollution sources are reduced. In Australia, the government is investing in the restoration of 30 percent of the country's lost shellfish reefs, which were abundant across tropical and temperate waters. Australia's most common shellfish reef species had declined critically by the start of the twentieth century due to pollution and overharvesting for human consumption and lime production. The native Sydney Rock Oyster (*Saccostrea glomerata*) is now found in only 10 percent of its historical locations and the endemic Southern Mud Oyster (*Ostrea angasi*) in less than 1 percent. It has been estimated that 1 hectare of restored shellfish reef in Australia will provide habitat for 100 marine species and each year will produce 375 kg (825 lb) of fish catch, filter 2.7 billion liters (700 million US gal) of water, and remove 225 kg (500 lb) of nitrogen and phosphates.

One of the techniques used to reinstate oyster reefs involves creating a reef base on the seabed with a mix of rock rubble and shells recycled from restaurant waste. The new substrate is seeded with juvenile oysters grown in local hatcheries to help colonization, which will attract more oysters and, over time, a range of other organisms.

▲ **Restoring**
A barge carrying granite rubble to form the base of oyster reefs in the Piankatank River, a tributary of Chesapeake Bay in the United States, where more than 200 ha (500 acres) of oyster reefs have been restored. It is estimated that every US$1 invested in oyster reef restoration here creates US$7 in benefits.

Coastal squeeze

Intertidal habitats, such as mangroves and saltmarshes, develop in areas that are inundated during rising tides and exposed at low tides. Mangroves and saltmarshes can keep up with moderate rates of sea-level rise by accumulating sediments, if they are available, and shifting inland to higher grounds. When sea levels rise, saltwater penetrates further inland and intertidal wetlands start colonizing new areas, replacing grasslands and other vegetation less adapted to saline waters. Through time, the maximum incursion of high tides can be restrained by coastal engineering structures, while the low-tide mark continues to be pushed landwards by rising seas. This human interference leads to a gradual but persistent reduction in intertidal habitats, known as coastal squeeze. Breaching or removing coastal protection structures can re-create intertidal space and offset the losses caused by coastal squeeze.

Re-creating intertidal space

In many coastal locations worldwide, upgrading or building new seawalls or earth embankments along estuaries or coasts is now constrained by high economic costs and undesired environmental impacts regulated by legislation. Reinstating the tides into farmland reclaimed from estuaries, rivers, or the open coast through the planned breaching or removal of flood protection structures is increasingly used in Europe to re-create intertidal space for the development of saltmarshes. This approach is known as managed realignment, depoldering, or de-embankment, and reflects a growing trend toward nature-based solutions that provide multiple environmental benefits, in addition to a more sustainable approach to managing flooding and erosion risks. Managed realignment has been implemented at more than 140 sites in Europe, particularly where maintaining flood protection would not be cost-effective and there is a legal requirement to offset coastal squeeze or compensate for the loss or degradation of habitats.

▼ **Breached**

Steart Marshes in southwest England on September 8, 2014, the day 200 m (650 ft) of earth embankments along the Parrett estuary were removed to allow tidal flow to re-create 477 ha (1,200 acres) of intertidal habitats lost through land reclamation. The area is managed by the Wildfowl and Wetlands Trust as a nature reserve.

SQUEEZED

The process of coastal squeeze, using saltmarshes as an example of an intertidal habitat. Saltmarsh species colonize lower or higher ground according to their tolerance to flooding and salinity. Given space, they gradually shift inland as a response to rising sea levels. When this shift is obstructed by the presence of coastal engineering structures such as seawalls, the intertidal area is squeezed as sea level rises through time, eventually leading to habitat loss due to inundation. Reinstating the tidal flow can re-establish the space for intertidal habitats to develop.

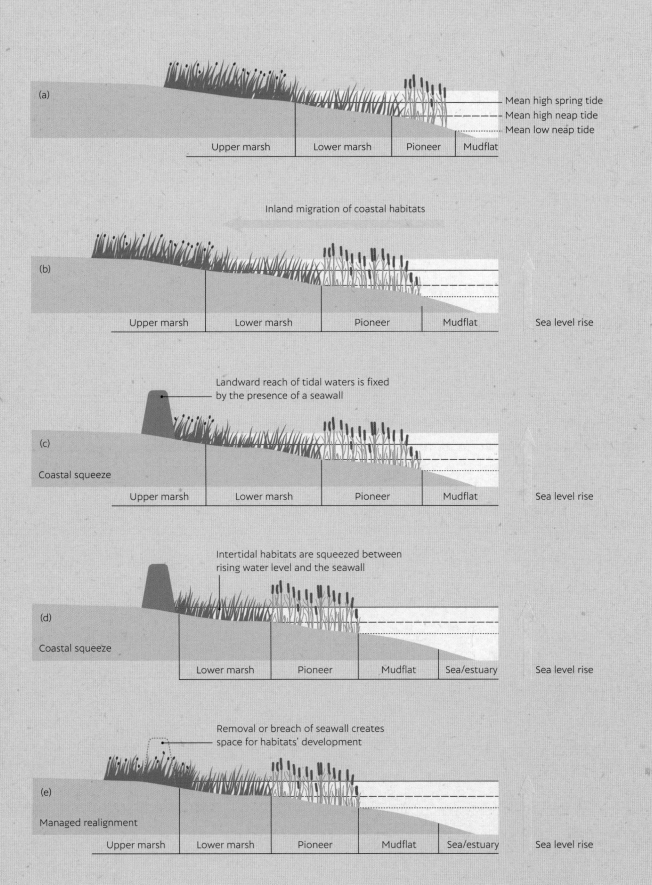

Saltmarshes to prevent flooding

Coastal squeeze and land reclamation are often cited as the main causes for the loss of intertidal habitats, and have contributed to a 50 percent reduction in the global extent of saltmarshes over the last century. Saltmarshes develop in temperate and subtropical climates, and it is now widely known that they provide important benefits to people and the functioning of healthy estuaries. They support fisheries and biodiversity, improve water quality, produce organic matter and create favorable conditions for the microbial life that recycles it, are very effective at storing carbon, and attenuate wave energy, reducing coastal erosion and flooding.

To benefit from these ecosystem services and offset or compensate for habitat loss, many countries are now investing in creating conditions for the development or restoration of saltmarshes. In Germany and the Netherlands, for example, saltmarshes are an integral part of a hybrid concept of more sustainable flood protection that combines "living" dikes and saltmarshes. Living dikes have gentler slopes to incorporate elements of nature and facilitate multiple uses, such as grazing areas and recreation. The development of saltmarshes seaward of dikes enhances flood protection and reduces maintenance costs over time, as they can naturally adjust to changing sea levels.

▼ **Protection**
Artificial fences and creeks stimulate saltmarsh colonization in front of dikes, helping to prevent flooding and the risk of overtopping along the Groningen coast, part of the Dutch Wadden Sea National Park.

Dwindling estuaries

Through time, many estuaries have been greatly reduced in size to make way for farmland, ports, urban areas, and other types of coastal development, leading to habitat and biodiversity loss and degradation of water quality. Former intertidal habitats have been drained, and tidal incursion prevented by the construction of embankments or other forms of flood control. These waterbodies have been transformed in ways that have restricted the space for water, paradoxically increasing the risk of flooding to the reclaimed land and other low-lying areas along the estuary and upriver.

Lake Hachirō, north of Akita in Japan, illustrates the intense changes experienced by estuarine systems worldwide. This freshwater lake of around 4,000 ha (15 square miles) is a remnant of a brackish lagoon that was more than five times larger but was transformed by extensive land reclamation. Postwar, helped by Dutch engineers and the World Bank, the Japanese government revived an old plan to transform the lagoon into rice paddies to resolve food shortage and unemployment. About 585 families selected from across the country settled on the new land, naming their village Ōgata, which means "big lagoon." To control flooding and reduce salinity, the lagoon was cut off from the sea by a tide gate and the polder was enclosed by embankments. The fishing communities opposed the project as local fisheries were lost, but rice production here is the most productive in the country, despite the low water quality and algal blooms that now affect the lake.

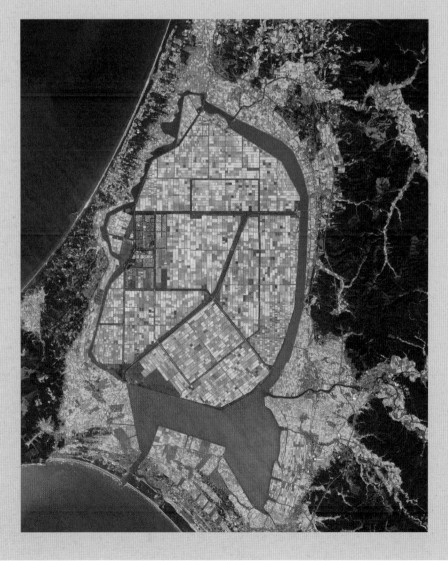

Reclaimed
This Landsat 8 satellite image, taken on September 27, 2014, shows the large polder covering more than 17,200 ha (66 square miles) that was built on Hachirōgata Lagoon between 1957 and 1977. The reclamation transformed the lagoon, which was the second-largest waterbody in Japan, into the much smaller Lake Hachirō.

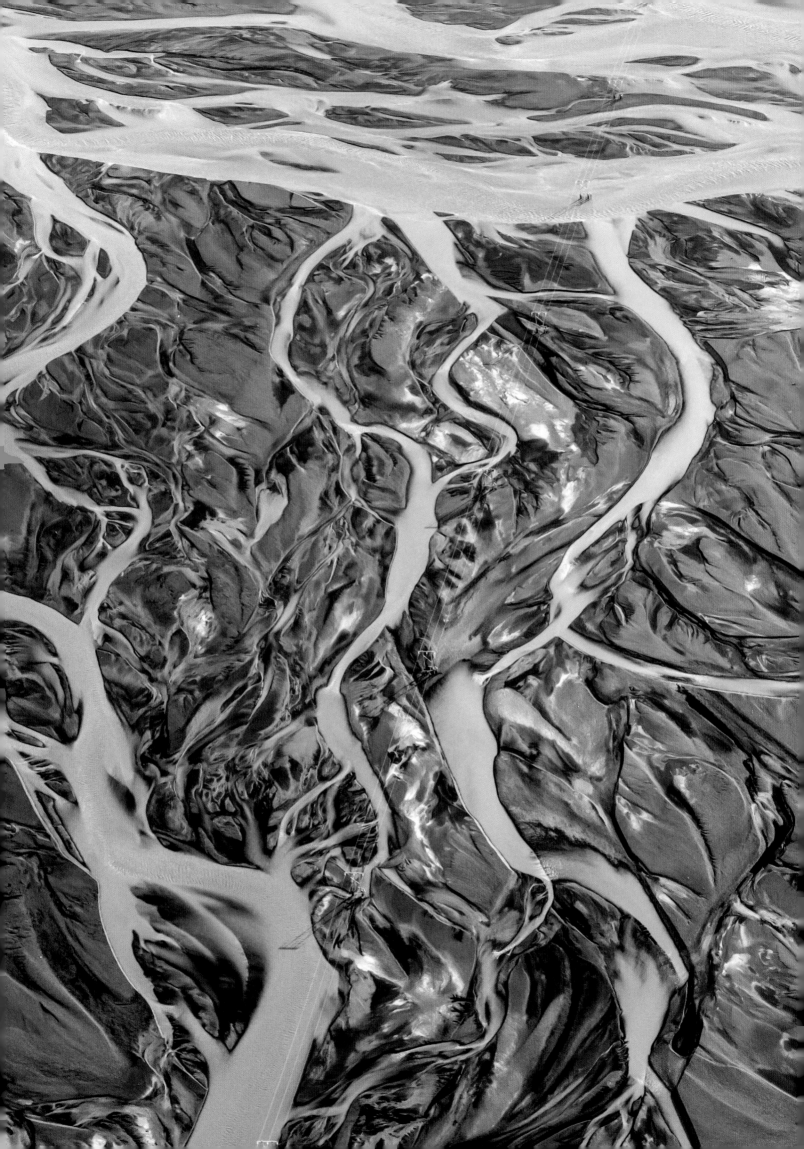

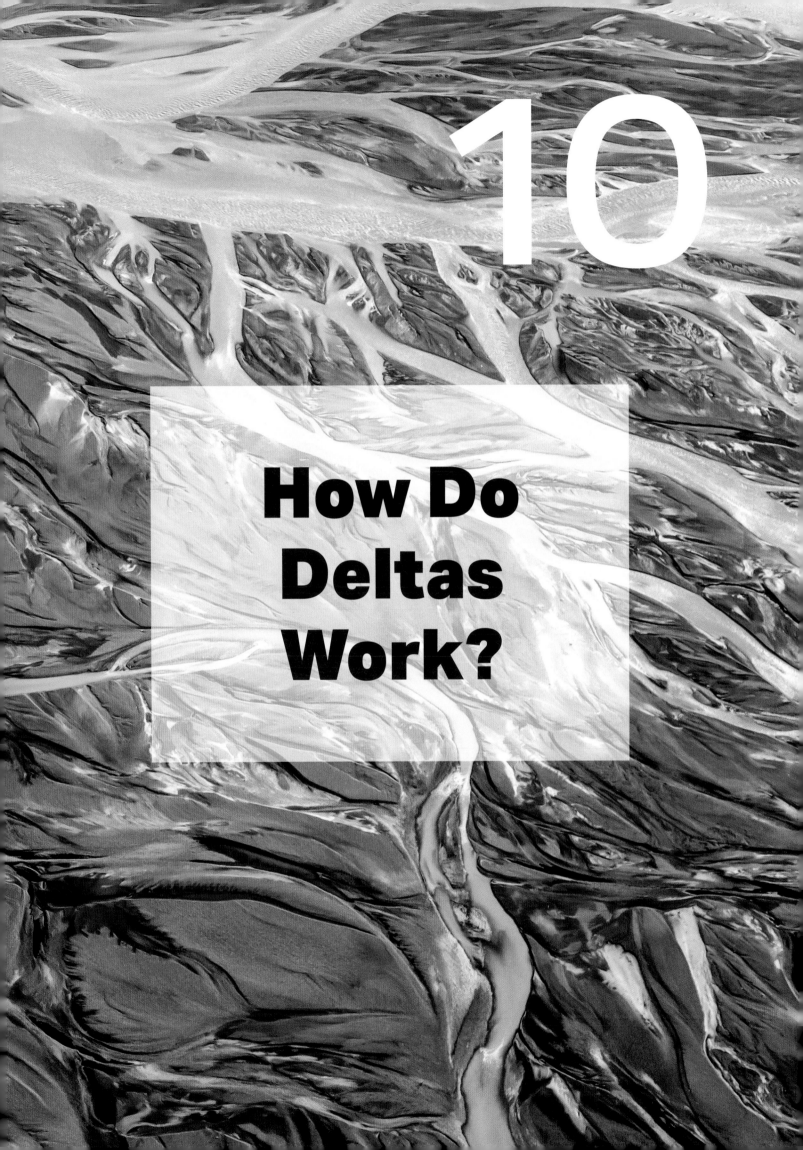

How Do Deltas Work?

What is a delta?

A delta is formed at the mouth of a river when it flows into a lake or the ocean and is carrying enough sediment to build a pronounced landmass. These features are created where sediment accumulates faster than it is submerged or removed by marine or lake processes. Deltas have a life cycle, where their growth is controlled by sediment deposition and organic production (plant growth), and their decline is controlled by erosion and subsidence of the deltaic deposits.

The ancient Greek geographer Herodotus (*c.* 484-*c.* 425 BCE) coined the term "delta" when describing the outlet of the Nile River into the Mediterranean Sea, where it forms a lush green oasis in the shape of the upper-case Greek letter Δ, or delta. Throughout the centuries of studying deltas, we now know that not all take the shape of a perfect triangle like the Nile's. This is because when a river reaches its terminus, it is affected not only by the terrestrial (landward) forces on the river such as climate, rainfall, slope, and vegetation along the shoreline, but also by marine (seaward) or lake processes, including tides, waves, coastal currents, and storms.

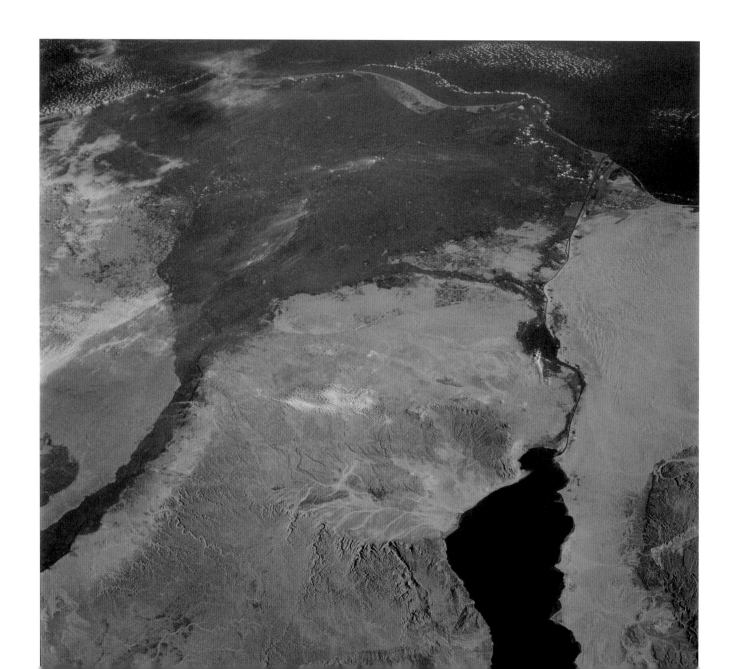

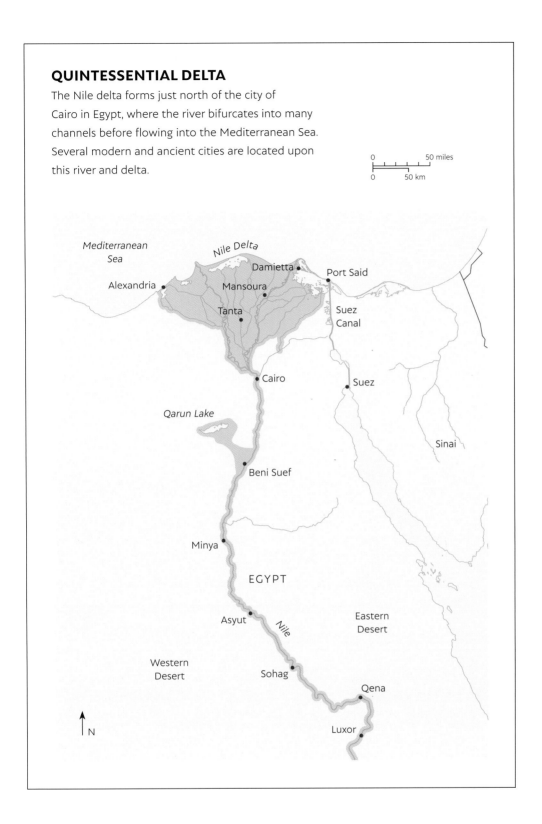

QUINTESSENTIAL DELTA

The Nile delta forms just north of the city of Cairo in Egypt, where the river bifurcates into many channels before flowing into the Mediterranean Sea. Several modern and ancient cities are located upon this river and delta.

When stepping foot on a delta, it is easy to imagine that it operates like a living and breathing organism. River and tidal waters that flow through the channels are like the lifeblood coursing through its circulatory system, the trees and the wetland plants are the breathing (respiratory) system, and the river and the tides are the heart, pumping and regulating the whole delta. In this analogy, there are many components that must work together to create a functioning organism. Not only is the delta landform itself like a living organism, but many living species of flora and fauna also use deltas as their home, either on a temporary basis or permanently.

◀ **Delta oasis**

The Nile River delta is an important ecological oasis in northern Africa. It dissects the Sahara Desert and provides vital agricultural land for the 50 million or so people who live on the delta.

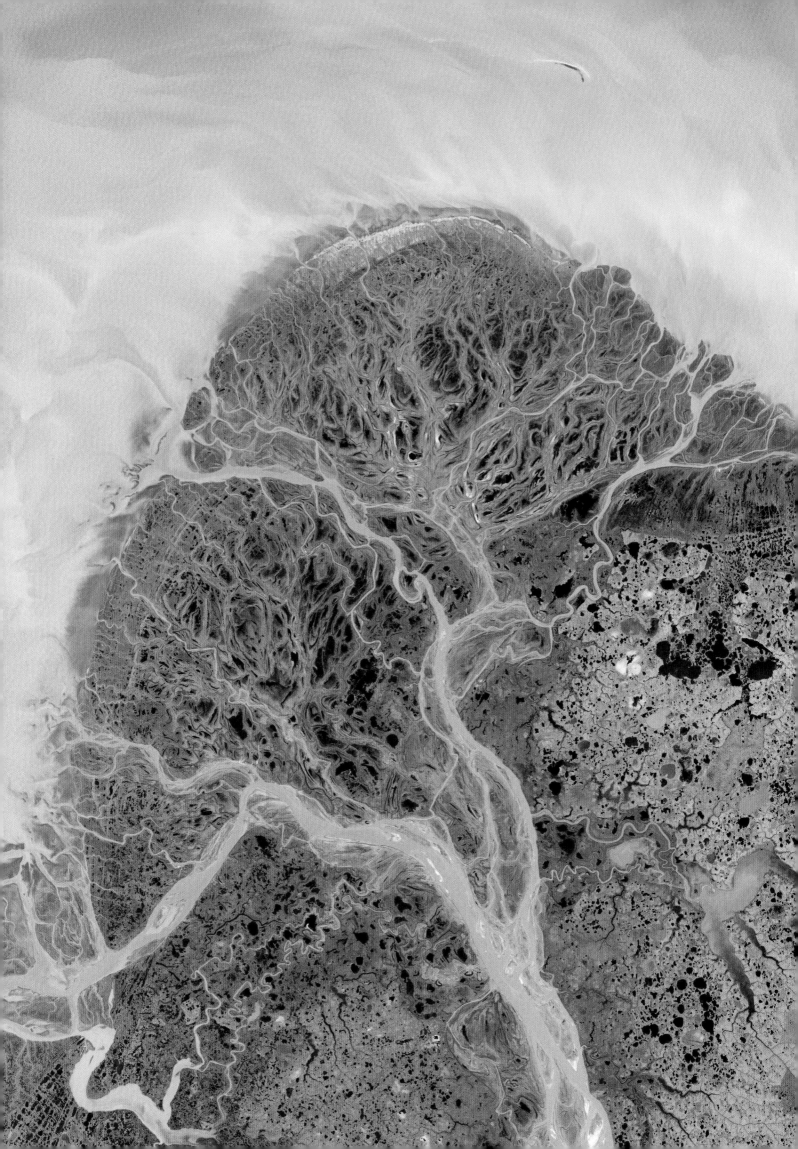

The birth of new land

Deltas form where a river discharges into a body of water and deposits sediment. Wetland vegetation can then take hold and help with the subaerial growth and productivity of the delta. Tiny deltas form in puddles or lakes, whereas very large deltas can form where a great river discharges into an ocean. The amount of water and sediment supplied to the river mouth dictates this myriad of delta sizes on Earth. Not only do deltas vary in size, but they can also vary in location, existing anywhere from lush tropical zones to deserts and polar landscapes. They serve as important nursery grounds for ecologically sensitive fish, birds, reptiles, and mammals, and larger deltas—such as the Indus in Pakistan, the Ganges-Brahmaputra in India and Bangladesh, and the Mekong in Vietnam—are food baskets for large numbers of people in the countries in which they are situated.

◥ **Delta inhabitants**
The Okavango delta forms in a lake in Botswana and is home to some large and charismatic animals.

▶ **Snowfall on the Selenga**
The Selenga delta in Russia builds out into Lake Baikal, the deepest lake in the world.

▶ **Seen from afar**
Astronauts aboard the International Space Station snapped this picture of two river deltas along the southwestern shoreline of Lake Ayakum in Tibet. River deltas can exist in lush tropical regions, as well as deserts and polar landscapes.

◀ **Plumes of sediment**
The sediment plumes around the Yukon delta in Alaska indicate where new land may form.

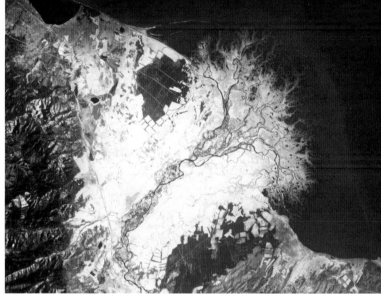

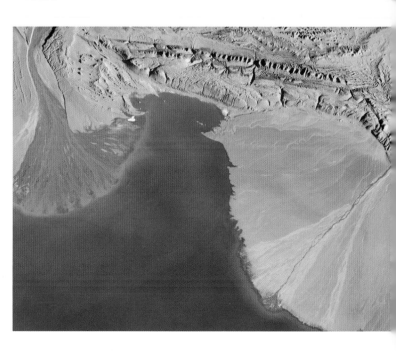

Delta classification

Why are some deltas triangle-shaped and others not? Why do some deltas exhibit a very pronounced protuberance of land, while for others the shoreline is quite smooth? What controls the shape of deltas? Deltas are typically classified as either river-dominated, wave-dominated, or tide-dominated, depending on the relative strength of the primary processes moving sediment around. Different delta shapes emerge as a result of these varying processes.

River-dominated deltas

The Nile delta is an example of a river-dominated delta, where the landmass that builds out into the body of water is shaped like a triangle, and the river channels split and partition water and sediment into smaller offtakes, called distributary channels. In river-dominated deltas, the unidirectional pulsing of the river dictates when it will flood, overtop its banks, and build up land. Such transport of sediment typically happens seasonally, when warming spring temperatures cause snow to melt or summer monsoons deliver enormous quantities of precipitation, as occurs in Southeast Asia.

Tide-dominated deltas

In tide-dominated deltas, not only is unidirectional seasonal river flow important, but the delta is also impacted by tides at the ocean outlet. Tides are long-period waves of water caused by the gravitational pull of the moon and sun, and in coastal environments this causes the water levels to rise and fall periodically. Deltas are low-lying, so this rise and fall in water level causes two-way, or bidirectional, flows, where water is transported landward for several hours while the tide rises (or "floods"), and seaward for several hours while the tide falls (or "ebbs"). This bidirectional movement redistributes sediment at the river mouth in both landward and seaward directions. As a result, the land between the distributary channels becomes more elongated and finger-like than in river-dominated deltas, and is oriented in a landward-to-seaward direction. In addition, the distributary channels in tide-dominated deltas tend to flare outward. Thus, the planform of a tide-dominated delta in bird's-eye view is different from that of a river-dominated delta.

▼ Flared channels

Tides at the mouth of the Indus River, Pakistan, form tidal channels that flare seaward, separated by elongated islands.

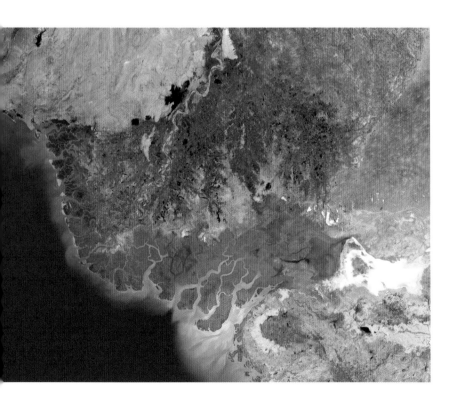

DELTA CLASSIFICATION

The world's deltas can be classified within a triangular diagram of river, tide, or wave dominance, depending on the relative balance of power occurring at the river mouth.

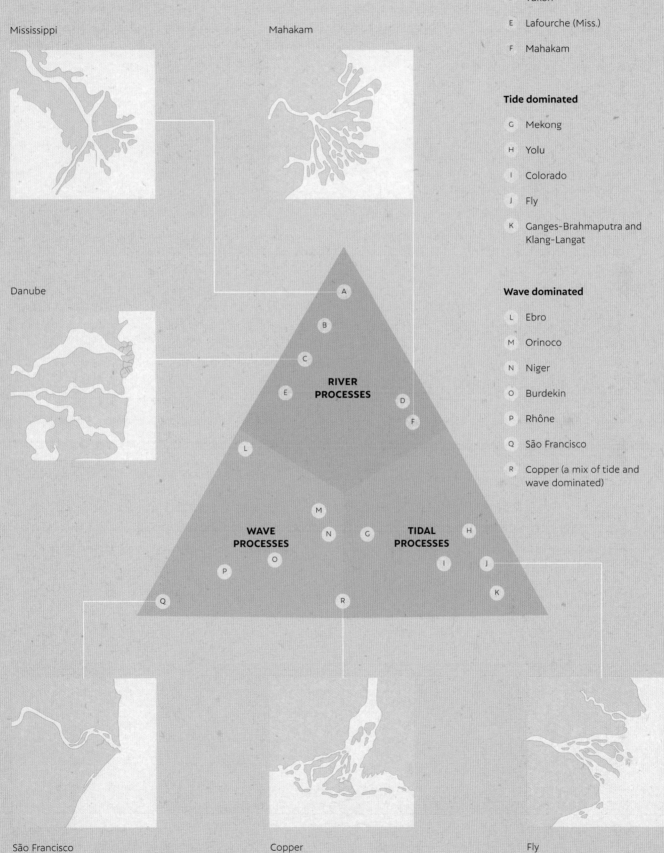

Mississippi

Mahakam

Danube

São Francisco

Copper

Fly

River dominated

A   Modern Mississippi

B   St. Bernard (Miss.)

C   Danube

D   Yukon

E   Lafourche (Miss.)

F   Mahakam

Tide dominated

G   Mekong

H   Yolu

I   Colorado

J   Fly

K   Ganges-Brahmaputra and Klang-Langat

Wave dominated

L   Ebro

M   Orinoco

N   Niger

O   Burdekin

P   Rhône

Q   São Francisco

R   Copper (a mix of tide and wave dominated)

RIVER PROCESSES

WAVE PROCESSES

TIDAL PROCESSES

► **Wave-straightened shoreline**

Waves along the Camargue in the Rhône delta, France, form the characteristic ridges and swales of a wave-dominated delta.

Wave-dominated deltas

These deltas form when a river enters a body of water with extensive waves that constantly redistribute sediment deposited at the river mouth. Wave action tends to form a current parallel to the shoreline, called a longshore current, which transports sediment along the shore. As a result, wave-dominated deltas have their own distinctive planform patterns, which exhibit ridges of sand parallel to the shore, called beach ridges, separated by lower-elevation swales. Sediment reworking along the shore also disrupts the "normal" distributary pattern seen in other deltas: typically, wave-dominated deltas maintain only one outlet distributary at a time.

WORLD DISTRIBUTION AND CLASSIFICATION OF DELTAS

River-, tide-, and wave-dominated deltas exist all over the world, from the lush tropics to the frozen poles. An analysis of the energies present at each river mouth showed more wave-dominated conditions exist around the world, but tide- and river-dominated deltas exhibit more freshwater discharge and sediment flux, as shown in the pie diagrams below.

△ Wave dominated
▲ Tide dominated
◁ River dominated

Number of deltas
13%
10%
77%

Discharge
16%
30%
54%

Sediment flux
11%
43%
46%

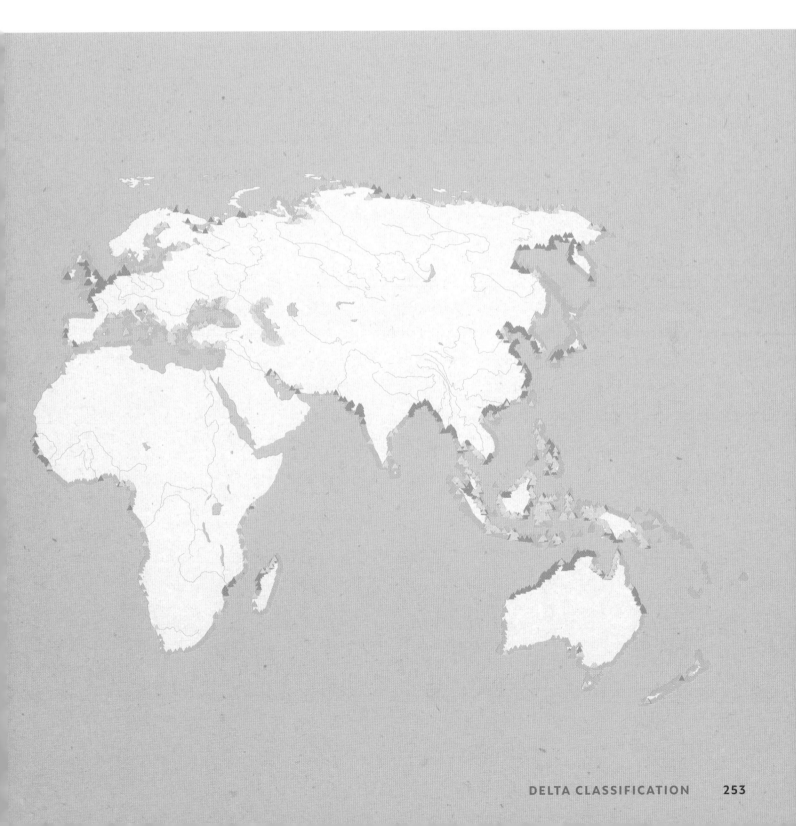

Diminishing flows

When a river discharges into a body of water, water and sediment that was once confined to distributary channels flows into a much larger (and unconfined) space, where it can spread out. When this happens, the current slows down and the river water mixes with that of the receiving basin, be it a puddle, lake, estuary, or ocean basin. But what happens to all the water that has been carried by the river—and the sediment, nutrients, and organic debris that have hitched a ride?

Homopycnal flow

The difference in density between two waters dictates what will happen when they mix. When the density of the river water is the same as that of the receiving basin, and mixing throughout the water column occurs at the river mouth, a homopycnal flow is created. In this scenario, sediment that is moving as bedload along the bottom of the channel slows and is deposited, usually quite close to the river outlet, creating accumulations, or bars, at the river mouth. These bars are typically quite sandy, and over time they can build up to the point that vegetation can colonize the surface. As the bars grow, they start to split the flow at the mouth, and two distributary channels are formed.

Sediment that is in suspension in the water can be deposited on the bed, but only when current velocities slow down enough for this to happen. At this stage, finer-grained (muddy) material is deposited far from the river mouth in deeper waters, or at the very crests of the bars at the end of flood pulses. These different types of deposits form discrete habitats that support a variety of organisms. Some plants and animals are well adapted to living in rapid flows and sandy conditions, while others prefer the quieter conditions of deeper waters.

Hyperpycnal flow

If the density of the river water is greater than that of the receiving basin, a hyperpycnal flow is created, where the river water sinks underneath the basin water as a density current. These currents can flow as "underwater rivers" for great distances. High-density currents are very common where rivers carry large volumes of sediment into their receiving basin, such that the additional mass of sediment increases the density of the water-sediment mixture. Glacial outwash rivers that flow into lakes, and rivers that flow out of steep mountain ranges into lakes or oceans, commonly generate hyperpycnal flows.

Hypopycnal flow

Where the density of the river water is less than that of the receiving basin, a hypopycnal flow is generated and the river water-sediment plume floats on top of the basin water and spreads out. This is very similar to what happens when oil is mixed with water: the oil floats on top because it is less dense. Hypopycnal flows are common where rivers carrying little sediment enter saline oceanic waters, which have a greater density than the fresh water. When this happens in a river delta, the river (aloft) and receiving (basal) waters remain vertically stacked for great distances from the river mouth, until waves and oceanic currents mix them over time.

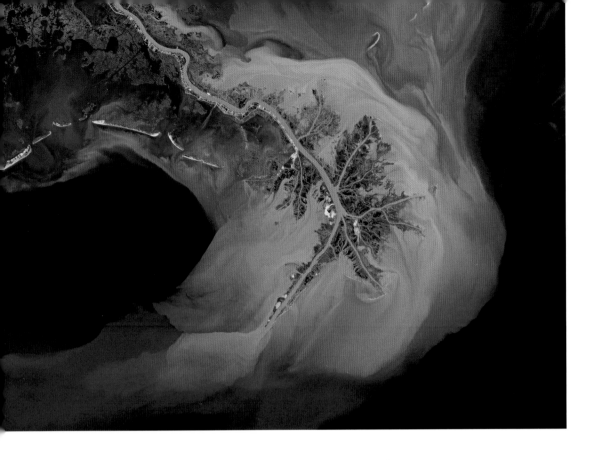

Hypopycnal plume

The buoyant plume of the Mississippi River in the northern Gulf of Mexico is easy to see with the naked eye in this satellite image. The plume is caused by the muddy sediment contained in the freshwater discharge as it enters the saltwater. The Mississippi plume can extend more than 100 km (60 miles) from the river's mouth.

FLOW MIXING

Differences in water density dictate whether the river water at the mouth of a delta will readily mix, sink below receiving basin water as an underwater river, or float on top.

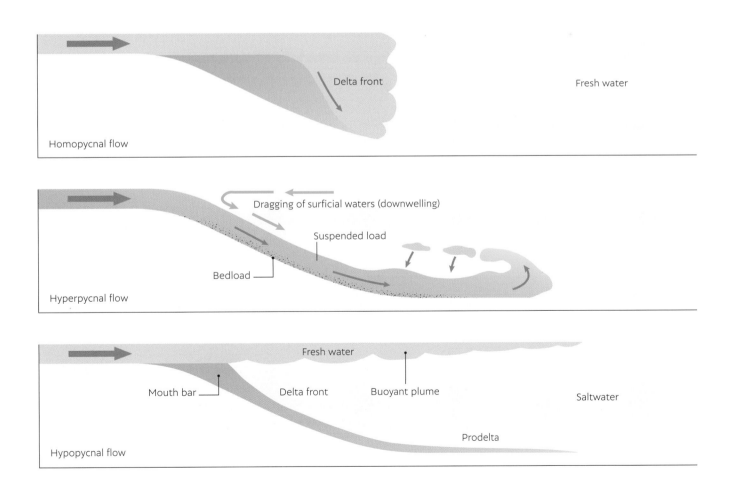

Sediment sinks

When a river enters a body of water, its velocity decreases and the sediment it carries is eventually deposited. Sometimes, the quantity of this sediment deposition is enormous—indeed, Earth scientists call deltas "sediment sinks" because so much sediment may have built up below their surface. The Ganges and Brahmaputra, two of the rivers that form the world's largest delta, carry the greatest sediment load on Earth: approximately 1 billion tonnes (1.1 billion US tons) of sediment flow to the delta each year! But what happens to this sediment, and where is it ultimately deposited?

WHERE DOES ALL THE SEDIMENT GO?

Contemporary sediment partitioning in the Ganges, Brahmaputra, and Meghna Rivers and delta. Note the offshore centers of deposition: a subaqueous delta on the continental shelf and a deep-sea fan delta.

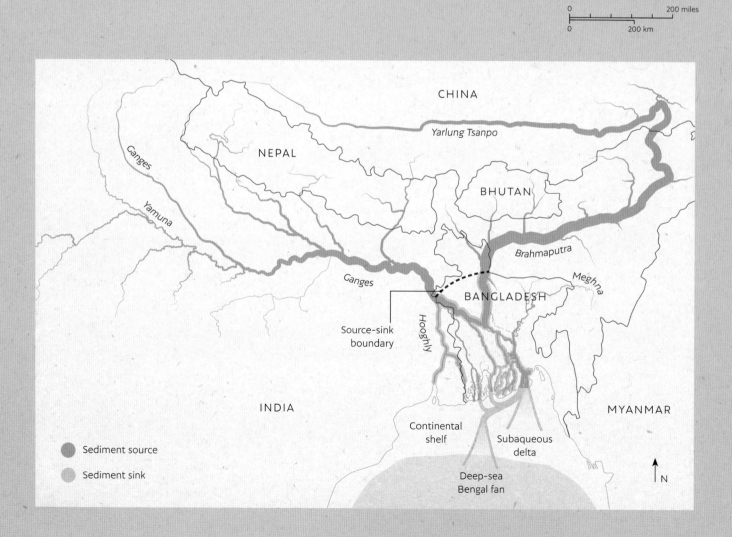

Layers of sediment

Research has shown that sediment carried by rivers is partitioned into several regions of the delta. Some is deposited on the subaerial (i.e., terrestrial) delta, while some is sequestered in subaqueous (i.e., underwater) regions. In the case of the Ganges–Brahmaputra–Meghna delta, approximately 20 percent of the sediment carried by these rivers is deposited on the delta floodplain, 30 percent is deposited at the river mouth (including along the channels), 10 percent is transported back onshore by the tides and sustains the Sundarbans (the largest mangrove preserve in the world and home to the Bengal Tiger, *Panthera tigris*), and 40 percent is deposited offshore on the continental shelf and into the deep-sea environment. This indicates that a substantial amount of sediment storage occurs in subaqueous regions, which are difficult to observe. The delta has been forming at the head of the Bay of Bengal for around 40 million years. Assuming 1 billion tonnes (1.1 billion US tons) of sediment has been transported to the delta each year over 40 million years, a vast quantity has clearly built up there. In fact, using geophysical surveying techniques that can "see" below the surface, we now know that sediments to a depth of about 20 km (12 miles) are stored below the surface of this delta!

▶ **What is a sediment sink?**

Deltas are sediment sinks, where vast quantities of sediment are deposited and stored below the surface. The sediment-laden, nutrient-rich waters of deltas support abundant wildlife, like this Saltwater Crocodile (Crocodylus porosus), and human populations that live along the nexus of the terrestrial and subaqueous environments.

Vast repositories

Other deltas likewise store vast thicknesses of sediment beneath their surface. The Amazon delta in Brazil and the Mississippi delta in the United States both store sediment to a depth of about 10 km (6 miles), while the Volga delta in Russia has 6-8 km (4-5 miles) of sediment deposits beneath its surface.

What causes so much sediment storage? Part of the reason is that delta deposition into ocean basins puts a significant load on the Earth's crust. Deltas are also subjected to a process called subsidence, or lowering of the land surface over time. It is both the inorganic (sedimentary) and organic (life-sustaining) nature of deltas that makes them such excellent sediment traps. Climate and tectonics also play a role in the delivery of sediment to rivers and thus deltas worldwide—and we have to recall that these are not static forces that have stayed the same over geological timescales.

CLIMATE RECORDS

Kilometer-thick sediment packages are found beneath the surface of large-scale deltas, as seen in this illustration of the Volga delta in Russia. This is due to the process of subsidence, which allows thick accumulations of sediment to build up over geological timescales (thousands to millions of years). Seismic data from the Volga delta reveal layers of sediment building out into the Caspian Sea. These data help scientists understand how long deltas have been active in a particular area and allow them to reconstruct the paleogeography and paleoenvironmental conditions of the region, which in turn help elucidate the paleoclimate.

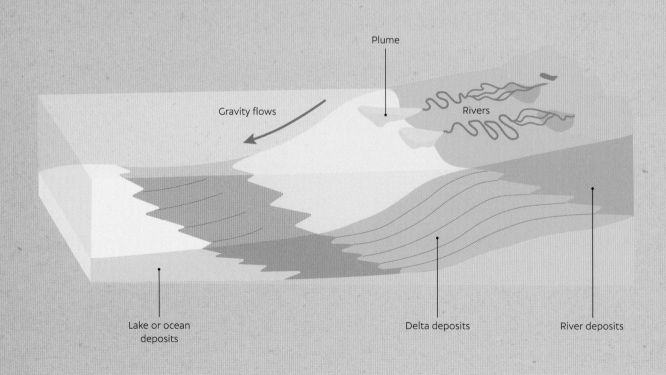

Plume

Gravity flows

Rivers

Lake or ocean deposits

Delta deposits

River deposits

Tectonic forces that build mountain ranges also change river courses and patterns, sometimes enhancing sediment delivery to deltas and at other times diminishing sediment supply. For example, the Rhine-Meuse delta in the Netherlands is located in a rift zone, where geological forces have torn the Earth apart. The delta receives sediment from the Alps, which have been building for millions of years. However, glaciers have eroded the land surfaces of many regions across Europe, and as a result the Rhine-Meuse delta stores only around 500 m (1,600 ft) of sediment below its surface. Since deltas are such excellent records of sediment erosion and delivery over time, recent investigations into these landforms are helping unravel the climate history of Earth.

SEDIMENT THICKNESS

The sediments in the Huang He (Yellow River) delta in China have been deposited over the past 11,000 years. Notice that the sediment thickness increases seaward as the delta builds into its receiving basin.

Sediment thickness

Low 4 m   High 16 m
(13 ft)   (52.5 ft)

– – – – – – – – –
Coastline in 1855

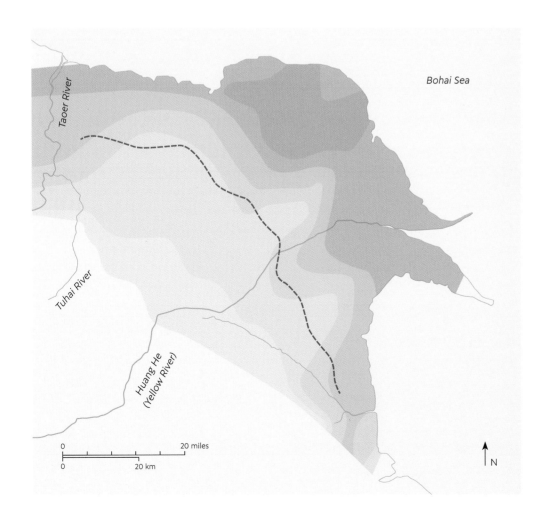

Grain size

Stepping foot on a coastline, you may wonder about the sediment grains that squeeze between your toes. Why is one area nice and sandy, while another is quite squishy, causing your feet to sink in? Deltas around the world are predominantly composed of gravel, sand, or mud (a generic term for silt and clay), depending on the carrying capacity of the river that supplies them. The river's gradient and length, the characteristics of precipitation in its catchment, and the rate of erosion all dictate the sizes of sediment grains that ultimately make it to the coast and form a delta.

  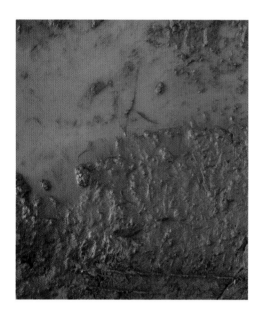

▲ **Grain size**

Grain sizes most commonly observed in deltas vary from gravel (larger than 2 cm/¾ in), sand (from 2 cm/¾ in to 63 μm), and mud (smaller than 63 μm). Individual mud grains are difficult to see with the naked eye. 1 micron (μm) is one thousandth of a millimeter.

Gravel-dominated deltas

Gravel-dominated deltas usually occur where glacial meltwater flows into lakes or fjords, or steep rivers empty abruptly from mountains into bodies of water. The key here is that there is a relatively steep gradient and also sufficient water discharge, causing the flow to be quite rapid and able to transport large sediment grains. For instance, the Yallahs River in Jamaica drains the Blue Mountains southeast of the capital, Kingston. This short tropical river is only around 40 km (25 miles) in length, but it drops 1,500 m (5,000 ft) in elevation down the rocky slopes. This steep river gradient (approximately 40 m/km, or 200 ft/mile), combined with the rainy tropical environment, means the flow of the Yallahs is very rapid, and the delta sediment where the river discharges into the Caribbean Sea is predominantly gravel-sized.

Sand-dominated deltas

The Niger River in Africa has a much gentler gradient (approximately 10 cm/km, or 6 in/mile). It has two deltas: an inland delta that builds wetlands into the lakes of central Mali; and a coastal delta, where the river finally terminates at the Atlantic Ocean to the south. Both of these deltas are composed predominantly of sand. This is due to the sediment sources and semi-arid environment in the drainage basin, and the river gradient, which is too shallow to transport much gravel to the deltas.

Mud-dominated deltas

The Huang He (Yellow River) is the second-longest river in China and drains mountains in Southeast Asia that are more than 4,800 m (15,700 ft) high. But the river is very long—more than 5,000 km (3,100 miles)—and travels across long, flat plains before it reaches its terminus in the East China Sea. The Huang He is notable for the large amount of silt it carries from erosion of the Loess Plateau. As a result, the mature landmass that we associate with its delta is predominantly mud.

▲ **Mud-rich delta**

The Huang He (Yellow River) drains some amazing landscapes in China, including the Loess Plateau. The eroded loess (silt-sized sediments) causes the river's waters to be very muddy and its delta to be predominantly composed of deposits of mud.

Sediment compaction with burial

When sediment is first deposited, it is often relatively "fluffy," or full of water between the individual grains. Over time, pressure from the overlying sediment squeezes water out of the spaces between the grains (called the "pore space"), consolidating the material. This process is called compaction, and results in a reduction in volume of the sediment due to dewatering (loss of water) and reorganization of the sediment grains to fit within a smaller, tighter space.

The influence of grain size

Grain size dictates how much sediment compaction can take place. For example, freshly deposited sand has 20-50 percent pore space, which means the remaining 50-80 percent of the volume is taken up by the sand grains themselves. By the time this sand deposit is buried to depths commonly observed in deltas (several kilometers), its pore space is reduced to 10-30 percent. When freshly deposited, grains of silt and clay have a much higher pore space than sand, at 50-90 percent. The plate-like grains reorganize substantially once buried, resulting in extensive dewatering and a reduction to 10-20 percent pore space.

Organic deposits

Deltas not only contain inorganic sediments, but they also accumulate thick packages of organic material from living organisms—in this case, wetland plants. When plants die, they can decompose in an oxygenated environment or may be buried and accumulate at depth. Wetland soils are typically oxygen-poor within a few centimeters of the surface, so decomposition rates are much slower than in oxygenated environments, and thus wetlands tend to accumulate organic material. Deposits with more than 60 percent organic material form a peat.

Like inorganic sediment, organic-rich soils are highly susceptible to compaction, and most of the volume loss takes place within 10 m (30 ft) of the surface. This is because organic material is composed primarily of water and void space—by weight, wetland soils are commonly 70-95 percent water. Thus, deltas that have extensive organic-rich soils are highly susceptible to compaction. For example, the Rhine-Meuse delta in the Netherlands has experienced extensive compaction of its organic-rich wetland soils. This has been exacerbated by embanking and poldering for agriculture, subjecting the area to oxygenated conditions. As a result, many parts of the Rhine-Meuse delta are now several meters below sea level, and expensive walls and pumped drainage are required to maintain dry conditions.

Compaction of peat

Organic-rich wetland soils in deltas can accumulate thick layers of peat. These are highly susceptible to compaction due to the high water volume of the organic material, which is pressed out as it is buried. Humans have used peat soils as an energy resource throughout the centuries, as it can be burned—as seen here in the Rhine–Meuse delta. Modern infrastructure built on organic-rich deltaic soils can suffer devastating effects from settling over time, as seen in the lower photo of buildings in Amsterdam, Netherlands.

Alive: Fluid within cell pushes against cell wall

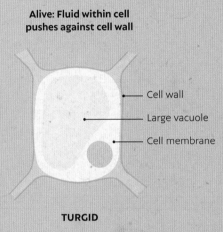

Cell wall

Large vacuole

Cell membrane

TURGID

Dead: Cell membrane has no turgor pressure

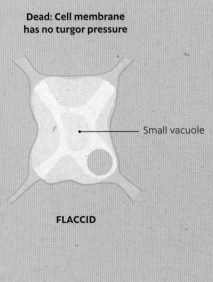

Small vacuole

FLACCID

SALT DIAPIRS

Diapirs are columns of sediment that are squeezed upward and sideways due to the pressure of overlying sediments, and typically form when a less dense substrate is overlain by a higher-density material. In deltas, this may occur where coarser sediments overlie finer-grained, more deformable muds, creating mud diapirs. In special circumstances, layers of low-density salt (called evaporite deposits) are overlain by higher-density particles such as sand, silt, and clay, forming salt diapirs. The greater the overburden pressure, the more salt can be squeezed vertically upward. In the Mississippi delta in the United States, the opening of the Gulf of Mexico about 170 million years ago created a shallow evaporating sea where very thick salt deposits formed. Over time, the gulf became wider and deeper, turning into a true oceanic basin, and the paleo-Mississippi River (along with other rivers draining North America) began depositing sediment on top of this salt. To date, sediment in the delta has accumulated to a depth of around 10 km (6 miles) and thousands of salt diapirs have been documented. These salt diapirs are very important traps for natural resources, such as oil and gas, and they form curious topographic highs on the delta top, such as Avery Island.

160 million years ago

North

South

Today

Salt diapirs

Mississippi salt diapirs
Low-density salt diapirs can be up to 10 km (6 miles) thick. Several of them have been extensively excavated to create vast caverns that are used for storing petroleum reserves. More importantly, diapirs generate important oil and gas traps for conventional drilling in the northern Gulf of Mexico.

- Seawater
- Salt
- River and delta deposits
- Marine deposits
- Basement rock

Height (miles)

Height (km)

0 — 100 miles

0 — 100 km

▶ A geological oddity

Avery Island (circled) is the top of a salt diapir in the Mississippi delta (right). It rises dramatically above the surrounding low, flat wetlands to a height of 50 m (165 ft) above mean sea level, making it the highest point on the northern Gulf of Mexico shoreline. Tabasco hot sauce is produced on the site using the local salt in its recipe. Indigenous people living in the area first discovered the source of the salt, and it was later turned into one of the first salt mines in the United States (below right). Collapses and concern for safe working conditions resulted in the closure of the commercial mine in 2022.

Subsidence in deltas

With compaction common in deltaic sediments, a process called subsidence (a lowering of the land surface) occurs frequently. While other processes, such as tectonics, can also drive subsidence, compaction is the dominant force causing subsidence in many deltas worldwide.

Subsidence in deltas is critically important because it causes the elevation of the land surface to decrease, or become lower relative to sea level. If new sediment is delivered to the region from the river, is redistributed from storms, or increases in situ from the accumulation of peat from wetland plants, this subsidence can be compensated for, and the land surface elevation remains constant. However, if no new material fills the void space, the surface elevation will decrease and submergence is possible.

Factors influencing subsidence

Subsidence in deltas varies over different temporal and spatial scales in complex ways. Importantly, the processes that contribute to compaction also contribute to subsidence. However, the partitioning of sediment is not uniform across a delta: coarser-grained sediments (sand and gravel) tend to be deposited in the distributary channels and mouth bars, while finer-grained sediments (mud) are deposited in deeper waters and interdistributary regions (the areas between distributary channels). These muds and organic particles compact more than sand and gravel, and thus greater subsidence is likely in these regions. Over time, we can imagine that freshly deposited sediment houses more water within its pore space, and thus can compact more than material that was deposited previously. In this process, the sediment therefore loses some of its pore water, especially if it is buried and compacted for a long time.

The canary in the coal mine

Globally, sea levels are rising as a result of climate change (eustatic sea level rise) at an average rate of about 2–3 mm (1/8 in) per year. Deltas are highly susceptible to rising water levels because subsidence compounds the situation. In deltas, subsidence and global mean sea-level rise combine to drive relative sea-level rise, which is the change in land surface elevation relative to sea level. This is what matters to the people who live on deltas. In fact, subsidence rates in many deltas are much greater than global mean sea-level rise, and so it is this process that dominates relative sea-level rise here. Only the vertical accretion of new material in the space created by the combined rising sea level and subsidence can offset this process and prevent the drowning of delta surfaces. If there is insufficient accumulation of sediment, inundation inevitably occurs.

The natural phenomenon of subsidence is one of the reasons delta landscapes exhibit a life cycle. It is also touted as the reason why deltas are called the "canary in the coal mine" for global sea-level rise. What does this mean? Deltas are low-lying coastal features that are close to sea level, and because of subsidence they exhibit larger relative rates of sea-level rise. If global warming patterns continue over the next few decades, more and more coastal areas will become prone to the flooding and issues of coastal erosion that are already plaguing deltas. These landforms are therefore the canary in the global-warming coal mine.

DELTA SURFACES GOING DOWN

Subsidence is a natural phenomenon in deltas, resulting from consolidation and compaction of sediment. If new river sediment is delivered to the area and healthy wetlands continue to grow, this elevation loss can be compensated. If not, the delta surface is doomed to drown.

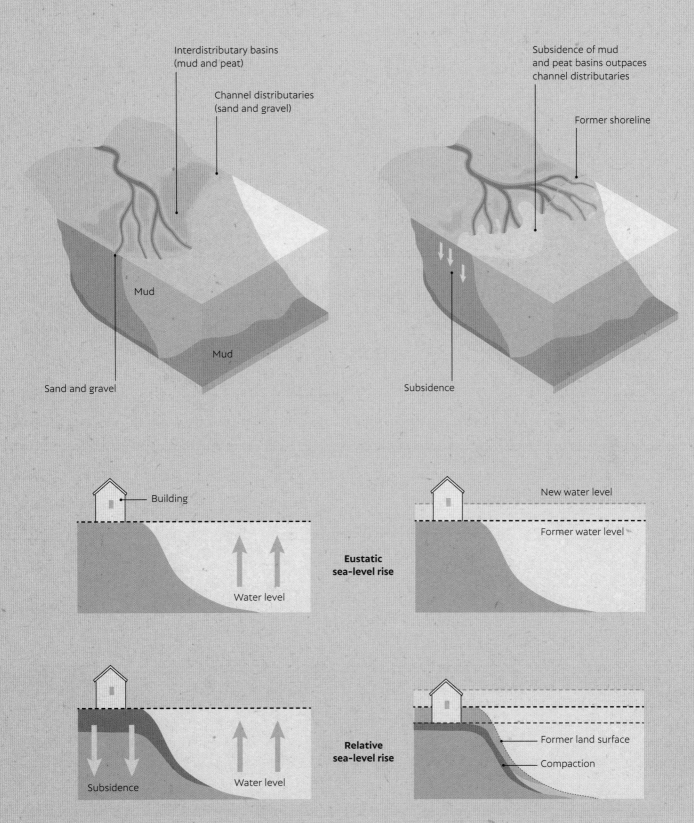

Interdistributary basins
(mud and peat)

Channel distributaries
(sand and gravel)

Mud

Mud

Sand and gravel

Subsidence of mud
and peat basins outpaces
channel distributaries

Former shoreline

Subsidence

Building

Eustatic
sea-level rise

Water level

New water level

Former water level

Subsidence

Relative
sea-level rise

Water level

Former land surface

Compaction

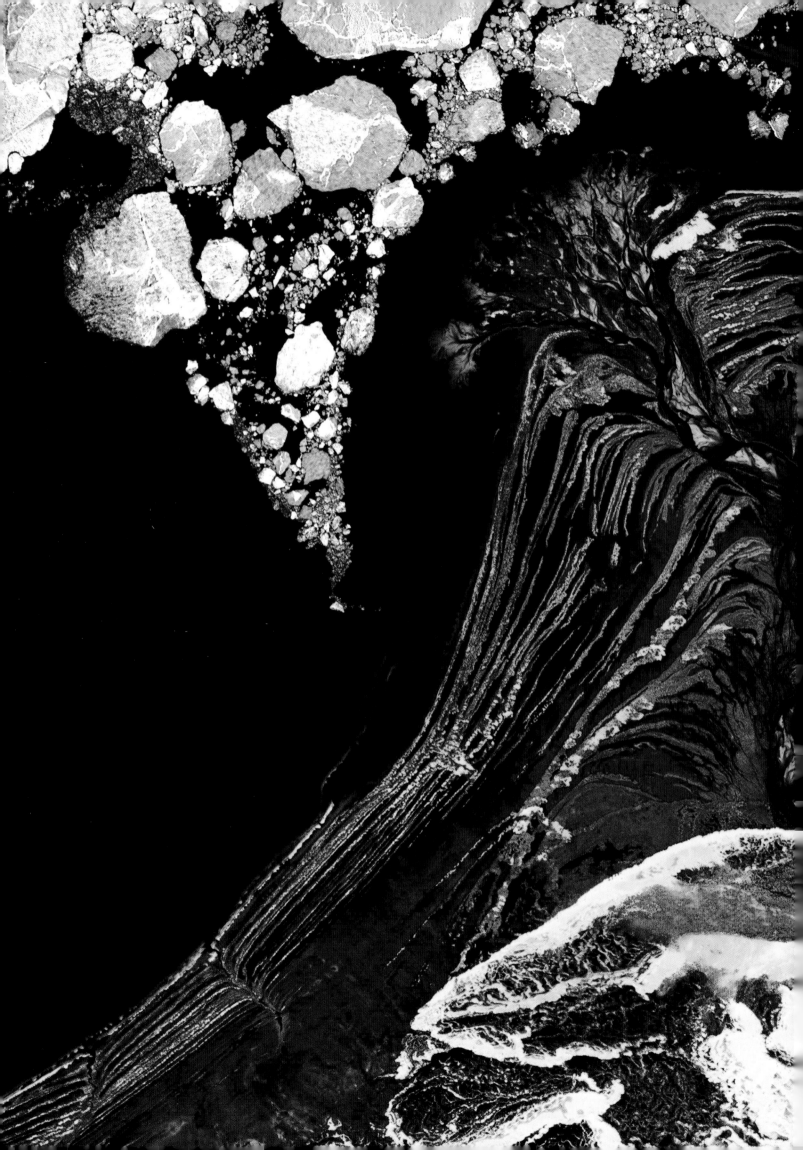

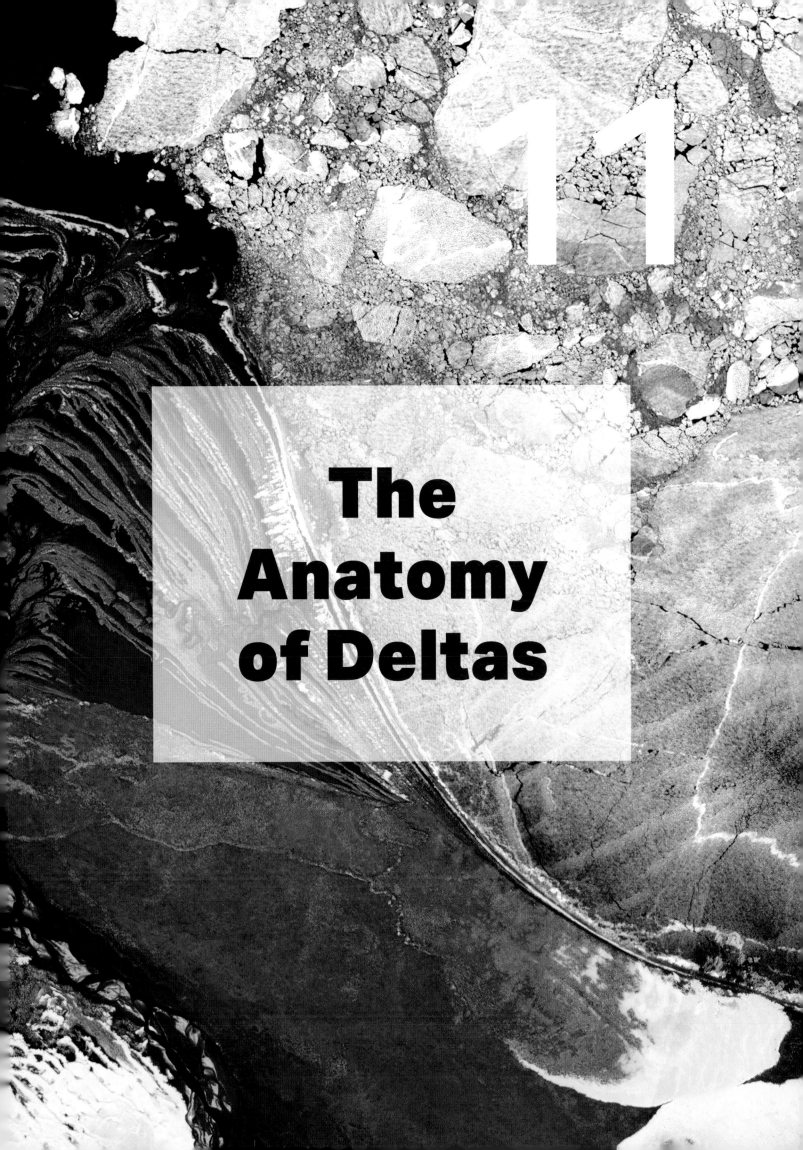

11

The Anatomy of Deltas

The anatomy of a delta

If a delta is like a living organism, where are its head, body, and tail? Morphologically, deltas comprise diverse regions where varying hydrodynamic processes dominate, and these are categorized into different geomorphic zones. Each delta also has a life cycle, whereby a new landmass can form, spread out, and mature, then senesce and retreat.

A bird's-eye view of the deltaplain—the terrestrial portion of the delta—shows that the river bifurcates (splits) into its distributary channels on its journey to its outlet, and also decreases in elevation as it approaches the water level of its receiving basin. Some rivers finally end in inland lakes, while others journey all the way to the ocean. Regardless, we can divide deltas into different regions along their upstream to downstream pathway.

The upper deltaplain

The upper deltaplain comprises the upstream regions of the delta where the slope of the land surface is relatively steep (even though it is still rather flat!). In this zone, the slope of the water surface is identical to that of the channel bed; geoscientists and engineers call this normal flow conditions. River channels are highly mobile here and can sweep across the deltaplain over time. Tides do not penetrate these upstream reaches of the delta, and the flow is unidirectional. Flora and fauna are highly diverse here as the freshwater conditions facilitate productivity.

THE PARTS OF A DELTA

Deltas can be divided into upper and lower deltaplains, depending on how far upstream the distributary channels, backwater-influenced flow, and tides extend. These interactions establish different ecological zones, which support widely varying flora and fauna.

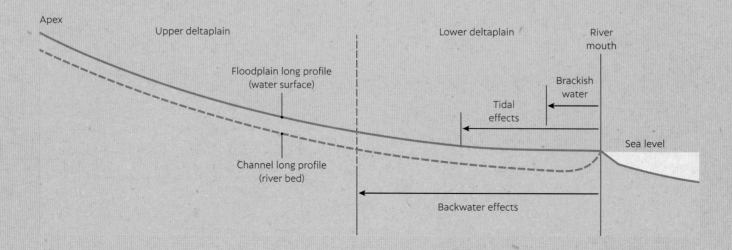

The lower deltaplain

Downstream of the upper deltaplain, the delta surface slope flattens greatly; this region is called the lower deltaplain. The hydrodynamics of the river change here because the water surface begins to back up and "feel" the effects of the receiving basin. In this zone, termed the backwater zone, the river experiences a deceleration in its current velocity. Many fish species from both inland and marine sources interact in the backwater region of the delta, as the tranquil flows create numerous pockets that are suitable for feeding and spawning. The head of the backwater zone is considered by some to be the apex of the delta, as many deltas develop distributary channels downstream of this point and nodes for avulsion (channel switching) are common here.

Further downstream, where tides can impact the river, the flow can shift from unidirectional to bidirectional. Closer to the coast, a salinity gradient is set up, and the flora and fauna living here must be highly specialized to withstand varying grades of salinity.

▲ **Upstream deltaplains**
Swamplands are common in upstream deltaplain regions, as seen here in the Mississippi delta in the United States.

▶ **Delta inhabitants**
A colorful Hoatzin (Opisthocomus hoazin) perched in the lush green Amazon delta in Brazil.

The delta cycle

Each delta has a life cycle, from the birth of the new river delta, through growth controlled by sediment deposition and organic production, and finally, over time, to senescence, when it reaches its maximum extent, becomes abandoned, and dies. Geoscientists call this process the delta cycle.

▼ **Building out**

When deltas are young and actively receiving sediment, the channels bifurcate (split) and the land builds out, like this delta in Costa Rica.

Spatial and temporal changes

The delta cycle is a little more complicated than the process of a single delta building out and then retreating. There is a hierarchy of geomorphic features that make up the deltaplain, each scaled according to size and with different lifespans. Every geomorphic feature goes through its own life cycle, the only difference between them being the spatial scale of the new land that builds out and the duration it takes to go through its life cycle. For example, a crevasse splay (see page 274) is a small delta formed where a channel breaks its levee and deposits sediments in the swampy

interdistributary area (the area between distributary channels). In a delta like the Mississippi, crevasse splays may be active for about 10-20 years and typically build around 2 km² (0.8 square miles) of new land before deposition ceases. A larger-scale delta lobe can build much larger tracts of land (100-1,000 km²/40-400 square miles) and may be active for 100-500 years before the river distributary feeding the lobe becomes inactive as sedimentation switches elsewhere in the landscape. An entire delta complex could be active for 1,000-2,000 years and build up to 10,000 km² (4,000 square miles) of land before a drastic change in river direction occurs, and the river switches course in a process called avulsion.

Varying processes

A delta that is growing and building is very different from one that is in its senescence. Both stages are part of the life cycle of the delta and reflect the major processes controlling deposition at the time. A delta that is growing is physically building land out into a body of water—a process called progradation—and here riverine processes are dominant. A delta that is senescing is receiving diminishing river water, and marine processes start to take over and the land retreats.

GROWTH AND SENESCENCE

The delta cycle includes growing and dying landscapes. A river building into a body of water starts the delta cycle, and senescence occurs when it reaches its maximum extent, becomes abandoned, and dies.

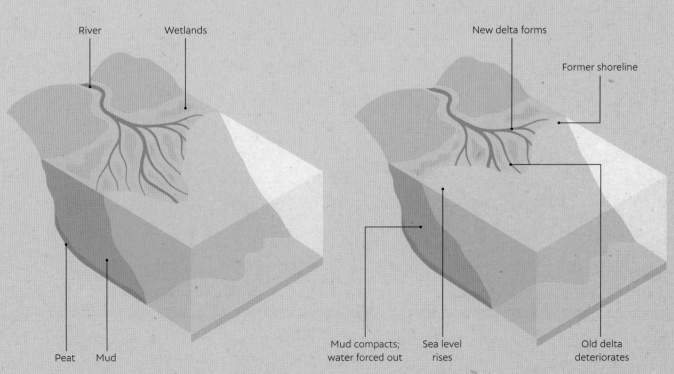

DELTA BUILDING

River   Wetlands

Peat   Mud

DELTA CHANGING COURSE

New delta forms

Former shoreline

Mud compacts; water forced out   Sea level rises   Old delta deteriorates

DELTA DEVELOPMENT

Each delta has a life cycle, from the smallest crevasse splay that forms on the margins of a distributary channel and is active for only a few years, to larger-scale delta lobes that are active for hundreds to thousands of years.

Ancient Greek and Roman settlements along the Black Sea (yellow circles) were affected by the delta cycle of the Danube River, and several of these sites are now underwater.

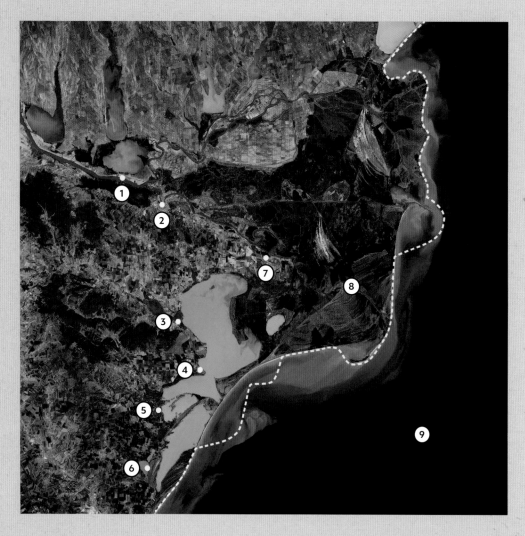

- - - - - - - - -
Former delta coastline

```
0          10 miles
0     10 km
```

1 Noviodunum
2 Aegyssus
3 Enisala
4 Orgame
5 Caraburun
6 Istros
7 Halmyris
8 Modern Danube River
9 Black Sea

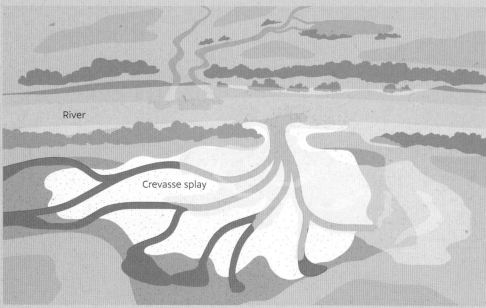

Small-scale delta
Crevasse splays are small deltas that form along the margins of rivers when they overflow their channels.

River

Crevasse splay

The Mississippi River delta cycle

One of the best-documented delta cycles is that of the Mississippi delta in the United States. The Mississippi River has been building land out onto the continental shelf of the Gulf of Mexico for the past 8,000 years, producing a landmass that covers more than 10,000 km² (4,000 square miles). When humans first colonized the region between around 10,000 and 7,000 years ago, the river delta looked very different than it does today. Extensive dating of the sediments of the modern delta has revealed that the river has occupied six different regions and effectively built out six delta complexes through progradation. Each of these delta complexes had a lifespan of around 1,000-2,000 years. When a river changes course in a process called avulsion, its existing delta complex no longer receives enough sediment to offset natural subsidence and relative sea-level rise, and the lobe begins to retreat. Submergence is the ultimate stage of the delta cycle, when a few sandy shoals are all that remain.

The newest delta complex to form in the Mississippi delta consists of the Atchafalaya and Wax Lake delta lobes in the southwest region, which have been building for 100 years. They first became subaerial in the 1970s and have since grown to cover more than 100 km² (40 square miles). Modern civilization and infrastructure require management of the river water that flows to these lobes to prevent delta avulsion, a process that is inevitable under natural conditions. On the other side of the delta, two deteriorating delta complexes (the St. Bernard and the Lafourche) built the landscape that modern New Orleans currently occupies. The mainstem outlet of the Mississippi River (Balize delta) is commonly called the Bird's Foot delta because of the distinct shape of its three distributary channels. This delta has prograded so far into the ocean that most of the sediment delivered by the river is transported offshore, as opposed to sustaining the deltaplain.

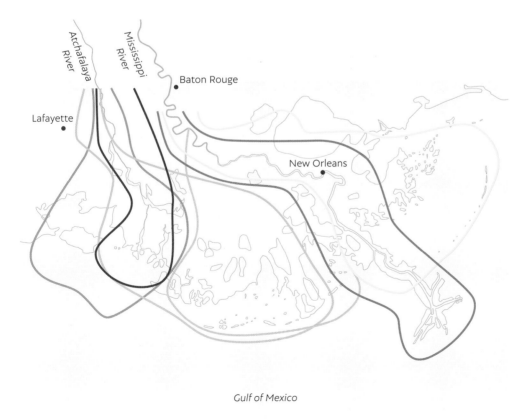

The Mississippi delta cycle
The Mississippi River has been building into the northern Gulf of Mexico for 8,000 years in a process called the delta cycle, whereby the river changes course over time and deposits sediment to different regions.

Maringouin

Teche

St. Bernard

Lafourche

Plaquemines/Balize

Atchafalaya/Wax Lake

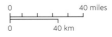

Evidence of delta building

The birth of a delta creates new landmass. Whether the river is building into a shallow lake or a giant ocean basin, it slows down, allowing deposition of the sediment it is carrying.

Early deposition

When a river is building a delta, the sediment it carries must first fill the subaqueous portions of the basin into which it is discharging. This process is hard to see because it occurs beneath the water's surface, but delta geomorphologists can confirm it is taking place by recording the bathymetry (water depth) of the basin, which will reveal that shoaling is taking place. Usually, the coarsest grains of sediment are deposited near the river mouth, while finer material is deposited further seaward. As sediment is deposited at the mouth of the river, bars form here. Over time, these build up enough height to become emergent, at which point vegetation will start to take hold. This signals a whole new stage of growth, as vegetation helps trap sediment on top of the bars: the plant stalks further slow down the river currents, facilitating the process of deposition.

The role of vegetation

The vegetation that establishes on the river mouth bars is subaqueous, residing mostly underwater and with only the tops of the plants visible. Once the mouth bars aggrade further, grassy vegetation takes hold, forming marshes. Marsh vegetation is capable of withstanding inundation for relatively long periods of time, typically within the daily tidal fluctuations. In tropical regions, mangrove vegetation soon succeeds marshland. Mangroves can be small and shrub-like, or grow as trees up to 15 m (50 ft) tall, and they can dominate tidal regions of the delta.

Close to the river mouth, the bars can aggrade even further as the river floods, with each inundation bringing a new blanket of sediment. Eventually, the bars will build up vertically to surpass the tidal range, and the vegetation will be succeeded by terrestrial shrubs and trees. When this happens, areas that were aquatic become transitional and finally fully terrestrial.

Progradation

Not only does the river delta build up vertically, but it also builds out horizontally, a process called progradation. As this happens, channel distributaries bifurcate and the dendritic pattern of the delta begins to emerge. Channel mouth bars extend and elongate, and eventually amalgamate as the mouth of the river builds further seaward. In some deltas, this process can happen quite rapidly, with hundreds of meters of land built out each year. In others, the process is slower, with change barely noticeable over human timescales. However, deltas are dynamic features, and their landscape constantly shifts over their life cycle as the power of the river battles with tides and waves.

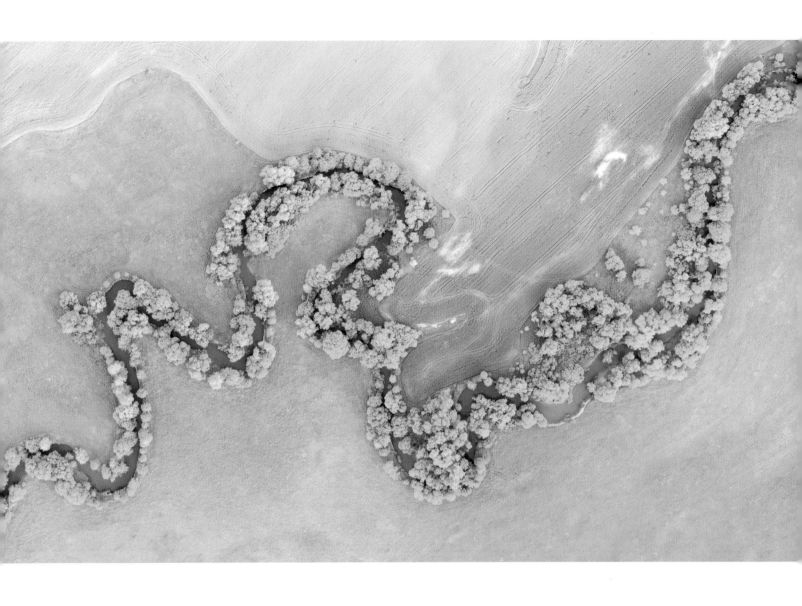

Delta surface aggradation

When a delta floods, lower-lying land is inundated first. Over time, the land surface aggrades, and taller vegetation such as trees begins to colonize the new land.

▶ **Delta progradation**

The islands of Dimer Char (left arrow) and Dhal Char (right arrow) in Bangladesh's Ganges–Brahmaputra delta are the newest of a series of islands building out into the Bay of Bengal.

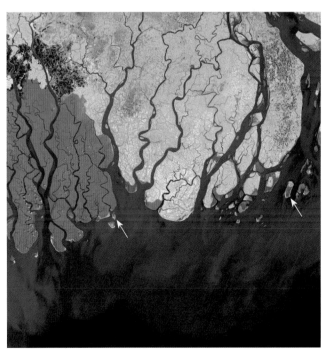

Delta channel abandonment

A river always seeks the steepest and shortest path to the ocean. During a delta cycle, the landmass of the delta extends into its receiving basin, but over time, the channels become so elongated that the flow is no longer hydrodynamically favorable. At this point, the river will find another, steeper pathway and the old channel is abandoned.

The abandonment of a delta channel typically happens when the river is in flood and much of the delta is inundated. While flooded, a river can "find" a steeper descent and begin carving a path in that direction. The newer river channel will start to take more and more of the river flow, while the older one loses its lifeblood and begins to senesce and close. The old course eventually becomes abandoned, and the river switches its course entirely to the new, steeper path.

Such river avulsion is not an instantaneous process, but it can be observed over human lifespans. Many ancient civilizations have built their foundations alongside riverbanks within deltas and then watched the river close up and die because it has avulsed and started flowing down an alternate path.

Cutting off the lifeblood

Since rivers act as vital networks for food, transportation, and commerce, many ancient cities built along them have themselves died with the drying up of the waterway. New civilizations then become established on the new river course. In ancient Greek and Roman times, the Danube delta became a hotspot for the circulation of both goods and ideas, playing a major role in Balkan culture. Over time, however, the delta channels changed course and these settlements were abandoned. On the Indus and Ganges deltas, several centuries-old Hindu temples still stand where civilizations thrived in areas that are now uninhabited. The Huang He (Yellow River) delta in China has also experienced major river avulsions, even in the past few decades.

▶ **Channel abandonment**

Channel abandonment begins when a river finds a new, steeper pathway to its receiving basin, whether this is a lake or an ocean. Abandoned channels can quickly become silted in with sediment and vegetation, as seen here in the Okavango delta, Botswana.

THE IMPACT OF CHANNEL SWITCHING

Several historical courses and older deltas have been identified
on the Huang He (Yellow River) in China. These changes have
had a great impact on past settlements in the region.

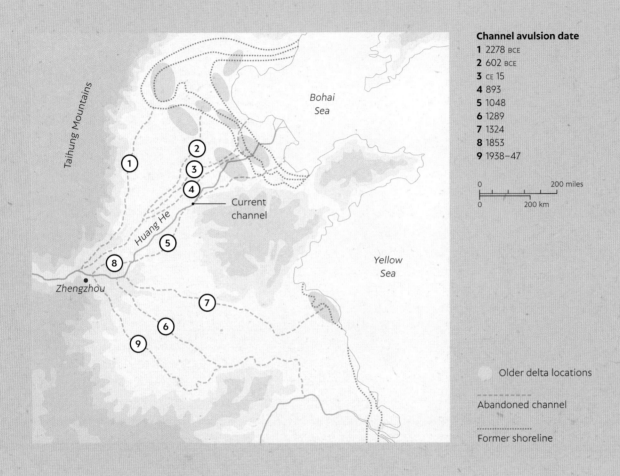

Channel avulsion date
1 2278 BCE
2 602 BCE
3 CE 15
4 893
5 1048
6 1289
7 1324
8 1853
9 1938–47

Older delta locations

Abandoned channel

Former shoreline

Today, many communities on deltas around the world are working hard to
prevent avulsion events from happening. For example, the Mississippi River has
built its delta so far out into the Gulf of Mexico that it is discharging at the edge
of the continental shelf and is primed for an avulsion. Enormous mitigation efforts
were put in place in the 1960s to prevent this from occurring because the city of
New Orleans obtains its drinking water from the river and the economic value
of the ports in lower Louisiana is substantial.

Bayous and backwaters

When channel distributaries become abandoned, they no longer receive any
active river flow. In some cases, they become quiescent bayous, remaining deep and
open. In other cases, however, the channels silt in and close up entirely. This has
significant implications for fish and other aquatic organisms that rely on the river
water and nutrient lifeblood to maintain a sustainable habitat. When a channel
is abandoned, change is imminent.

Delta retreat

When river avulsion occurs, what happens to the delta lobe that is abandoned? When a river abandons its course, its water and sediment are diverted to the new pathway, causing the old pathway to senesce. The process is similar to how a succulent plant reacts when it becomes water-stressed: it initially retracts and becomes smaller, but over time it may disappear entirely.

Delta retreat occurs when waves, tides, and storms begin to take over the major redistribution of sediment because the river is dying. Seen in bird's-eye view, a delta will start to retreat due to wave attack, and the protuberance that had built up will begin to erode. Sand that was originally delivered by the distributaries at the river mouth will be reworked along the shore into barrier islands. These barrier islands will at first provide critical protection to the fragile wetlands lying between them and the coast, such as marshes and mangroves. Over time, however, relative sea-level rise will continue, and the landmasses that no longer receive riverine sediment will begin to submerge. This means that many areas that were once aquatic and subsequently became terrestrial during delta growth will revert to interior bay habitat, and eventually open-ocean marine conditions.

The timescale over which delta retreat occurs can be in the order of decades to thousands of years. The rate of relative sea-level rise dictates this timescale to some extent, but the health of the vegetation that takes hold on the deltaplain is also important. Recent findings show that when the inorganic supply of sediment wanes, some tidal mangrove and marsh areas are able to keep pace with relative sea-level rise via organic production below ground, through the accumulation of roots, rhizomes, and other organic material collectively called peat. However, these regions are typically found in quiescent areas, and thus require the protection of the barrier islands or offshore sediment shoals to minimize wave attack. When marine conditions begin to dominate, or rates of relative sea-level rise are great, it is merely a matter of time before an entire delta lobe is washed away or subsides and submerges, at which point only a portion of it might be preserved in the geological record.

▼ **Beach formation**
The beaches in the Parc naturel régional de Camargue in France formed from reworking of Rhône delta sediment. The region is home to the famous wild gray Camargue horse.

POST-ABANDONMENT

When delta abandonment takes place in river-dominated deltas, they go through a transition from erosional headland to transgressive barrier island arcs, and finally to submerged shoals.

- ● Abandoned distributary channel
- ● Active distributary channel
- ● Channel belt/alluvial
- ○ Distributary mouth bar
- ○ Sand sheet
- ○ Salt marsh

Active delta
Progradational deltaic headland

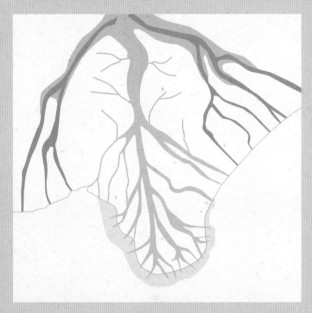

Stage 1
Erosional headland: flanking barriers

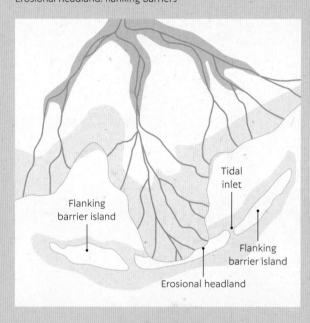

ABANDONMENT

REOCCUPATION

SUBMERGENCE

SUBMERGENCE

Stage 3
Shoreline retreat: inner shelf sand shoal

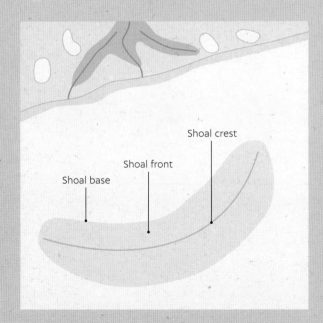

Stage 2
Shoreline retreat: transgressive barrier arc

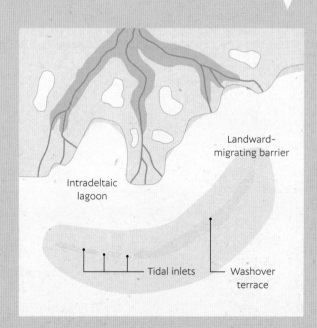

Deltas in deep time

Geologists use the rock record to reconstruct Earth's history over millions of years. This includes unraveling what paleoenvironments existed where and for how long, and what organisms thrived in these ancient settings. Today, the spectacular remains of paleodeltas (old deltas that are no longer active) can be seen in rock outcrops.

▼ **The Book Cliffs**
The Book Cliffs rise above Grand Valley in Colorado and Utah in the United States. They offer some of the most spectacular three-dimensional and cross-sectional views of river and delta formations preserved in rocks.

Paleodeltas contain the deposits of old river deltas that carved across the landscape, debouching into paleolakes and ancient oceans. Over time, tectonic activity has changed the landscape, shifting the course of rivers and creating mountains and even entire continents. Paleodelta deposits can be moved along with the new landscape, like luggage on a conveyor belt. Some of these ancient deposits have been uplifted from sea level into mountain ranges, while others have subsided beneath the surface and are now visible only using advanced scientific equipment that can "see" under thick sediment and rock layers. Countless paleodeltas exist in the rock record around the world, but the Book Cliffs in the United States and the paleo-Volga delta in the Caspian Sea are two of the best studied.

The Book Cliffs

Millions of years ago, the Earth's continents looked very different than they do today. After the massive supercontinent of Pangaea broke apart some 200 million years ago, North America began a westward migration that continues today. Around 100 million years ago along the western side of the continent, a new mountain range—called the Sevier Mountains—began to form as the old oceanic Farallon Plate moved underneath North America in the process of subduction. This buckled up the Earth's crust, building mountains, and created a large subsiding basin on the eastern side of this range. An extensive interior seaway, called the Western Interior Seaway, then flooded the North American continent, to the extent that it was like an ocean between the Sevier and Appalachian Mountains. Erosion of the Sevier Mountains yielded sediments that were then deposited to the east of the range as deltas in the Western Interior Seaway.

Over millions of years, the Sevier Mountains completely eroded away, and other tectonic forces impacted the western margin of the North American continent. These forces produced a completely new and younger mountain range, the modern-day Rockies, which are around 35-50 million years old.

ANCIENT ENVIRONMENTS

The Book Cliffs provide an amazing opportunity for geologists and hydrocarbon engineers to study river and deltaic deposits. These deposits record paleoenvironmental changes, such as sea-level rise and fall, and help geologists understand the structure of potential subsurface reservoirs in other paleodelta environments.

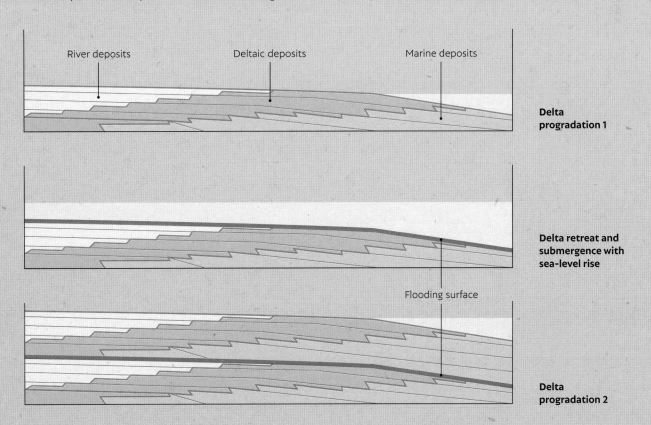

River deposits   Deltaic deposits   Marine deposits

Delta progradation 1

Delta retreat and submergence with sea-level rise

Flooding surface

Delta progradation 2

How do we know this happened? The Book Cliffs in modern-day Utah and Colorado offer a unique window into the geological past of the region—a bit like stepping into a time machine to see the environments that existed there millions of years ago. The Book Cliffs extend for 350 km (215 miles) and feature spectacularly exposed sandstone rocks dating from the Cretaceous period around 100 million years ago. The rock ledges that cap the cliffs look like a shelf of books—hence their name. Geologists have discovered that the rocks within the Book Cliffs contain a complete record of the river and delta deposits that formed when the Sevier Mountains existed and rivers flowed from them into the Western Interior Seaway. These indicate that the rivers were flowing east and southeast, and analyses help identify where paleoshorelines once existed and how far inland tides propagated.

The Book Cliffs are some of the best-studied examples of river and delta geology because the exposures are so clean, and they provide unique opportunities to see in three dimensions how subaqueous delta deposits are structured. This information is necessary to create accurate paleoenvironmental reconstructions and unravel sea-level histories, and has been used to help elucidate oil and gas reservoir potential in other paleodelta environments around the world.

Paleo-Volga delta in the Caspian Sea

Today, the Caspian Sea is an enclosed basin in temperate Eurasia, but the climate in the region has fluctuated extensively over the past several million years, between glacial and interglacial periods and wet and dry conditions, causing the shoreline to expand and contract like a rubber band. When this happens, the outlets of the rivers in the region similarly march across the landscape, either landward or seaward. Ancient deltaic deposits in the rock record can be used to help reconstruct this climate history.

There have also been extensive tectonic forces at play in the region as the Arabian Plate collides with the Eurasian Plate. This has forced several mountain ranges to rise, while other regions have been affected by subsidence. Research suggests that the Caspian Sea may have been connected to the Black and Mediterranean Seas to the west during certain episodes of these tectonic movements and when relative sea levels were high.

The Volga delta currently builds out into the Caspian Sea along its northern shoreline. However, around 5 million years ago, the extent of the Caspian Sea was much smaller (20 percent of its present-day size), incorporating only the southern section, called the South Caspian Basin. This basin was quite deep because accelerated subsidence was taking place in the region as a result of the tectonic collision between the Arabian and Eurasian Plates. The paleo-Volga River flowed into the basin, and a depth of up to 8 km (5 miles) of fluviodeltaic sediments accumulated there. Over time, the regional climate brought more precipitation to the region, which shifted the shoreline of the Caspian Sea—and the more modern Volga delta—to its present-day northern position.

The paleo-Volga delta is extremely important today thanks to the huge reservoirs of oil and gas contained deep within its sediments. Some of the world's first oil wells were drilled here in the mid-nineteenth century, and modern-day nations along the margins of the Caspian Sea, such as Azerbaijan, rely on these conventional oil resources for energy.

THE PALEO-VOLGA DELTA

The paleo-Volga delta built out into the South Caspian
Basin around 5 million years ago, when the extent of the
Caspian Sea was quite small. Over time, the water level
in the Caspian Sea has risen, and the modern Volga delta
now discharges along its northern shoreline in Russia.
The sedimentary deposits of the paleo-Volga delta are the
epicenter of oil and gas fields in modern Azerbaijan.

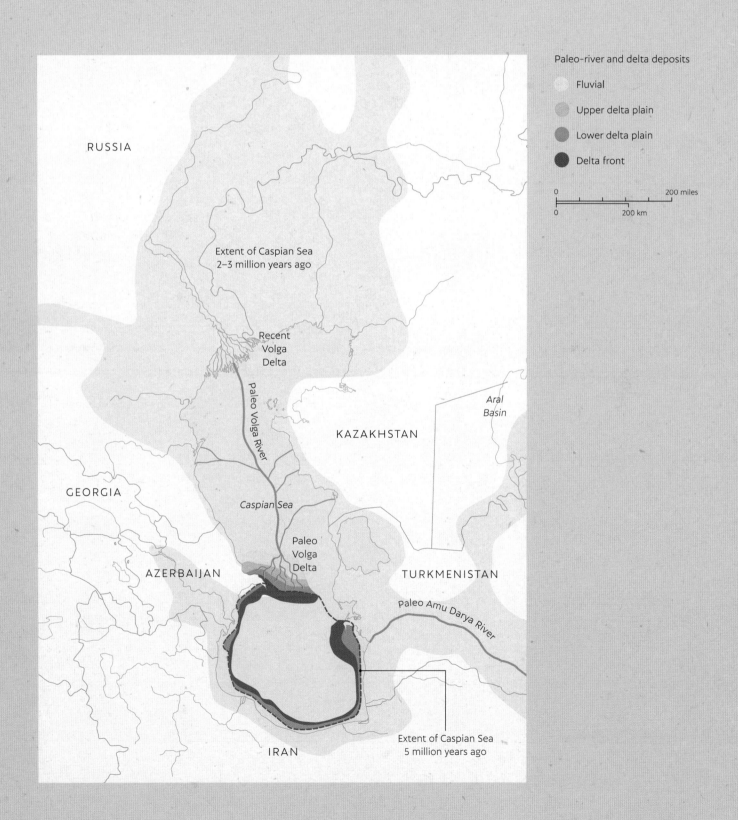

Paleo-river and delta deposits

Fluvial

Upper delta plain

Lower delta plain

Delta front

0   200 miles
0   200 km

RUSSIA

Extent of Caspian Sea
2–3 million years ago

Recent
Volga
Delta

Paleo Volga River

Aral
Basin

KAZAKHSTAN

GEORGIA

Caspian Sea

Paleo
Volga
Delta

AZERBAIJAN

TURKMENISTAN

Paleo Amu Darya River

IRAN

Extent of Caspian Sea
5 million years ago

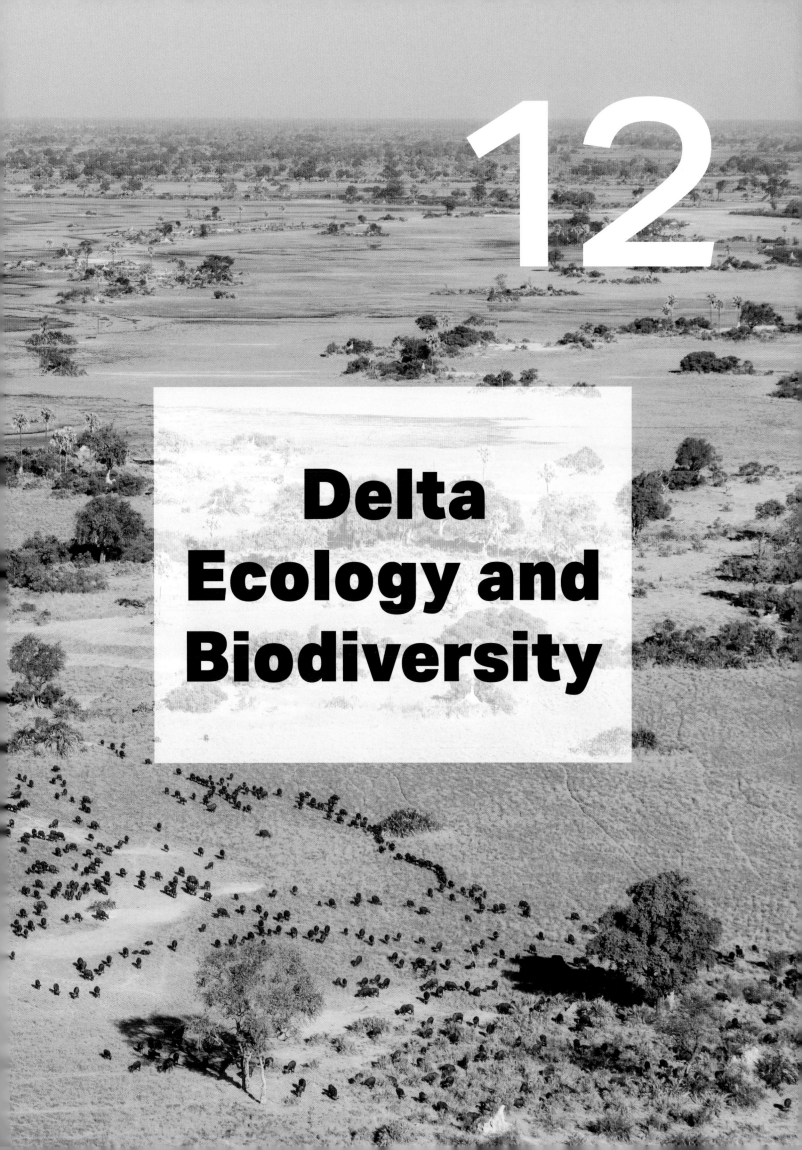

Delta Ecology and Biodiversity

12

Biodiverse landscapes

Like rivers and estuaries, deltas harbor abundant life, from the smallest microscopic algae floating in the water column, to the giant mammals and reptiles that swim in its channels and march along its shores. The climate and adaptation capabilities of different organisms determine which species call deltas their home.

Arctic delta inhabitants

Imagine living in a landscape where the average summer temperature does not rise above 10 °C (50 °F) and the average winter temperature is -40 °C (-40 °F), well below freezing. Polar regions reach these temperatures regularly, and deltas north of the Arctic Circle are unique in the face of these cold challenges: trees are replaced by shrubs and lichens, and for much of the year they are covered by ice and snow. Despite these extreme landscapes, Arctic deltas offer unique habitats for a diversity of life-forms. For example, river channel and tundra habitats in the Yukon delta in Alaska are home to one of the largest aggregations of waterbirds in the world, critical spawning and rearing habitat for Pacific salmon, and uplands that support Brown Bear (*Ursus arctos*), American Black Bear (*U. americanus*), Caribou (*Rangifer tarandus*), Moose (*Alces alces*), Wolves (*Canis lupus*), Arctic Foxes (*Vulpes lagopus*), and Muskox (*Ovibos moschatus*). Along the coast, the Bering Sea hosts various marine mammals. The Yukon is the ancestral home of the Yup'ik, Cup'ik, and Deg Xit'an people, and the delta facilitates their active subsistence lifestyle.

Temperate delta inhabitants

Deltas in temperate climate zones are typically dominated by wet grasslands, called marshes, and are teeming with life within both the channels and the intertidal platforms. The charismatic wildlife found here includes Lions (*Panthera leo*), Cheetahs (*Acinonyx jubatus*), elephants, rhinos, deer, ox, apes, Chimpanzees (*Pan troglodytes*), weasels, otters, muskrat, alligator, pelicans, egrets, snails, crabs, and catfish, to name just a few.

The Manning River delta on the east coast of Australia, north of Sydney, supports large populations of oysters. The Sydney Rock Oyster (*Saccostrea glomerata*) and Australian Flat Oyster (*Ostrea angasi*) are endemic to Australia, while the Pacific Oyster (*Magallana gigas*) was introduced from Japan in the 1940s. This delta is also famous for its captivating marine life, including dolphins, whales, and sharks, which frequent the channels leading in from the Pacific Ocean.

▶ **Yukon caribou**
*Caribou, also called Reindeer (*Rangifer tarandus*), are well adapted to living in harsh Arctic climates, with thick coats of fur for warmth and digestive systems that can break down lichen and moss.*

Tropical delta inhabitants

The tropics are regions where temperatures rarely drop below freezing (0 °C/32 °F), and whose deltas are dominated by mangrove forest vegetation. The warm temperatures are more favorable to amphibians and reptiles due to the cold-blooded nature of these animals, and birds thrive on the abundant fish and insect life. You do not have to look far to find mudskippers crawling along the channels at low tide, or Saltwater or Estuarine Crocodiles (*Crocodylus porosus*) sunning themselves along the banks. The Orinoco delta in Venezuela is home to the famous Orinoco Crocodile (*Crocodylus intermedius*), Orinoco River Dolphin (*Inia geoffrensis humboldtiana*), Giant River Otter (*Pteronura brasiliensis*), Green Anaconda (*Eunectes murinus*), and more than 1,000 species of fish. It is also a hub for birdlife, including flamingos, colorful parrots, scarlet ibis (*Eudocimus ruber*), and toucans.

◀ **Pretty in pink**

American Flamingos (Phoenicopterus ruber) can gather in the thousands. Scientists estimate that there are more than 200,000 American Flamingos in the neotropics of North and South America. Their distinctive pink plumage results from their diet of algae and brine shrimp that contain carotenoids, the natural pigments that give carrots their orange color and turn ripe tomatoes red.

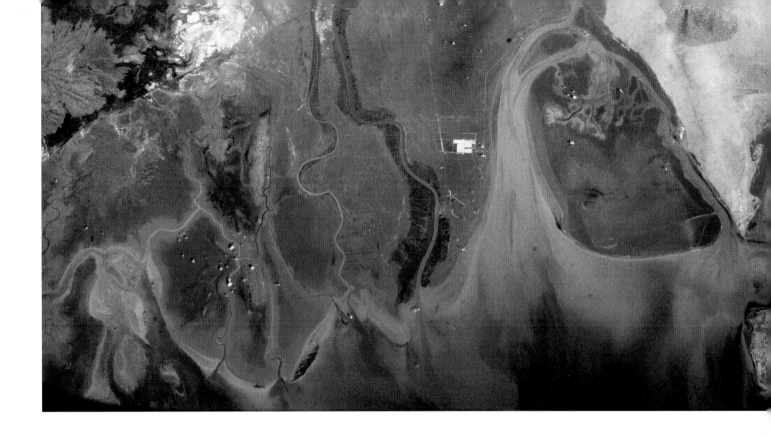

▶ **Honey collectors**

Natural products such as honey are extracted from lush delta lands. These honey collectors in the Sundarbans mangrove forest of Bangladesh use smoke to pacify the bees during harvest

▲ **Date Palm kingdom**

The Shatt al-Arab River, formed by the confluence of the Tigris and Euphrates Rivers, bisects the southern Iraq desert, as seen in this Space Shuttle image from 2000. Its delta was once home to one of

the largest Date Palm (Phoenix dactylifera; right) forests in the world, so famous it was even depicted on the royal coat of arms for the Kingdom of Iraq between 1924 and 1958.

Desert delta inhabitants

Unlike the iceboxes of polar regions, deltas in desert environments are unique oasis habitats. The Okavango delta brings water and life to the inland Kalahari basin of Africa, and has become home to thousands of species. Many of these animals are endangered, including the African Bush Elephant (*Loxodonta africana*), Lion, Cheetah, Southern White Rhinoceros (*Ceratotherium simum simum*), Black Rhinoceros (*Diceros bicornis*), and African Wild Dog (*Lycaon pictus*). (For more on the Okavango, see pages 300–301.)

The Shatt al-Arab delta, formed at the confluence of the Euphrates and Tigris Rivers in Iraq, is another oasis surrounded by desert sand, where temperatures regularly exceed 40 °C (104 °F) in the summer (May to September). In the 1970s, the delta's waters supported the largest Date Palm (*Phoenix dactylifera*) forest in the world, with 17–18 million date palm trees, or an estimated 20 percent of the world total at the time. Today, overextraction of fresh water for human consumption and irrigation has caused saltwater intrusion and decimation of the forest, and is also threatening critical fisheries in the area and the northern Arabian Gulf.

Wetland vegetation in deltas

The lush vegetation that exists in deltas is subject to frequent inundation by water, be it from rivers, tides, waves, or storm surges. What species exist where is determined by the climate and salinity, and the frequency and duration of inundation.

Coastal marshes

Marshes are dominated by grasses and thrive where periods of inundation are modulated by the tides. Vegetation can exist in three zones—subtidal, intertidal, or supratidal. Hydroperiod (the frequency and duration of inundation) and salinity control species variation. For example, in the Danube delta of Romania, the dominant vegetation comprises Common Reed (*Phragmites australis*), cattails (*Typha latifolia* and *T. angustifolia*), bulrushes (*Scirpus radicans* and *S. lacustris*), sedges (*Carex pseudocyperus*, *C. dioica*, and *C. stricta*), horsetails (*Equisetum* species), Arrowhead (*Sagittaria sagittifolia*), Yellow Iris (*Iris pseudacorus*), and Grey Sallow (*Salix cinerea*), each occupying their preferred ecological niche, where they outcompete other vegetation. Since inundation is highly tied to elevation and salinity, location and sedimentation dynamics can strongly control the spatial extent of vegetation in deltas. In the newly formed Atchafalaya delta in Louisiana, willows and cattails are found on the older and higher islands of the delta, while Arrowhead is found at lower elevations in the interior. As the islands develop and build up over time, a succession of species can be expected across the landscape, from low to high marsh.

ELEVATION CONTROL

Transverse cross section through one of the delta islands of the Atchafalaya delta in Louisiana shows how wetland vegetation is controlled strongly by the length and frequency of inundation to which it is subjected and hence the elevation at which it grows. Lower elevations are inundated regularly by the tides, while higher elevations are flooded only seasonally by the river and harbor freshwater plant species.

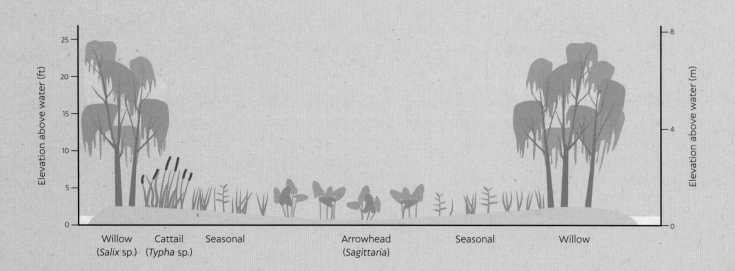

Willow (*Salix* sp.) Cattail (*Typha* sp.) Seasonal Arrowhead (*Sagittaria*) Seasonal Willow

Tropical mangroves

Mangroves dominate deltas in the tropics. They are affected by seasonal and tidal variation, but are very sensitive to temperature and cannot thrive at latitudes that experience freezing for long periods of time. In the Fly delta in Papua New Guinea, more than 29 mangrove species thrive in forests that cover more than 900 km² (330 square miles). In the most marine-influenced areas, Tall-stilt Mangrove (*Rhizophora apiculata*) and *Bruguiera parviflora* predominate as they are well adapted for water salinity that exceeds 10 parts per thousand (ppt). Where salinity ranges between fresh water and saltwater (2-10 ppt), Nipa Palm (*Nypa fruticans*; the only palm adapted to mangrove forests) dominates, while the freshest reaches closest to the river mouth are dominated by *Sonneratia lanceolata* and Gray Mangrove (*Avicennia marina*). Coastal equatorial regions in West Africa and South America stand out as mangrove hotspots, with some of the tallest trees (usually more than 20 m/65 ft in height), although the tallest mangrove trees on Earth, reaching up to 65 m (215 ft) tall, are found in the Gabon estuary, West Africa.

▲ **Swamp giants**

Bald Cypress (Taxodium distichum) *trees live in river and delta floodplains that are seasonally inundated with fresh water, such as Bolieu Pond in the Atchafalaya Basin shown here. Ancient tree trunks more than 50,000 years old were found in 18 m (60 ft) of water off the coast of Alabama and thrived there when the sea level was lower.*

Mangrove sediment trapping and salt extraction

In deltas that drain into the ocean and are impacted by tides, fresh water from rivers mixes with the saltwater of the ocean, creating a salinity gradient across these landscapes. How do mangroves cope with these challenging conditions?

Pneumatophores

Mangrove platforms receive tidal water during the flood tide but are drained when the ebb tide follows. Mangrove trees have developed special roots called pneumatophores, which are adapted to fluctuating currents, periods of inundation, and oxygen-poor sediment. These portions of the mangrove root system stick out above ground and facilitate the aeration necessary for root respiration. The pneumatophores help the mangrove "breathe" when underwater and in oxygen-poor conditions, similar to a snorkel.

Depending on the mangrove species, pneumatophores can be a mini forest of single stalks that stick out of the mud (sometimes up to 50 within an area of 1 m²/10 ft², each 5–300 cm/2–120 in high), or they can be a tangled web of roots that crisscross the mudflat. A 2–3 m-tall (6½–10 ft) Gray Mangrove tree usually has more than 10,000 pneumatophores. An important function of mangrove pneumatophores is that they slow tidal currents, helping to trap sediment on the surface and thereby building new land.

MANGROVE SPECIES DISTRIBUTION

Mangroves dominate in the tropics and at low latitudes. Southeast Asia is home to a plethora of mangrove species.

Mangrove species

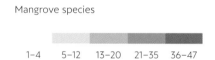

1–4   5–12   13–20   21–35   36–47

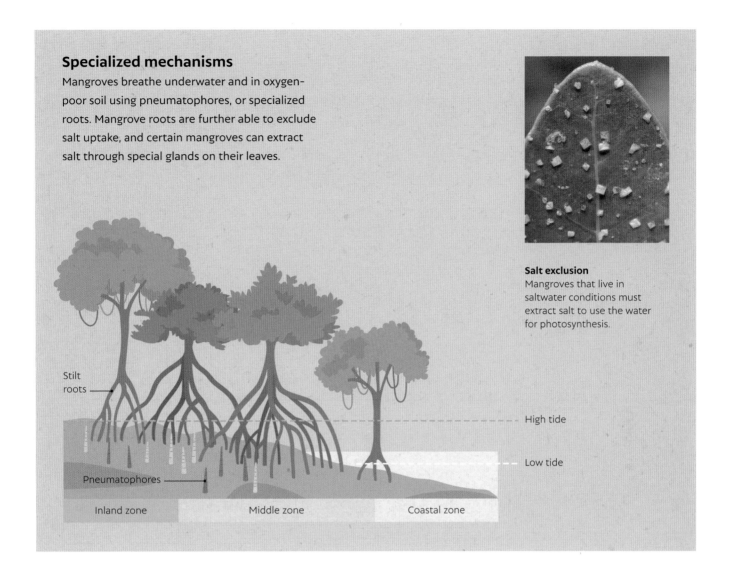

Specialized mechanisms

Mangroves breathe underwater and in oxygen-poor soil using pneumatophores, or specialized roots. Mangrove roots are further able to exclude salt uptake, and certain mangroves can extract salt through special glands on their leaves.

Salt exclusion
Mangroves that live in saltwater conditions must extract salt to use the water for photosynthesis.

Stilt roots

High tide

Low tide

Pneumatophores

Inland zone

Middle zone

Coastal zone

Salt exclusion

Saline conditions are difficult for organisms to manage because although salt is a necessary nutrient, too much of it causes water imbalance and cellular stress, particularly through the absorption of other nutrients. Mangroves cope with these challenges by filtering salt out of the seawater surrounding their roots, allowing only purer water to pass within. Some mangroves remove more than 90 percent of salt in seawater this way. A second line of defense is excreting salt through glands on the underside of their leaves, visible as a white dusting (see above photo).

A third defense involves concentrating salt into older leaves or bark. Since many mangrove species are deciduous, they can get rid of excess salt as they drop their leaves or shed their bark in the "fall." The Mahakam delta in Indonesia is dominated by the species Looking-glass Mangrove (*Heritiera littoralis*), *Avicennia*, *Sonneratia alba*, Nipa Palm, *Rhizophora*, and *Bruguiera*. Of these, *Avicennia* and *Rhizophora* are facultative halophytes, meaning they can tolerate saline conditions. They inhibit salt absorption by their roots and excrete salt from their leaves, so they are most commonly found at the seaward limits of the delta. Leaves that drop from mangroves are soon broken down by bacteria and fungi, forming detritus in the water column that provides a food source for marine life, including economically important shrimp, crab, and fish.

Temperate marshes

Marshes are wetland landscapes that are dominated by grasses. They teem with life and form some of the most ecologically diverse natural habitats in the world. Historically thought of as stagnant cesspools and breeding grounds for disease, these soggy landscapes are now recognized as crucial for their roles in filtering pollutants and providing nursery grounds for wildlife.

Cesspools?

A casual onlooker might consider a marsh wetland to be a quagmire or cesspool, a sloppy, slimy, muddy realm where monsters reign. This is hardly surprising—before the advent of modern medicine and engineering practices, these waterlogged bogs harbored mosquitoes and biting insects that spread malaria and other infectious diseases. They flood, and alligators, crocodiles, and venomous snakes slither along their shorelines and in the water. When describing the coast along Europe's Rhine-Meuse delta, the Greek geographer Pytheas (c. 325 BCE) wrote, "more people died in the struggle against water than in the struggle against men." To say the least, marshes were not considered productive or desirable land where people could live. In France in 1599, malaria and yellow fever were such a problem that King Henri IV issued an edict allowing the drainage of all lakes and wetlands to make them more productive and less deadly. Over the next 60 years, the French reclaimed land on a large scale, backed by Dutch financing and engineering, and large portions of the Rhône delta, which forms the Camargue region near Arles in southern France, were embanked and drained.

The large-scale drainage of wetlands and marshes can be traced back to 3000 BCE by Pharaoh Menes on the Nile delta, and as recently as 1950-1990 in the Mesopotamian marshes in the Shatt al-Arab delta in Iraq. Whether for economic, health, or political reasons, people have been trying to control the natural state of wetlands for millennia, especially those in delta regions where large populations thrive. So, when did opinions and perceptions about the value of wetlands change?

▼ **Drained marshes**

The wetlands near Aiges-Mortes on the western side of the Camargue in the Rhône delta were historically drained for agriculture. Today extensive salt ponds are utilized to harvest salt.

Altered marshland

Civilizations since Roman times have relied on the extensive salt pans of the Camargue (Rhône) delta for harvesting this essential mineral. From the sixteenth century, numerous embankments and canals were built on the delta to drain the marsh "cesspools" and convert them to "more productive" agricultural land. This 2022 Copernicus Sentinel-2 satellite image of the largest river delta in western Europe shows how much of the delta is now agricultural land, with only 950 km² (370 square miles) of natural marsh remaining.

1 Salt pans
2 Petit Rhône
3 Vaccarès Lagoon
4 Mediterranean Sea
5 Grand Rhône

Or ecological gems?

Early Egyptian and Chinese writings and Indigenous oral histories extolled the beauty of nature along the river's waterways, and some even deified rivers. Hapi was the ancient Egyptian river god who personified the Nile and its flooding, while in Aotearoa New Zealand, Māori regarded the Whanganui River as a spiritual mentor.

> *Whistling wind, in a driving autumn rain*
> *Clacking stones, stumbling on the shore*
> *Waves leaping, crashing into each other*
> *White egret startles, recovers, descends.*

> *—Wang Wei (740 CE)*

In the 1800s, perceptions of wetlands, in particular, gradually changed toward more positive values of beauty, fertility, variety, and utility. American poet Emily Dickinson (1830-1886) depicts some of the conflicting sentiments people faced:

> Sweet is the swamp with its secrets,
> Until we meet a snake;
> 'Tis then we sigh for houses,
> And our departure take.

Early writings by Scottish-American naturalist John Muir (1838-1914) considered swamps and marshes as beautiful landscapes, even after he contracted malaria in 1867, and in 1923 the American author Willa Cather (1873-1947) wrote about the "idleness and silvery beauty" of a marsh located near a character's farm in the novel *A Lost Lady*. Aldo Leopold (1887-1948), considered by many to be the father of American wildlife ecology, expounded on the idea of the "land ethic," an appeal in 1949 for society to not only treat one another with respect, but also to expand that sentiment to the natural world of soils, waters, plants, and animals.

Since 2009, Aboriginal oral histories revealing the deep cultural significance of wetlands have helped shape preservation efforts in the Murray-Darling Basin of Australia. Today, many people across the world consider wetlands as "ecological gems" because of the natural resources and ecosystem services they provide.

▶ **Value of wetlands**
Egyptian hieroglyphs and paintings dating back as far as 3000 BCE depict marshes as beautiful and fertile. This artwork from the tomb of Nebamun shows him hunting birds in the marshes with his wife and daughter.

Wetland ecosystem services

The term ecosystem services (see also Chapter 4) refers to the benefits people gain from ecosystems, including wetlands. These may be natural resources such as ecological havens for recreational and commercially important plants, fish, birds, and other wildlife; landscape modulation such as flood control, groundwater replenishment, storm protection, shoreline stabilization, and water purification; or economic and social advantages such as tourism and cultural value.

One 2019 study estimated that global wetlands are worth the equivalent of more than US$47 trillion a year. Coastal marshes and wetlands, in particular, stabilize shorelines and have an estimated value of US$164-761 per linear meter (US$50-232 per linear foot) in helping to protect inland communities. And in terms of filtering pollutants, the Mississippi delta's water quality enhancement has an estimated value of US$245-13,710 per hectare (US$99-5,551 per acre).

Marshes provide abundant nursery grounds for birds and fish (see Chapters 4 and 8). They also sequester massive quantities of carbon as organic plant material accumulates and is buried (see Chapters 2 and 10). Unfortunately, as marshlands are converted to agriculture or urban centers, their ecosystem services are diminished.

▲ **As far as the eye can see**

The low grasses of marshlands create spectacular landscapes teeming with life, where the horizon stretches for miles, and are highly valued for the ecosystem services they provide.

INTERIOR DELTA: THE SPECTACULAR OKAVANGO

Established as a UNESCO World Heritage Site in 2014, the Okavango delta is one of the world's largest landlocked inland deltas. It is located in northwestern Botswana, and its waters drain into the desert sands of the Kalahari Basin in south-central Africa. Juxtaposed against a dusty brown landscape, this vibrant delta receives floodwaters during the peak of Botswana's dry winter season (June/July), when it supports an incredible ecological oasis. The 20,000 km² (8,000 square miles) of the delta harbor abundant wildlife in and along its permanent and seasonal rivers and lagoons, swamps, grasslands, riparian forest, deciduous woodlands, and islands.

The annual floods trigger spectacular wildlife displays, including large herds of African Bush Elephants (*Loxodonta africana*), African Buffalo (*Syncerus caffer*), Red Lechwe (*Kobus lechwe*), and Burchell's Zebras (*Equus quagga burchellii*). The animals splash and drink the clear waters of the Okavango River in celebration of their survival during the dry season or the weeks-long migration they made across the Kalahari Desert. Botswana is currently home to more elephants than any other African country—295,000, or approximately 70 percent of the worldwide population of this endangered species—and the Okavango delta is vital to their survival. The females of this largest land mammal remain together in family units that are led by the oldest matriarch, and they all assist in raising and protecting the calves.

Lion prides as large as 30 strong feast on African Buffalo, antelopes, Spotted Hyenas (*Crocuta crocuta*), and African Wild Dogs in the delta, and they are agile swimmers and climbers. One of the iconic animals of the Okavango is the Hippopotamus (*Hippopotamus amphibius*); these rotund pachyderms are often heard, their deep honking laughs and snorts carrying across the delta's lakes, channels, and lagoons. The Nile Crocodile (*Crocodylus niloticus*) is frequently seen swimming in the channels or sunning itself along the shores—the reptiles even play hide-and-seek within intricate underwater tunnels beneath floating islands of papyrus. Flocks of birds number in the thousands, with spectacular displays of Red-billed Quelea (*Quelea quelea*), flamingos, Slaty Egret (*Egretta vinaceigula*), kingfishers, and Wattled Crane (*Bugeranus carunculata*) seen on a seasonal basis.

Ecotourism is a major boon to the Okavango delta and Botswana, with tourism contributing approximately US$2.2 billion or 13 percent of the country's GDP in 2019. Ecotourism promotes travel to areas of natural and cultural heritage, with the objectives of maximizing the involvement in, and the equitable distribution of economic benefits to, host communities; maximizing revenues for conservation; educating visitors and local people; and minimizing any negative social, cultural, and environmental impacts.

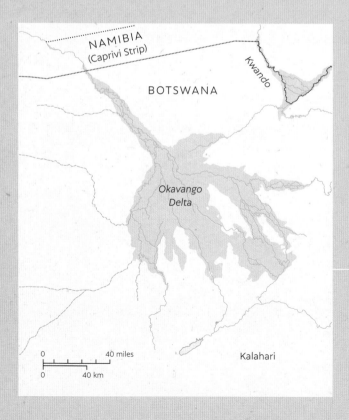

Inland delta
The Okavango that drains into the Kalahari Desert brings life to the charismatic wildlife of the region.

Life-giving water
Red Lechwe antelope (*Kobus leche*) run across the flooded Okavango delta in Botswana.

Elephant sanctuary
The Okavango is a haven for the vulnerable African Bush Elephant (*Loxodonta africana*), thanks to the water and food resources the delta provides in the otherwise dry desert landscape.

Arctic deltas and peatlands

Deltas in the Arctic are unique due to the extreme temperatures, punctuated meltwater pulses, and a landscape that is frozen for much of the year. They also harbor a unique plant and animal wildlife.

ARCTIC RIVER WATERSHEDS

Arctic rivers and their deltas face extreme challenges as they flow northward into the Arctic Ocean. As climate change warms the planet, the extensive permafrost landscapes of many deltas in Canada, Alaska, and Russia are melting.

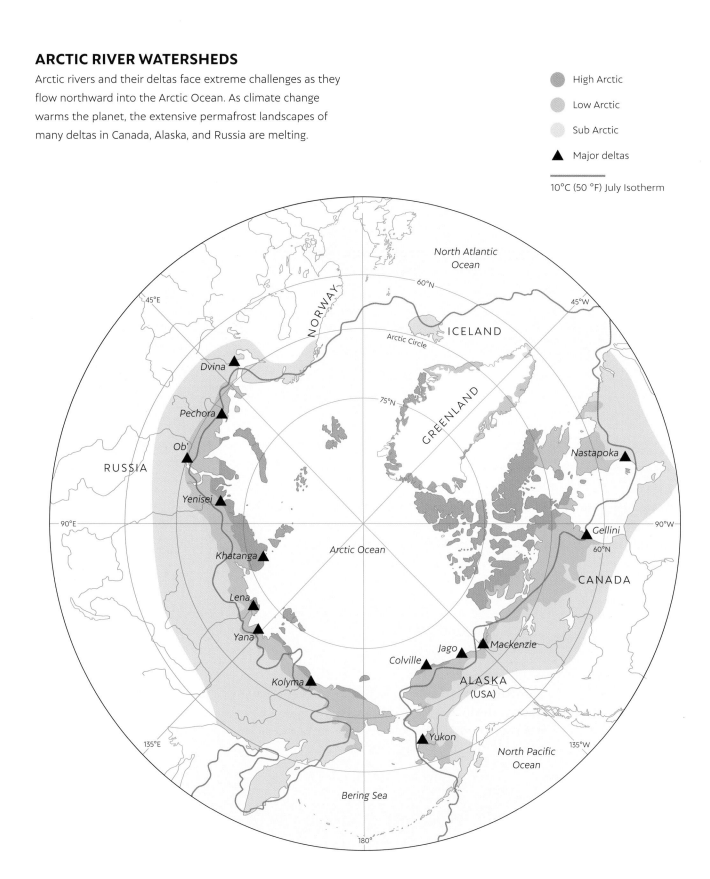

High Arctic

Low Arctic

Sub Arctic

▲ Major deltas

10°C (50 °F) July Isotherm

Permafrost

In polar regions, average summer temperatures do not rise above 10 °C (50 °F) and the average winter temperature is -40 °C (-40 °F). While the surface of the land can thaw in summer, a zone of permanently frozen rock and soil, called permafrost, remains underground. The permafrost has a strong influence on polar deltas, limiting channel migration and the depth at which organisms can live and obtain food. As a result, most Arctic regions are dominated by shrubs and lichens that have shallow roots and can survive dormant during the 7-9 months of below-freezing temperatures.

Permafrost is also an extremely important store for carbon. Recent estimates suggest that Arctic permafrost regions hold about 1,035 petagrams (1 petagram = 1×10^{12} kg, or 2.2 trillion lbs) of frozen carbon from the buildup of organic matter in their uppermost 3 m (10 ft), which is more than 30 times the amount of carbon present in the atmosphere before the industrial revolution in 1850. As our planet warms, however, this permafrost peatland is melting, exposing more and more of its stored organic matter to oxidation and release into the atmosphere.

▲ **Arctic delta**
Deltas in polar regions are frozen for much of the time, and subject to punctuated pulses of meltwater.

PERMAFROST MELTING

As temperatures warm, permafrost melts, forming bodies of water called thermokarst lakes. These waterbodies are important for waterfowl and fish that reside in the region. With warming global temperatures, the thermokarst landscape is expanding and releasing vast stores of carbon.

- Ice wedge
- Permafrost (CO_2 storage)
- Thermokarst lake formation
- Talik (unfrozen ground)

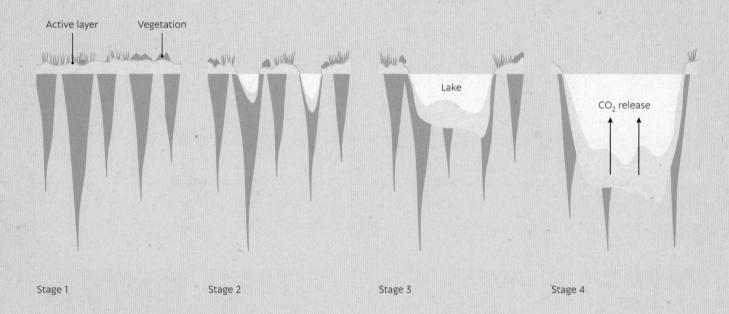

Active layer　Vegetation

Lake

CO_2 release

Stage 1　　　　　Stage 2　　　　　Stage 3　　　　　Stage 4

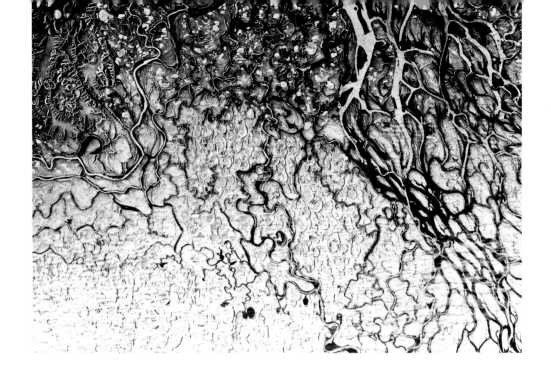

◀ **Frozen delta**
Like many Arctic rivers, the Lena River in Siberia flows northward, a phenomenon that means it melts inland in spring before it does so at the coast further to the north. This causes ice dams to form and increases inundation of the river's floodplains and wetlands.

Freezes and floods

More than 13 major rivers in the Arctic drain 14 million km^2 (5.4 million square miles) of northern permafrost terrain. The deltas that form at the mouths of these rivers are ice-dominated systems affected by land ice, permafrost, and sea ice for much of the year. When seasonal temperatures begin to rise in the Arctic, they warm first in the south, forcing meltwater to flow northward under ice and frequently resulting in punctuated meltwater pulses and ice-jam floods. As these floods occur, important biological and chemical exchange occurs between the river channels and bodies of water called thermokarst lakes, as fresh water, nutrients, and sediment are delivered to the lakes.

Polar deltas range in size from the giant Lena in Russia and the Mackenzie in Canada (see box overleaf), to tiny features that form in glacial lakes and have similar meltwater "pulses" and ice-dominated conditions for much of the year. The Lena River in Siberia discharges into the Arctic Ocean, forming a 400 km-wide (250 mile) delta that is a frozen tundra for seven months of the year. In May, the region transforms into a lush wetland for a few months, when it harbors abundant life.

Arctic delta inhabitants

Deltas in the Arctic provide important refuge and breeding grounds for many species of wildlife. The Lena Delta Wildlife Reserve is a protected wilderness area, home to Siberian wildlife such as the Polar Bear (*Ursus maritimus*), Caribou, Bighorn Sheep (*Ovis canadensis*), Kamchatka Marmot (*Marmota camtschatica camtschatica*), Arctic Fox, lemmings, Beluga Whale (*Delphinapterus leucas*), Narwhal (*Monodon monoceros*), seals, Walrus (*Odobenus rosmarus*), swans, geese, ducks, loons, shorebirds, raptors, and gulls. In addition, it is an important fish spawning site for Siberian Sturgeon (*Acipenser baerii*) and 12 different species of salmon. The delta has also supported species that are now extinct but have been preserved in the frozen ground: in 1799, the body of a frozen mammoth approximately 36,000 years old was discovered here. Many of the modern species that are found in Arctic deltas are temporary summer migrants that are unable to cope with the freezing winter temperatures, but others have adapted to the icy conditions and can remain year-round.

The Mackenzie delta

At 4,241 km (2,635 miles) in length, including its tributaries, and discharging into the Arctic Ocean in the Northwest Territories, the Mackenzie River is the longest river system in Canada. One theory, heavily debated, is that the Mackenzie valley was the path taken by peoples migrating from Asia to North America more than 10,000 years ago. Today, the delta is a frozen tundra for approximately eight months of the year, with permafrost extending down to a staggering depth of 700 m (2,300 ft).

The inland portions of the Mackenzie delta are covered by subarctic boreal forest of White and Black Spruce (*Picea glauca* and *P. mariana*), while the coastal areas are a tundra of low willow, birch, and alder shrubs and sedge wetlands. With such environmental diversity, it is an ecological oasis that supports Beluga Whales (*Delphinapterus leucas*), Bowhead Whales (*Balaena mysticetus*), Ringed Seals (*Pusa hispida*), Bearded Seals (*Erignathus barbatus*), Polar Bears, Arctic Foxes, Caribou (*Rangifer tarandus*), North American Beavers (*Castor canadensis*), Muskrats (*Ondatra zibethicus*), waterfowl, Northern Pike (*Esox eliac*), Arctic and Least Cisco

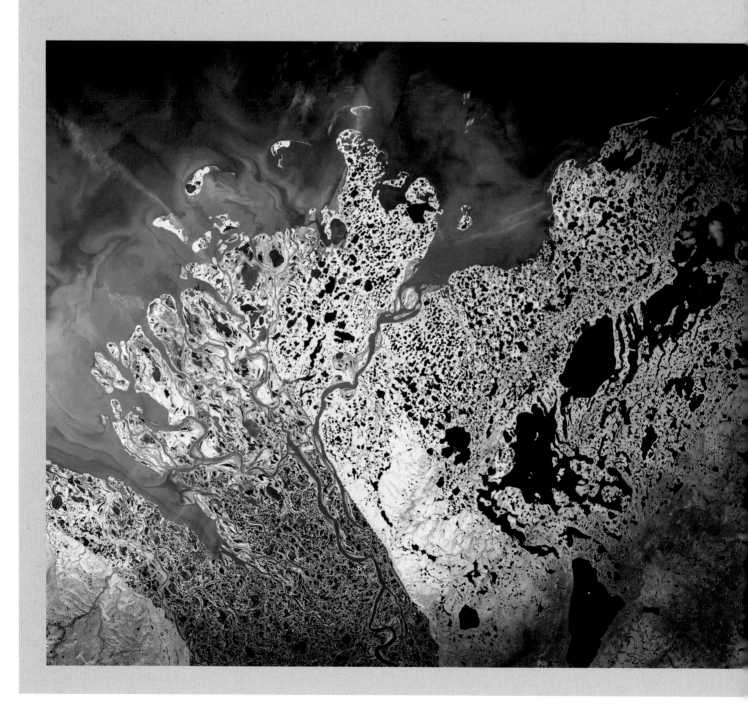

(*Coregonus autumnalis* and *C. sardinella*), Lake and Broad Whitefish (*Coregonus clupeaformis* and *C. nasus*), Arctic Grayling (*Thymallus arcticus*), and Arctic Char (*Salvelinus alpinus*).

With the coastal ocean covered in ice for much of the year, Bowhead Whales are only summer visitors to the delta. Their numbers depend on the ice and upwelling conditions of the Beaufort Sea: heavy ice moves them closer to the delta, but they prefer to feed offshore where blooms of krill and copepods occur. These whales, once greatly valued for their baleen (whalebone) and oil, were nearly hunted to extinction by American fishers between the seventeenth and nineteenth centuries. Commercial whaling of Bowheads near the delta ended in 1914, and since then numbers of the western Arctic population of this endangered species have recovered to around 4,400.

Today, the Mackenzie delta is home to the western Canadian Inuvialuit and Iñupiat peoples, who have hunted and harvested Caribou and Beluga Whales for centuries.

Hunted nearly to extinction
The Bowhead Whale (*Balaena mysticetus*), like this individual swimming off the Mackenzie delta, was valued for its oil and baleen (the filter-feeding plates in its mouth made of keratin), and was hunted nearly to extinction.

End of the road
With icy conditions much of the year, local western Canadian Inuvialuit and Iñupiat peoples must find alternate transportation for hunting, fishing, and recreation.

Deltas and the food web

Like rivers and estuaries, deltas harbor complex food webs that hang in a delicate balance. They straddle the transition zone between land and sea, and the nutrients and sediments they provide create some of the most productive waters in the world.

What is a food web?

Organisms obtain their energy either from the sun (autotrophs), from breaking down chemical compounds and not using sunlight (chemotrophs), or from eating other organisms (heterotrophs). Like the food webs of rivers and estuaries, those that exist in deltas are complex networks. The base of a delta food web comprises photosynthesizing organisms that live in the water column or on intertidal mudflats, such as phytoplankton and algae. Zooplankton are microscopic animals that feed on these smaller life forms, with the juvenile fish, mollusks, and crustaceans also occupying this niche. In turn, more complex, larger animals feed on these organisms. Humans are part of this higher trophic level, as we feed on the fish, mollusks, crustaceans, reptiles, waterfowl, and small mammals that call deltas their home.

Food webs are precariously balanced systems, such that impacts on one species can have ramifications that manifest up and down the trophic levels. For example, overfishing and pollution in the estuaries and deltas of the East Coast of the United States have decimated Blue Crab (*Callinectes sapidus*) populations for decades. As a result, the smaller Purple Marsh Crab (*Sesarma reticulatum*), normally consumed by Blue Crabs, has become more numerous. The Purple Marsh Crabs are notorious herbivores and burrowers, causing destruction of marsh plants such as Smooth Cordgrass (*Spartina alterniflora*) that has been well documented from Georgia to the Santee delta in South Carolina (the largest delta on the East Coast) and north to Cape Cod in Massachusetts.

▶ **Delta food webs**
*Smaller organisms are commonly preyed upon by larger ones, as seen with this Bald Eagle (*Haliaeetus leucocephalus*) catching a fish in a river. These birds can fly at speeds up to 50 km/h (30 mph) and can dive at speeds up to 160 km/h (100 mph).*

SALTMARSH DIEBACK

Blue Crab (*Callinectes sapidus*) populations in the Chesapeake Bay region of the United States have been on the decline for decades. As a result, one of their prey, the smaller Purple Marsh Crab (*Sesarma reticulatum*), has been wreaking havoc on West Atlantic saltmarshes, causing extensive dieback in the marshes of the East Coast, United States.

Blue Crab (*Callinectes sapidus*)

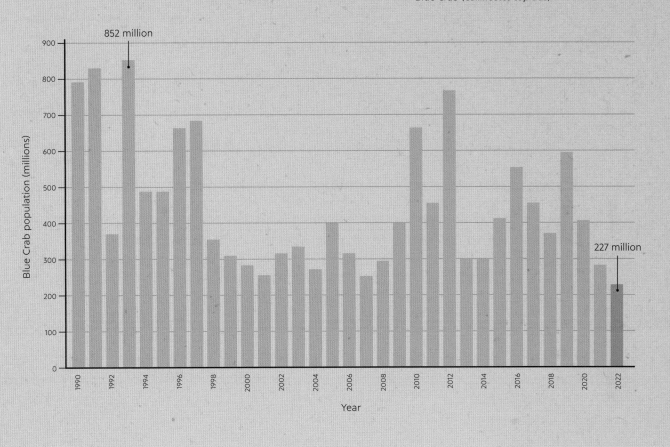

Seafood basket

Each coastal country has an exclusive economic zone (EEZ) that extends 320 km (200 miles) from its shoreline, and 99 percent of the global annual catch for commercial fisheries comes from the combined area of these zones. On the Orinoco delta in Venezuela, 40,000–45,000 tonnes (44,000–50,000 US tons) of fish are caught each year, bringing nourishment and economic prosperity to local inhabitants. In the Po delta in Italy, around 1,800 fishers annually harvest about 15,000 tonnes (16,500 US tons) of Manila Clams (*Ruditapes philippinarum*), more than half of the country's total clam production. And in the United States, 23 percent of seafood landings—335,000 tonnes (370 US tons)—comes from the Mississippi delta, amounting to US$263 million in 2020 alone. The total annual economic impact to the Mississippi delta region—factoring in jobs, commercial fisheries, and recreational tourism—is nearly US$2 billion.

However, climate change and anthropogenic impacts place a strain on fisheries around the world. In the Mekong basin, the operation of hydropower dams upstream has left waterbodies in An Giang, a deltaic wetland in Vietnam, high and dry, devastating fish stocks and threatening the livelihood of local communities that have relied on fishing for generations.

Wild-caught or farmed?

As the world's population grows, the need for reliable, affordable, and sustainable protein also grows (see pages 125 and 216-19). Deltas are important nurseries for juvenile fish, shrimp, crab, and mollusks, and thus historically have been key hunting grounds for wild-caught food. But as environmental degradation and land-use practices have destroyed many natural delta landscapes, wild shellfish and finfish fisheries are becoming increasingly threatened, necessitating expensive environmental regulations and limits on harvest sizes to prevent total collapse.

Areas in many deltas—including the Zhu Jiang (Pearl), Huang He (Yellow), Chang Jiang (Yangtze), Mekong, Ayeyarwady, Ganges, Indus, Nile, Po, Danube, and Mississippi—have been converted to inland fisheries and aquaculture. Here, fish, shrimp, mussels, and clams are isolated within containment ponds or confined to nets in the natural channels of the delta. The farms help feed burgeoning populations and provide valuable export products—for example, much of the Mekong delta floodplain in Vietnam has been converted to aquaculture. The annual yields from these extensive shrimp and fish ponds are estimated to be an astonishing 2.23 million tonnes (2.45 million US tons), enough to sustain national consumption at 12-60 kg of protein per person per year and allow export around the world. Likewise, the Chang Jiang (Yangtze) delta in China produces more than 60 percent of the national production of Giant Freshwater Prawn (*Macrobrachium rosenbergii*).

However, as aquaculture becomes more commonplace, societies are faced with incorporating complex—and often expensive—management strategies to mitigate disease transfer within farmed and natural populations, environmental contamination from effluent exchange, and the degradation of natural mangrove and marsh environments due to land-use conversion.

▶ **Amazon fish market**
*The Tiger Shovelnose Catfish, or Surubi (*Pseudoplatystoma tigrinum*), is a popular fish that haunts the rivers of the Amazon delta. It can grow to enormous size, reaching 1.3 m (4 ft) in length and weighing more than 100 kg (220 lb).*

IMPORTANCE OF FISHERIES IN DELTAS

Fish, mollusks, and crustaceans provide an important
source of protein for the world's burgeoning population,
and many deltas have the necessary nutrients and nursery
grounds to support coastal fisheries. As human populations
in delta regions rise, reliance on marine and inland fisheries
and aquaculture is on the rise, as seen in the historical catch
data of the Ganges–Brahmaputra–Meghna (Bangladesh),
Mahanadi (India), and Volta (Ghana) deltas.

Fishing yield (fish, mollusks, crustaceans)
in tonnes per km² per year

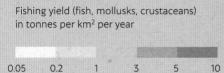

0.05   0.2   1   3   5   10

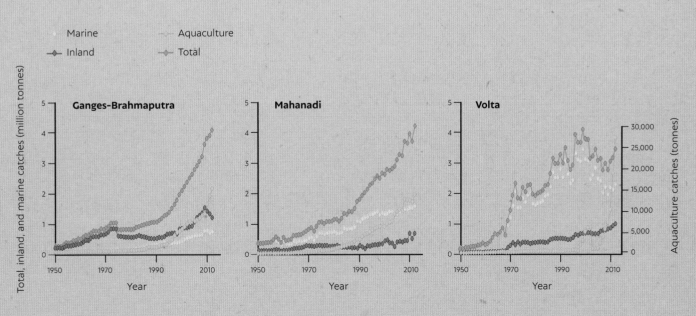

Bird sanctuaries

Birds are some of the most stunning delta inhabitants, with often fantastic plumage and behaviors ranging from the dramatic to the elusive. They can be seen nesting on shorelines, plunging into the waters for a snack, or merely flying in unison in great V-formations overhead. Of the world's 50 billion birds, a large proportion reside along rivers and in estuaries and deltas. Unfortunately, declining habitats due to deforestation, overfishing, and pollution threaten their survival.

What is a bird sanctuary?

Birds not only provide a food source and recreational viewing pleasure for humans, but they also play many roles as ecosystem engineers, including acting as predators, pollinators, scavengers, vegetation modulators, and seed dispersers. Bird sanctuaries are areas of land in which birds are protected and encouraged to breed.

The Kızılırmak (previously the Halys and Alis) delta on the Black Sea in Türkiye is a bird sanctuary and UNESCO World Heritage Site. Over the course of a year, millions of birds pass over the delta as part of the famous Black Sea–Mediterranean Flyway. Some 352 bird species have been recorded here, including rare species such as the Lesser Yellowlegs (*Tringa flavipes*), Arctic Warbler (*Phylloscopus borealis*), Dusky Warbler (*Phylloscopus fuscatus*), and Black-throated Accentor (*Prunella atrogularis*). Several endangered species also visit the delta, including the White-headed Duck (*Oxyura leucocephala*), Purple Gallinule (*Porphyrio porphyrio*), Red-breasted Goose (*Branta ruficollis*), Eastern Imperial Eagle (*Aquila heliaca*), and Egyptian Vulture (*Neophron percnopterus*). The Eastern Imperial Eagle is one of the rarest birds of prey in the world, with only 1,600 remaining.

▼ **Pelican colony**
*Sanctuaries provide protection to bird species such as Great White Pelicans (*Pelecanus onocrotalus*), seen here in the Danube delta, Romania. The delta has the largest colony of these pelicans outside of Africa.*

▶ **Lone aviator**
*The Eastern Imperial Eagle (*Aquila heliaca*) from eastern Europe and Asia is one of the rarest birds of prey in the world. Deltas provide critical nesting and feeding grounds for this endangered species, as well as many others.*

▶ **Nesting sites**
*Bird sanctuaries provide protected land where nesting and breeding is encouraged. The Black-headed Gull (*Chroicocephalus ridibundus*) breeds in much of Eurasia and the Mediterranean Basin, and in coastal eastern Canada.*

Migratory bird flyways

Several major flyways exist across the world's continents and oceans. A flyway is a geographical area covered by a migratory bird over the course of its annual cycle, encompassing breeding and wintering grounds and the connecting migration route. These "avian superhighways" are typically oriented north-south as birds move to milder climate zones (i.e., low latitudes) during their nonbreeding season (winter). Some birds travel up to 25,000 km (16,000 miles) during their migration, and fly as fast as 50 kph (30 mph). Their entire journey can take several months. Due to the energetic costs associated with migration, birds take the shortest route possible, but they also modify their routes based on weather patterns and food resources: flyways often follow mountain ranges, watercourses and coastlines, avoid large bodies of open water, and take advantage of prevailing wind patterns and updrafts. Thus, rivers, estuaries, and deltas provide excellent stopovers along these major annual journeys.

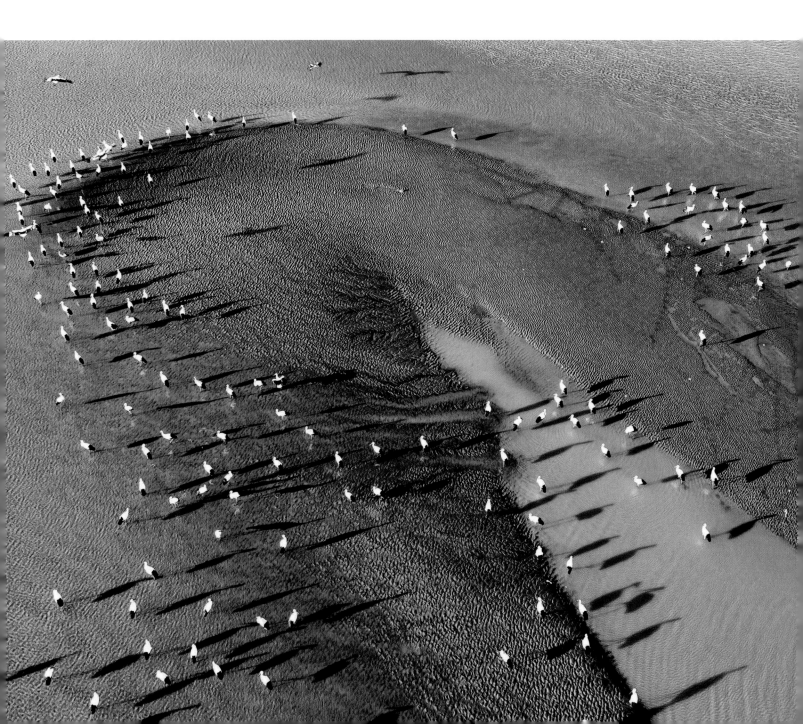

GLOBAL FLYWAY PATHS FOR BIRDS

Migratory birds undertake regular biannual movements between distinct breeding and nonbreeding areas, usually in a north–south direction. Some species travel tens of thousands of miles over the course of several months. The eight flyways that have been recognized are shown here.

Pacific Americas

Central Americas

Atlantic Americas

East Atlantic

Black Sea/Mediterranean

East Asia/East Africa

Central Asia

East Asia/Australasia

▲ Major delta

■ Major estuary

〰 Rivers and lakes

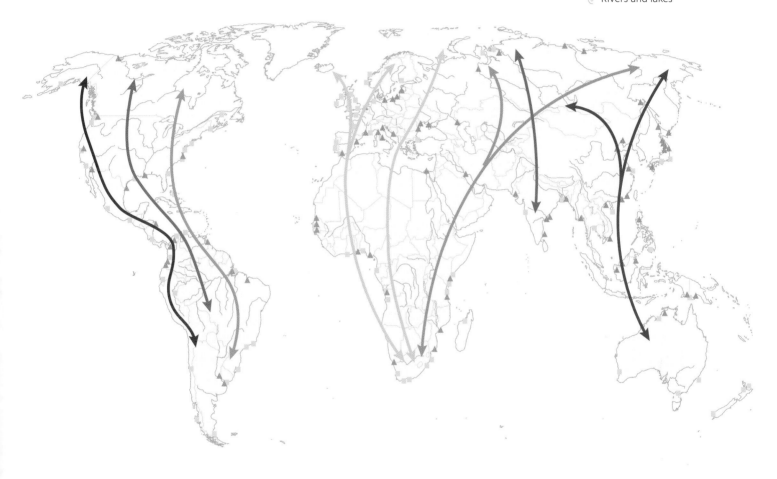

Effective conservation of a migratory species requires coordinated efforts along the entire length of the flyway. Safeguarding a migrant's breeding ground is insufficient if its wintering sites are lost, or if heavy hunting or loss of food resources resulting from overfishing occurs along its migration route. Fortunately, international conservation is gaining traction. The United States Migratory Bird Treaty Act of 1918 implements international conservation efforts with Canada, Mexico, and Russia, and is intended to ensure the sustainability of protected migratory bird species across these political borders. The Oriental Stork historically migrated between southeastern China, North and South Korea, Japan, and eastern Russia, but owing to habitat loss and hunting along its flyway it is currently classed as Endangered on the International Union for Conservation of Nature's Red List of Threatened Species. Conservation efforts in the Yellow River Delta National Nature Reserve in China, which was created in 1992, have facilitated the birth of 2,200 Oriental Storks since 2003, a substantial fraction of the current estimated world population of 3,000.

13

Deltas and Us

Deltas under stress

With human population growth, coastal areas are burgeoning with new inhabitants. Deltas attract a large portion of this population explosion thanks to their opportunities for employment, their bountiful food resources, and their beautiful natural landscapes. However, a growing human population places stress on the landscape.

Population centers

The world's human population remained below 500 million until around 1500 CE, after which medical advances, improvements in sanitation, industrialization, and mechanization of agriculture saw it increase to a billion by around 1830, then double in 100 years and double again in just 50 years by 1970. Today, 500 million people out of a global population of 8 billion live on or within 25 km (15 miles) of the world's largest deltas, several of which support megacities, huge urban centers with populations exceeding 10 million.

Some of the world's deltas—including the Tigris-Euphrates (also called the Shatt al-Arab delta), Nile, Indus, Huang He (Yellow), and Chang Jiang (Yangtze)—are considered cradles of civilization, with population centers dating back 5,000 years. The modern city of Guangzhou on the flanks of the Zhu Jiang (Pearl) delta in southern China has been an important trade port since the third century CE. Much earlier, during the Qin (221-206 BCE) and Han (206 BCE-220 CE) dynasties, the city was located slightly north of the modern urban center, but as the Zhu Jiang prograded seaward and the channel narrowed through the deposition of sand and silt, the city grew southward. Today, an estimated 18 million people live within the megacity of Guangzhou, with 57 million in the entire Zhu Jiang delta region.

Changes to delta lands

With human population growth, there is ever-increasing pressure to convert natural landscapes, such as wetlands and forests, to agricultural land. As a result, several deltas worldwide have seen rapid deforestation, shifts from rich forests to agricultural fields, and conversion of rural land to urban centers. Early civilizations

▼ **Deltas as urban centers**
Today, little remains of the natural Zhu Jiang (Pearl) delta in Guangzhou, southern China.

DEFORESTATION TO MAKE WAY FOR AGRICULTURE

As with many deltas around the world, the Ayeyarwady delta in Myanmar has lost much of its original vegetation cover, removed to make way for agricultural fields. Only through the establishment of wildlife preserves—such as the green island that is the Meinmahla Kyun Wildlife Sanctuary—can human habitation and extensive land-use change be prevented in these sensitive areas.

The purple, red, and yellow colors here indicate previous forest tracts that have been lost.

1978
1989
2000
2011

0   20 miles

0   20 km

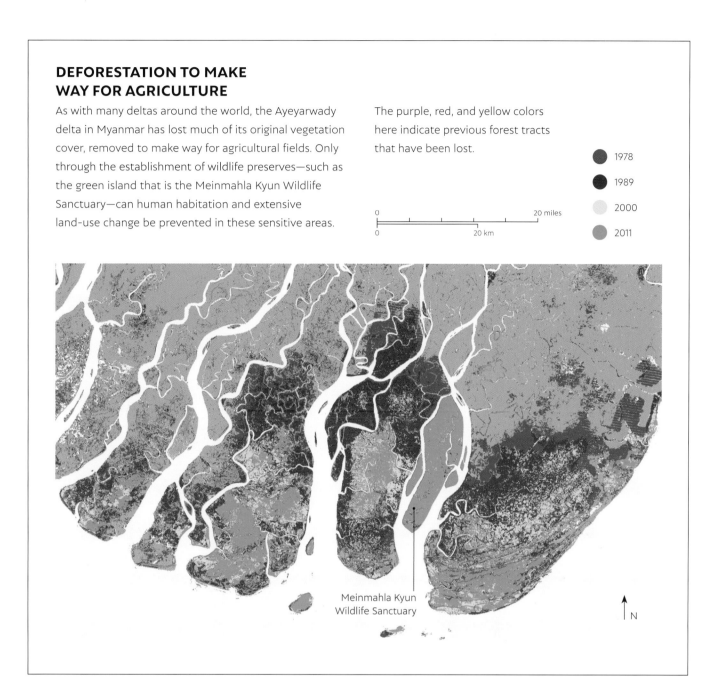

Meinmahla Kyun
Wildlife Sanctuary

N

living on the Tigris-Euphrates delta in modern-day Iraq converted the natural landscape to croplands 5,000 years ago, and even installed reed-woven banks to protect their agricultural fields from flooding. Windmills in the Rhine-Meuse delta of the Netherlands have been operational for several hundred years, draining the land and facilitating its conversion to agriculture. Deforestation is especially rampant in deltas, as exemplified by the loss of critical mangrove forests to rice and shrimp paddies in the Mekong delta—some estimate that as much as 100,000 km^2 (40,000 square miles) of forest has been converted, of which 500 km^2 (200 square miles) was mangrove prior to 1965.

Urbanization involves the replacement of other forms of land use, such as agricultural fields, by cities. Alexandria, founded in 331 BCE at the mouth of Egypt's Nile delta, is one of the oldest examples of urbanization. Today, more than 5 million people call the city home, while just 200 km (125 miles) away a further 21 million live in the metropolitan area of Cairo at the apex of the delta.

Calculating risk in deltas worldwide

With the increasing numbers of people inhabiting delta lands comes the challenge of protecting lives and infrastructure against the natural flood risk. Deltas in lower latitudes are also subject to seasonal and tropical storms that form in warm ocean

Estimates of relative rate of change in risk

RISK TO RESIDENTS

Shown here is the relative risk faced by the human residents of 48 deltas around the world. Deltas such as the Krishna, Ganges–Brahmaputra, and Indus have higher rankings due to the high

basins and cause coastal storm surges in low-lying regions. A recent risk assessment for more than 50 of the world's deltas found that, while some deltas are only moderately susceptible to short-term hazardous events (river or storm flooding), many are increasingly at risk from relative sea-level rise.

socioeconomic vulnerability of people living within these regions. Deltas that have extensive flood control engineering, such as the Mississippi and Rhine-Meuse, rank lower.

▲ Major delta

Lena

Rhine-Meuse

Vistula

Dnieper

Amur

Po

Rhone

Danube

Han

Ebro

Huang He

Moulouya

Chang Jiang

Tone

Sebou

Shatt-el-Arab

Nile

Indus

Ganges-Brahmaputra

Zhu Jiang

Mahanadi/Brahmani

Hong

Krishna

Ayeyarwady

Senegal

Godavari

Chao Phraya

Volta

Mekong

Niger

Mahakam

Fly

Tana

Congo

Burdekin

Limpopo

Flood defenses

Humans have been striving to tame rivers for millennia. In fact, as far back as 3000 BCE people living along the Euphrates River in Mesopotamia (modern-day Iraq) carried out flood control measures to prevent their crops and homesteads being inundated. Many modern flood defenses are not much more advanced than the reed and earthen dams built on riverbanks and deltas in ancient times.

Poldering and levees

Deltas offer unique challenges for humans, including floodwaters from rivers overtopping houses, salinity intrusion jeopardizing agricultural fields, and coastal cyclones pushing storm surges inland. Many of the earliest flood defenses in deltas relied on earthen dams placed along the margins of river channels or the coast; these structures are called embankments or levees. If levees completely enclose an area alongside a river or in a delta, the embanked region is sometimes called a polder, a term first coined by the Dutch in the Rhine-Meuse delta. Earthen levees aim to channelize rivers and create literal walls of defense against floodwaters.

When the French founded New Orleans on the Mississippi River in 1718, they built 2 m-high (6 ft) levees along the banks using manual (often slave) labor and mules. Today, an extensive levee system surrounds the city, and much of the lower Mississippi delta is protected by more than 5,000 km (3,100 miles) of earthen levees that exceed 10 m (30 ft) in height in an effort to prevent flooding in coastal communities and across valuable agricultural land.

Most deltas worldwide have extensive flood protection measures such as levees and polders in place, especially where rich agricultural land and large populations are at risk. These include the Rhine-Meuse in the Netherlands, the Indus in India, the Ganges-Brahmaputra in India and Bangladesh, the Ayerawaddy in Myanmar, and the Mekong in Vietnam, to name just a few. With developments in mechanization over the last century, flood protection measures have become physically larger in size, and in some places concrete walls have replaced earlier earthen embankments.

Levee failure

Embankment or levee walls are not infallible, however. Numerous failures have occurred when river and storm surge floodwaters overtop the walls, through internal saturation and blowouts (where overpressurized soils blow out the side), and due to undermining caused by subsurface flow, or "piping", of water. More rarely, accidents from ship collisions have caused levee collapse. Many of these failures are associated with catastrophic loss of life and property. Flooding in the Niger delta in October 2022 was due to embankment overtopping and failure, which when combined with uncontrolled urban growth into flood-prone deltaic lands led to tragedy. More than 600 people were killed and 1.4 million displaced as a result of the flooding.

▲ **Early flood protection measures**

Woven reed and earthen dams such as this example constructed in 1944 on one of the Euphrates River channels in Iraq have been used since 3000 BCE to protect agricultural land and homes.

◀ **Inherent risk**

Earthen embankments often fail due to overtopping during flood events (river or storm surges), undermining, internal saturation and failure, or ship collision. This 1884 Harper's Weekly illustration shows a catastrophic levee failure on the Mississippi River.

One of the most notorious levee failures occurred in the embanked city of New Orleans during, and immediately after, the passage of Hurricane Katrina in 2005. Katrina created an extensive storm surge that overtopped the levees in outlying communities, and several levees failed following subsurface piping of water and ship collision. More than 3,000 people were killed or injured and over a million were displaced by the floodwaters. Many who left the city never returned.

Negative levee effects

Although levees are important for the protection of human life and property, they may cause numerous negative effects. For a start, levees encourage development in flood-prone areas, placing more people at risk should failures occur. Once constructed, levees need continual maintenance, which is often expensive. Levees also prevent the natural flooding that supplies sediment to delta surfaces, the primary mechanism that offsets the effects of subsidence and sea-level rise in these areas, leading to elevation deficits. In addition, levees restrict water supply to the natural wetland ecosystems that thrive in deltas and that protect coastal regions. Together, these effects compound flood risk to deltaic lands over time, negating the benefits that levees were originally intended to provide.

LOSING GROUND

Over time, levees constructed in deltas are counterproductive, causing elevation loss of embanked areas, as shown in the bottom diagram. Levees also prevent natural flooding and sediment delivery to delta wetlands, causing their deterioration, shown in the top diagram.

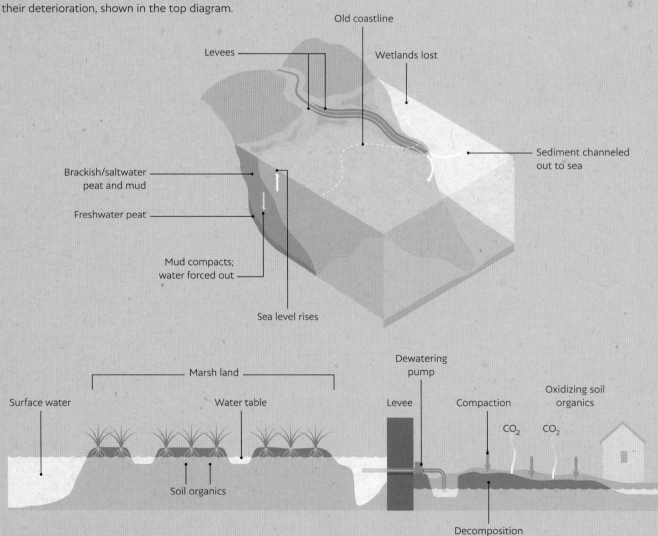

LIVING BELOW SEA LEVEL

Due to extensive poldering, there are now several areas within the Rhine–Meuse delta that are below sea level (highlighted in blue here). Only light green, yellow, and orange areas are well above the average sea level.

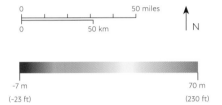

0   50 miles

0   50 km

N

-7 m   70 m
(-23 ft)   (230 ft)

Delta Plan, Netherlands
A combination of embankments, dams, sluices, canals (below), locks, and storm surge barriers (below right) is used to protect a large area of urban and agricultural land around the Rhine–Meuse delta from flooding caused by river and coastal storm surges.

Ports and navigation

Although our bipedal bodies are designed to walk on land, humans have been seafarers for several millennia. While navigation and trade have changed substantially over time, the principles of commerce and the benefits deltas offer have remained the same.

Early sea trade in deltas

Deltas are important hubs for seafaring activity because they provide protection from waves and storms, yet also proximity to the ocean. Some of the earliest maritime trade routes, active from around 1000-600 BCE, linked the Mediterranean Sea, Red Sea, Arabian Sea, and South Asia. These later formed the sea legs of the Silk Road and hence became known as the Maritime Silk Road. Goods such as grain, wine, ivory, gold and other precious metals, gemstones, cloth, silk, flowers, and spices were traded between deltas such as the Nile, Tigris-Euphrates, Indus, Huang He (Yellow), and Chang Jiang (Yangtze). Global colonization and conquests by the Portuguese, Dutch, French, and British between the fifteenth and nineteenth centuries further expanded maritime trade around the world, and ports such as Kolkata (formerly Calcutta and located in India on the Ganges-Brahmaputra delta) and Antwerp (in Belgium on the Scheldt delta) thrived.

Modern-day ports

Shipping and commerce have changed drastically since the early days of maritime trade. While food exports such as wheat, rice, and corn remain important commodities worldwide, goods have shifted from wine and lace to finished clothing, building materials, electronics, and petroleum products. Container ships have become colossal in size, requiring deep-water navigation ports to accommodate them. Of the top 20 busiest container ports in the world, seven are located on deltas: 1 Shanghai (Chang Jiang/Yangtze), 3 Ningbo-Zhoushan (Chang Jiang/Yangtze), 4 Shenzhen (Zhu Jiang/Pearl), 5 Guangzhou (Zhu Jiang/Pearl), 8 Hong Kong (Zhu Jiang/Pearl), 11 Rotterdam (Rhine), and 15 Antwerp (Scheldt). Port traffic helps the economy, industry, and commerce of the delta region and beyond.

Vancouver, the largest port in Canada, is located on the Fraser delta, which empties into the Strait of Georgia. It exports goods from the interior of Canada and imports products from around the world. The port of Bangkok on the Chao Phraya delta in Thailand did not open fully until 1947, but today it is one of the world's busiest container ports. Antwerp, on the Scheldt delta in Belgium, was originally constructed on the order of Emperor Napoleon Bonaparte in the early 1800s. The port is now the second largest in Europe, boasts state-of-the-art automated container terminals, and has plans to become the "green energy gateway" to Europe in the twenty-first century.

◤ Rush hour on the river
Boat traffic in the Mekong delta, Vietnam, carries people and goods to various destinations.

▼ Bright lights in Bangkok
Boat traffic lights up the already colorful channels of the Chao Phraya delta in Bangkok, Thailand.

Delta channels as navigation

With the increase in urban centers in many deltas around the world, their associated road and bridge infrastructures have to cope with hazardous flood conditions, highly mobile channels, and extensive wetland expanses. In delta lands in developing nations, road and bridge networks may remain poorly developed and connected, reaching limited communities across the landscape. In those locations, the distributary (river and tidal) channel network remains the primary means for navigation, whereby people and goods rely on boats for transport rather than roads. Boats ranging from two-person dugout canoes to large ferries carrying hundreds of passengers at a time can be seen crisscrossing delta channels, transporting people and goods to their various destinations. Some deltas, such as the Mekong in Vietnam, are the location for floating markets, where people gather on boats within the channels to conduct business.

◀ **Port of New Orleans**

New Orleans began as a French settlement on the banks (natural levees) of the Mississippi River. The urban area expanded rapidly into the organic-rich swamp and grew to become one of the largest cities and ports in North America. As seen in the 1884 chromolithograph (left), cotton, grain, and textiles were once the primary commodities passing through the port. Today (below left), annual exports from the Port of South Louisiana total more than US$63 billion.

SILK ROAD TERMINUS

Guangzhou, on the banks of the Zhu Jiang (Pearl) delta in southern China, has a rich 2,000-year history and was once a major terminus on the Silk Road. The city was founded in the third century CE, and during its early period was an important trade port in southern China, where, according to the History's Histories website, "hundreds of ships sat at anchor in its harbor while hundreds more smaller vessels shuttled trade goods to docks already piled high with merchandise. Because of its status as an open port for trade, Guangzhou was one of the most cosmopolitan of all Silk Road cities." Cross-cultural negotiations would take place between Persian, Arab, Chinese, and Indian sailors and businessmen, and the city had a flourishing community of Buddhist, Hindu, and Muslim settlers. During the Ming dynasty (1368–1644 CE) the city underwent considerable rebuilding and expansion, with several districts being combined into one walled city.

Over the centuries, the nationalities of the main traders shifted to Portuguese, Dutch, French, British, American, and finally Chinese companies. Today, the once modest-sized port has ballooned in importance to become one of the largest international shipping trade centers in the world. In 2020, Guangzhou port was ranked fifth largest in the world and transferred 23 million twenty-foot equivalent units (TEUs), closely behind its neighbor Shenzhen, located at the mouth of the Zhu Jiang (Pearl) delta (26 million TEUs), and Shanghai, located at the mouth of the Chang Jiang (Yangtze) delta (43 million TEUs)!

The Silk Road map
Deltas such as the Nile, Tigris–Euphrates, Indus, Ganges–Brahmaputra, Mekong, Zhu Jiang (Pearl), and Chang Jiang (Yangtze) were the locations of critical port cities and distribution centers along the important Silk Road trading route.

Silk Road

Other trade route (land)

Other trade route (sea)

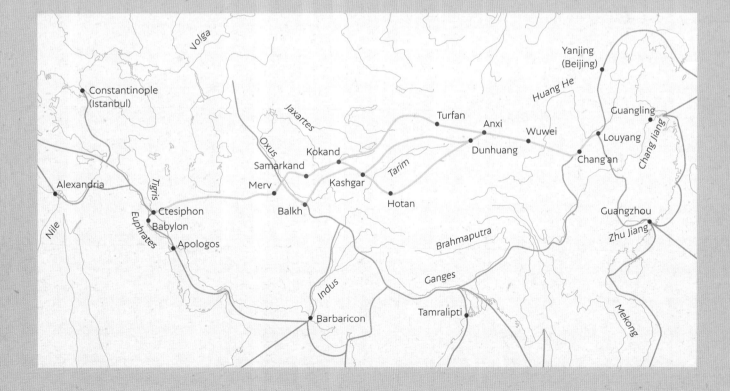

Dredged channels

Rivers naturally deposit and spread out sediment at their outlets. How can this possibly be a bad thing? Unfortunately, the sedimentation that occurs in delta channels causes many challenges that humans must manage in order to maintain navigation routes and open pathways for water to drain efficiently. How do we do this? The answer is dredging.

Why is dredging necessary?

As river flow enters into the low-lying reaches of a delta and spreads out into distributary channels, current velocities naturally decline and sediment that is being transported can no longer be carried along. At this point, these particles are deposited on the bed of the channels. In theory, particles that are deposited can be picked up and transported further downstream at a later time, perhaps during the next flood event. But during the life cycle of a delta, once a channel has overextended itself the river will search for a newer, shorter path to the receiving basin and the older channel will fill in and "die." So, sedimentation in channels is a very natural phenomenon.

▼ **Filling boatloads of sand**

Manual sand excavation is common when river levels are low, as seen here on the Beki River in India, which flows into the Ganges– Brahmaputra delta.

Unfortunately, sedimentation increases as a result of the modifications humans place along rivers. Damming in upstream reaches tends to weaken flood pulses, reducing the capacity of rivers to transport sediment downstream. Furthermore, artificial levees constructed to prevent floodwaters from damaging crops and infrastructure prevent the natural inundation of floodplain areas. This means that sediments carried by rivers no longer have "off ramps" from the water superhighway in which they are carried—i.e., more is moved downstream and can be deposited on channel beds. To keep the channels deep enough to ensure boat traffic can continue to operate, dredging is therefore required.

Dredging operations

There are two principal dredging methods used to maintain navigability. Small-scale projects on shallow channels can be carried out by hand, particularly in tidally dominated channels where the tidal range is large and it is therefore easy to dig at low tide when the channel bed is exposed. Larger-scale dredging requires more sophisticated machinery, such as bucket excavators placed either alongside the channel or on barges within the channel, or siphons that suck a slurry of water and sediment off the channel bed and place it into barges for removal. Because modern large-scale container ships require water depths of 10-15 m (30-50 ft), navigation channels typically require extensive and expensive dredging operations.

What is done with dredge material?

Material dredged from the bottom of channels, commonly referred to as dredge "spoil," can have various grain sizes (gravel, sand, silt, clay) and may be repurposed in many ways. In the Ganges-Brahmaputra delta in Bangladesh, for example, sediment derived from channel dredging is commonly used for construction materials (bricks, concrete). It is also used to build new land—an estimated 80-100 km² (23-40 square miles) of the megacity of Dhaka sits in an area that was once wetland and that has been filled in over the past few decades. Unfortunately, some of the highest subsidence rates, where the land surface is sinking by as much as 2 cm (⁴/₅ in) each year, occur in regions such as this, where fine-grained organic-rich substrates are loaded with new sediment and buildings are placed on top.

While much dredge spoil can be repurposed in beneficial ways in many deltas, some is polluted with toxic heavy metals, industrial waste products, sewage, and harmful microbes. Polluted dredge spoil should be considered a waste material and disposed of in an environmentally regulated manner; unfortunately, however, regulations vary widely between nations and deltas.

▲ **Suctioning sediment**

Gravel, sand, and mud suctioned from the riverbed in this channel in the Rhine-Meuse delta is delivered to the shoreline as a slurry and can be used to create new land.

Deltas and natural resources

While deltas are extremely important areas for natural wildlife, fisheries, and bird sanctuaries, humans have also benefited from many of their natural resources. Some may argue that this has turned into exploitation, especially over the last century with the growth of the human population and increased energy demands.

Wood

Wood is sourced from the wetlands (swamps, marshes, mangroves) and bottomland hardwood/rain forests that stretch across delta lands, and is used primarily for building materials and for fuel. While silviculture (the growing and cultivation of trees) ensures sustainable supply conditions, deforestation and conversion to agricultural or urban land is more common—for example, 5,000 km^2 (1,930 square miles) of forest was lost in the Niger delta between 1987 and 2013. Today, many coastal communities value the benefits provided by large, continuous forest tracts and vegetated wetlands. These include habitat for critical and endangered species, nursery grounds for fish and waterfowl, protection from storm surges, and offset of carbon emissions. As a result, many revegetation and reforestation efforts are under way in deltaplains worldwide.

Salt

Sodium chloride is a necessary nutrient for bodily function and diet, and is often added to food during cooking. In the Ebro delta in Spain, extensive ponds are used to evaporate salt from seawater. This tradition extends back to the Middle Ages, when the delta city of Tortosa was an important provider of salt to many places around the Mediterranean Sea. Similarly, salt extracted in brackish Lake Manzala in the Nile delta remains an important resource and export commodity today.

Water

Fresh, clean drinking water is perhaps the most important of all natural resources in the world, and the fresh water provided by rivers is critical for human settlement of delta lands. The Nile delta is an oasis in an otherwise hot, dry desert land, as is the Shatt al-Arab delta in Iraq, which receives water from the Tigris and Euphrates Rivers. Without the fresh water provided by these rivers, millions of delta dwellers could not survive or make their livelihoods there.

▲ **Wood resources**

Local wood and materials sourced from deltas are used to construct houses and other shelters, as seen here in the Mekong (top) and Danube (above) deltas.

▶ **Ebro delta saltworks**

The salt extracted from delta lagoons, such as these in the Ebro delta in Spain, is an important trade commodity.

However, freshwater supply in deltas around the world is not guaranteed and may even be described as endangered. Overextraction of water from the Colorado River to supply numerous large cities in the southwestern United States (including Las Vegas, Phoenix, and Los Angeles) and agricultural lands (such as California's Imperial Valley), coupled with recent drought, has significantly altered the Colorado delta in Mexico over the past century. Water discharge dropped exponentially from an annual supply of around 20 billion m^3 (5,300 billion US gal) in 1910 to less than a tenth of that in 2000. As a result, the delta and its former lush wetlands and lagoons have dried up. The supply of water to communities and environments upstream of the delta remains controversial to this day.

With a decrease in freshwater supply comes salinity intrusion, a problem that plagues many communities worldwide. Measures can be put in place to limit saltwater intrusion into delta lands, but these require extensive, and expensive, scientific and engineering research planning and implementation, along with multinational diplomacy if the rivers and deltas concerned cut across geopolitical boundaries. In the Mekong delta of Vietnam, upstream flow control and shifts in agricultural crop choice are being implemented to limit salinization and its impacts, with some success. In Australia, the Burdekin is the country's second most important river in economic terms. Communities living along its flanks have implemented an artificial recharge to its groundwater to store and supply irrigation water during the dry season, and to maintain pressure in the delta aquifer to withstand saltwater intrusion. As climates continue to shift through the effects of global warming, salinization of deltaic waters will continue to be a major concern.

▼ **Giant sentinels**

*Oil and gas platform
on the northern Gulf
of Mexico continental
shelf, off the Mississippi
delta in the southern
United States.*

Petroleum

Petroleum is a carbon-rich organic substance that occurs as liquid and gas accumulations in sedimentary rocks. Many paleo-delta regions contain vast amounts of sandstone and organic-rich shales, the original source for petroleum products. Organic-rich shales form when plant and animal matter accumulates in low-energy, muddy environments and is then buried and compacted from loose sediment into rock. With burial, the organic debris degrades into a product called kerogen, and with increasing temperature and pressure (greater than 100 °C/212 °F and depths of 1 km/3,300 ft) kerogen is converted to petroleum oil (liquid) and natural gas. These products are released from the organic-rich shale rocks through cracks and faults, and migrate to other rocks with greater pore space that store them, such as sandstones and carbonate rocks.

While civilizations have been using petroleum products for several thousand years, their use for combustion engines was a critical factor in the explosive growth of the petroleum industry around the turn of the twentieth century. Today, global oil production amounts to around 90 million barrels per day. Significant petroleum reserves have been discovered in at least 18 deltaic provinces worldwide, and a recent estimate proposes that these provinces could contain more than 50 billion barrels of oil and 10 trillion m³ (350 trillion ft³) of natural gas reserves. While these are astonishing volumes, oil and natural gas are finite resources. The extraction of fluids such as oil and natural gas in deltas can also have negative effects in these environments, including exacerbating subsidence and, in the case of oil spills, wreaking havoc on local wildlife and fisheries.

ENERGY GIANTS

Significant petroleum reserves have been discovered in at least 18 deltaic provinces worldwide, including the Niger delta. Note most of the present-day oil and gas fields are centered on the subaerial delta, but vast reserves of the petroleum system exist offshore in much deeper waters.

Bathymetric contour

● Center of oil or gas field

○ Maximum extent of petroleum system

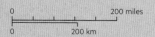

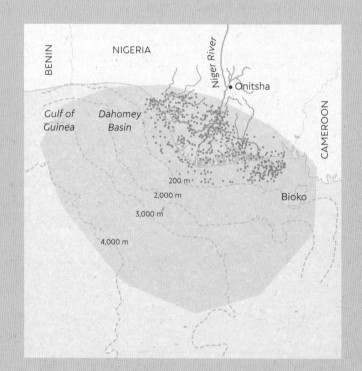

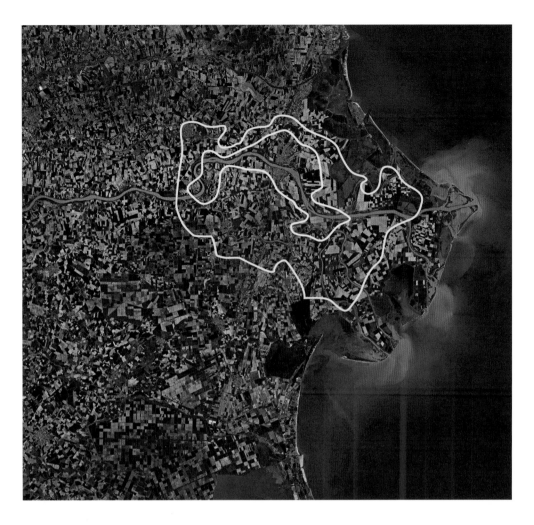

Impacts of extraction
Subsidence, or a lowering of the land surface, has occurred in several deltas worldwide as fluids such as oil, natural gas, and water have been overextracted from below ground. Left: Po delta, Italy. Below: Chao Phraya delta, Thailand.

1958–1962

Subsidence through gas mining

40mm/year (1.6in/year)

60mm/year (2.4in/year)

0   10 miles

0   10 km

Subsidence through water mining

50mm/year (2in/year)

100mm/year (4in/year)

0   10 miles

0   10 km

Changes in biodiversity

Due to rampant human-induced deforestation, land-use change, salinization, and pollution, the area and quality of delta habitats has drastically declined. As a result, countless mammals, birds, reptiles, amphibians, insects, and microbes have been pushed out of their native deltaic landscapes. Some remain within protected forest lands, but others have become significantly endangered or even extinct.

Changes in the Orinoco

The Orinoco delta is a large (45,000 km²/17,400 square miles) swamp forest in the tropical region of Venezuela where three distributary channels—the Rio Grande, Rio Caroni, and Caño Manamo—fan out into the Atlantic Ocean. Vast belts of mangroves fringe the coast, while the interior is seasonally to permanently flooded rain forest where the Indigenous Warao people live in more than 300 settlements. The swamp forests are home to species that include the Orinoco Crocodile (*Crocodylus intermedius*), Amazon River Dolphin (*Inia geoffrensis*), Jaguar (*Panthera onca*), Bush Dog (*Speothos venaticus*), Giant Otter (*Pteronura brasiliensis*), Orinoco Goose (*Neochen jubata*), and Harpy Eagle (*Harpia harpyja*).

In the 1960s, a flood control program dammed the Caño Manamo, which reduced water levels in the northwestern region of the delta. This changed the local hydrodynamics, and the region became increasingly impacted by tides and salinity intrusion. This had a drastic effect on the flora and fauna of the Orinoco delta swamp forests: freshwater mangroves became stressed or died, and were replaced by more salt-tolerant species, and freshwater fish (a staple of the Warao people) disappeared, causing high levels of malnutrition. The Orinoco River is now also heavily polluted due to oil and mining in its upstream reaches, and the Orinoco Crocodile, Giant Otter, Amazon River Dolphin, Bush Dog, and Harpy Eagle are considered threatened, as are the livelihoods of the Warao people.

Today, around 3,500 km² (1,350 square miles) of the delta is protected by the Delta del Orinoco National Park, which preserves the swamp forest and the delta ecosystem, and the way of life of the Warao people living here. Ecotourism has boomed in the region in recent years, bringing in valuable income.

Changes in the Mekong

Salinization is a major concern in the Mekong delta (see map overleaf). Upstream dams (11 in China, two in Laos, and a staggering 300 in tributary rivers) control floodwaters during the monsoon season (June to November), but they also diminish the amount of water released during the dry season (December to May). Some researchers estimate that there is a shortfall of as much as 10 billion m³ (2.6 trillion US gal) of water during the dry season compared to historical water discharge figures. As a result, saltwater from the ocean is penetrating around 60 km (35 miles) up delta channels and impacting natural mangrove landscapes and agricultural land (and hence human livelihoods).

The Mekong delta is home to two charismatic mammals that require conservation: the Hairy-nosed Otter (*Lutra sumatrana*) and the Dugong (*Dugong dugon*). There are also at least 37 species of birds and 470 species of fish living here that are classified as Vulnerable or Endangered.

To reduce impacts on human livelihoods, efforts are under way to improve crop irrigation, crops are being switched to saline-tolerant species, and there is a shift from growing three crops year-round to two crops a year. To protect local wildlife, UNESCO has designated the Can Gio Mangrove Biosphere Reserve and Mui Ca Mau Biosphere Reserve, which together cover around 4,000 km² (1,540 square miles) of the delta. These reserves are managed in a fashion that seeks to balance society and nature together, preserving the natural landscape and ecosystem. Despite these great efforts, local scientists maintain that no matter how much is done to adapt to saline intrusion in the delta, the root causes must be addressed or it will continue to plague the region, its people and its flora and fauna.

The largest mangrove forest in the world

The Sundarbans mangrove forest sits in the Ganges-Brahmaputra delta and is home to countless charismatic yet threatened species. Here, Bengal Tigers (*Panthera tigris tigris*) slink between Sundri (*Heritiera fomes*), the mangrove trees after which the forest is named (sundri means "beautiful"), while Saltwater Crocodiles (*Crocodylus porosus*) sunbathe on the muddy banks. Kingfishers call from the trees, Rhesus Macaques (*Macaca mulatta*) swoop overhead, Spotted Deer (*Axis axis*) munch on tree leaves with Red-tailed Bamboo Pit Vipers (*Trimeresurus erythrurus)* watching on, Barred Mudskippers (*Periophthalmus argentilineatus*) and fiddler crabs scurry along the mudflats, and Ganges River Dolphins (*Platanista gangetica*) and Hilsa Fish (*Tenualosa ilisha*) swim through the tidal channels.

The Sundarbans encompasses around 10,000 km² (3,860 square miles) of mangrove forest, and has been set aside as a preserve since 1875. It is currently a UNESCO World Heritage Site due to its natural biodiversity and cultural importance—humans have been living on the delta since about 400 BCE, and there are several Hindu and Buddhist archaeological temple sites scattered throughout.

Complex landscape
Tidal channels crisscross the mangrove forest of the Sundarbans, which is home to more than 1,000 species of flora and fauna including the elusive Bengal tiger (*Panthera tigris tigris*).

1 Matla River
2 Bidyadhari River
3 Sundarban National Park

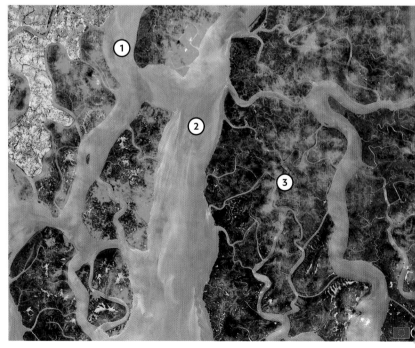

However, despite the mangrove forest's significant ecological value and status as a national icon, its area has been reduced by half through deforestation to create polders to house a burgeoning human population. The Bengal Tiger, a national symbol of pride for both India and Bangladesh, has been decimated by illegal poaching for its skin and for body parts used in traditional Asian medicine. As a result, the species has been brought to the brink of extinction, with only an estimated 500 individuals now remaining. Bangladesh and India are working diligently to protect the tiger and other important species in the Sundarbans. In 1973 the Indian government launched Project Tiger to save the species from extinction, and in May 2021 a man suspected of killing 70 endangered tigers was arrested after a 20-year search.

Local governments and forestry officials realize that conservation is key, but these efforts are often at odds with population pressures—around 15 million people live within 20 km (12 miles) of the forest borders. And while a new coal-burning power plant being built within 10 km (6 miles) of the forest will improve the quality of life of the people in this economically depressed region, it will put severe pollution pressures (acid rain, heavy metal contamination, petroleum leaks from boat traffic) on an already fragile ecosystem.

SALINITY INTRUSION

In the Mekong delta in Vietnam, salinity intrusion affects farming and fishing activities, as well as the health of the natural mangrove at the coast. Here (below), salt crystals can be seen dried on the surface of rice paddy mud.

Water salinity in delta

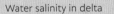

Severe   Mild

14

Future Rivers, Estuaries, and Deltas

Global challenges for the future

We know many of the large-scale changes that will affect global rivers, estuaries and deltas in our future world, along with the natural ecosystems and human communities they support. But how will these systems vary spatially and in time, and how might they interact?

▼ **Future populations**

Growing human populations will continue to place stresses on the world's rivers, estuaries, and deltas up to and beyond the estimated peak in the 2080s. Here tourists are rafting in Yuxi Grand Canyon, Henan Province, China.

Future stresses

In the future, we know that global human populations will rise to a peak of around 10.5 billion in the 2080s, thereby increasing stresses on many natural environments. Global temperatures are also forecast to increase, and will cause complex patterns of change in precipitation and drought. These changes will yield transformations in water, sediment, nutrient, and pollutant fluxes through the world's rivers, estuaries, and deltas that will impact the ecosystems and natural capital they provide. Global sea level will also continue to rise as ice sheets melt, threatening low-lying coastal regions, with estuaries and deltas being especially vulnerable. The geographical distribution of many species will shift as they adapt to changing global air and water temperatures, moving northward and/or to higher elevations.

FUTURE HEAT AND PRECIPITATION

Global maps of change in temperature and percentage change in precipitation by 2081–2100, relative to a 1961–1990 baseline. Warming is present across the globe, with high latitudes experiencing the greatest change. Some regions—including East Africa, the Middle East, India, and parts of Asia—will experience increased rainfall, whereas others—the southern United States and Central America, southern Europe, regions of North and South Africa, and parts of Australia—will suffer more periods of drought.

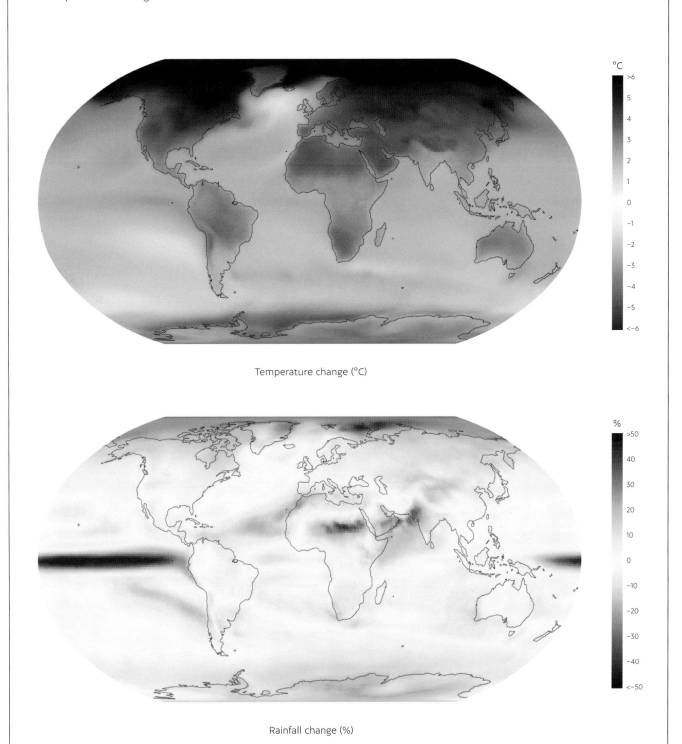

Temperature change (°C)

Rainfall change (%)

RISING SEA LEVELS

Projected global average of sea-level rise rates, relative to a 1995–2014 baseline, under different scenarios of projected socioeconomic global changes (Shared Socioeconomic Pathways [SSPs]; see lower diagram for definitions) up to 2150. For each SSP, the figures in italics show different Representative Concentration Pathways (RCPs) from very low (1.9) to very high (8.5) future greenhouse gas concentrations. The plot shows the mean (solid lines) and maximum (dashed lines) for each scenario. Under all scenarios, global mean sea level will rise at least 0.6 m (2 ft), but some projections show far more severe increases.

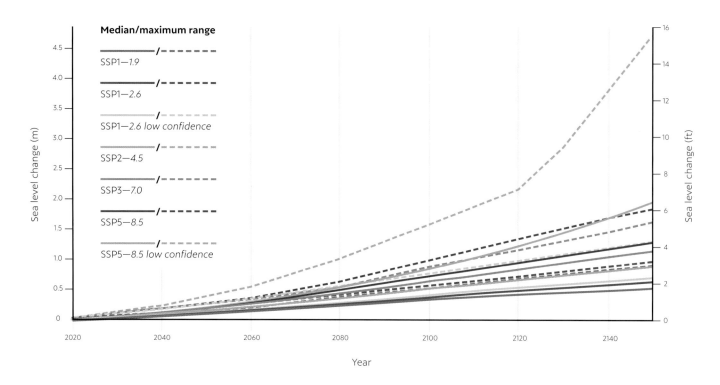

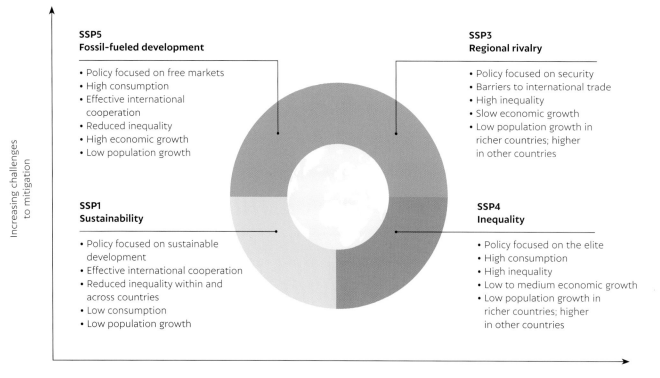

FUTURE PLANNING FOR RIVERS, ESTUARIES, AND DELTAS

In planning a sustainable future for the world's rivers, estuaries, and deltas, and improving or restoring ecosystem functioning, it is imperative to recognize and accommodate the links between the natural landscape, the ecosystems these waterways support, and the rural and urban landscapes that are superimposed upon them. Steering development toward a new balance between land use and ecosystem functioning is critical, and will demand coherent policies and governance from source to sink and linking of the physical landscape to management strategy. The geomorphology of rivers, estuaries, and deltas provides a template on which ecosystems and humans have become established. The conditions dictated by landscape morphology, water supply, and biophysical attributes must thus guide how we are to manage these environments in future.

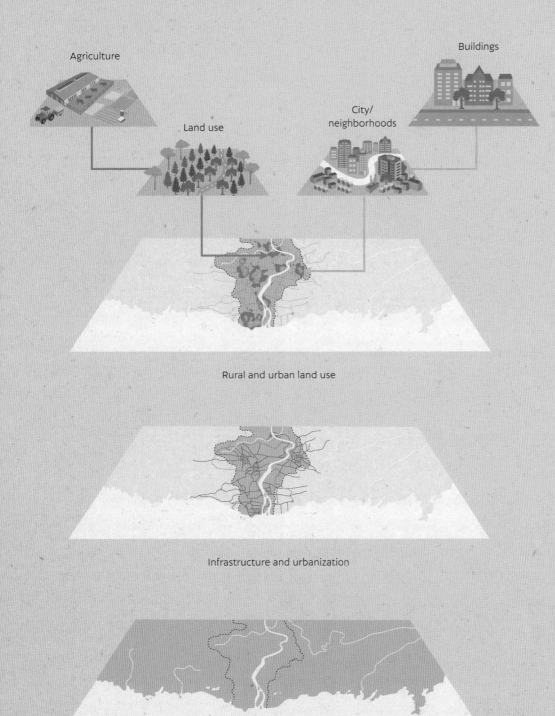

Agriculture

Land use

Buildings

City/ neighborhoods

- ● Urbanized areas
- ● Agriculture
- ○ Forest and scrubland
- ● Grassland

Rural and urban land use

Infrastructure and urbanization

Geomorphology, water, and soild systems

ACTIONS NEEDED FOR SUSTAINABLE RIVER BASINS

Local land and water use, building, and management
- Reduce water use
- Reduce emissions
- Improve water and soil management
- Implement zoning restrictions
- Water-proof buildings and infrastructure
- Apply nature-based solutions

Maintenance and improvement of critical infrastructure (dams, urban areas, electrical and communications networks, and transportation)
- Water-proof urban planning and infrastructural development
- Apply nature-based solutions and design

Integrated river-basin strategy and plan across borders
- Balancing development and conserving/restoring natural patterns, processes, buffers, and qualities
- Regulate water and sediment dynamics
- Regulate water use and polluting emissions
- Implement spatial zoning
- Draw up transboundary river-basin agreements

Future rivers

What do projected changes in the variables driving the behavior of rivers, estuaries, and deltas mean for how water, sediment, and nutrient flows across Earth's landscapes will alter in the future? The availability of a wide range of modeling tools allows such changes to be predicted with a good degree of confidence.

SEDIMENT FLOWS TO DELTAS

Percentage change in the mean annual sediment load supplied by major rivers to their deltas between 1990–2019 and 2070–2099, as simulated by a global model of sediment transport. The shaded outlines highlight the locations of the river basins feeding each delta.

Increase

Decrease

0%   50%   100%

1 Amazon	**13** Indus	**26** Niger	**38** Senegal
2 Amur	**14** Ayeyarwady	**27** Nile	**39** Tana
3 Burdekin	**15** Krishna	**28** Orinoco	**40** Tigris
4 Chao Phraya	**16** Lena	**29** Paraná	Euphrates
5 Colorado	**17** Limpopo	**30** Zhu Jiang	**41** Tone
6 Congo	**18** Mackenzie	(Pearl)	**42** Vistula
7 Ebro	**19** Magdalena	**31** Po	**43** Volta
8 Fly	**20** Mahakam	**32** Red	**44** Chang Jiang
9 Ganges-	**21** Mahanadi	**33** Rhine	(Yangtze)
Brahmaputra	**22** Mekong	**34** Rhône	**45** Huang He
10 Godavari	**23** Mississippi	**35** Rio Grande	(Yellow)
11 Grijalva	**24** Moulouya	**36** São Francisco	**46** Yukon
12 Han	**25** Murray	**37** Sebou	**47** Zambezi

Sediment load

Substantial changes in the future supply of river sediment to the world's major deltas are projected. For example, while increased river flows as a result of climate change tend to drive increases in future sediment transport in some deltas, many rivers—especially those in Asia, Africa, and parts of South America—are projected to likely experience a net decline in their sediment loads because major dam-building programs will prevent the transfer of sediment to the coast.

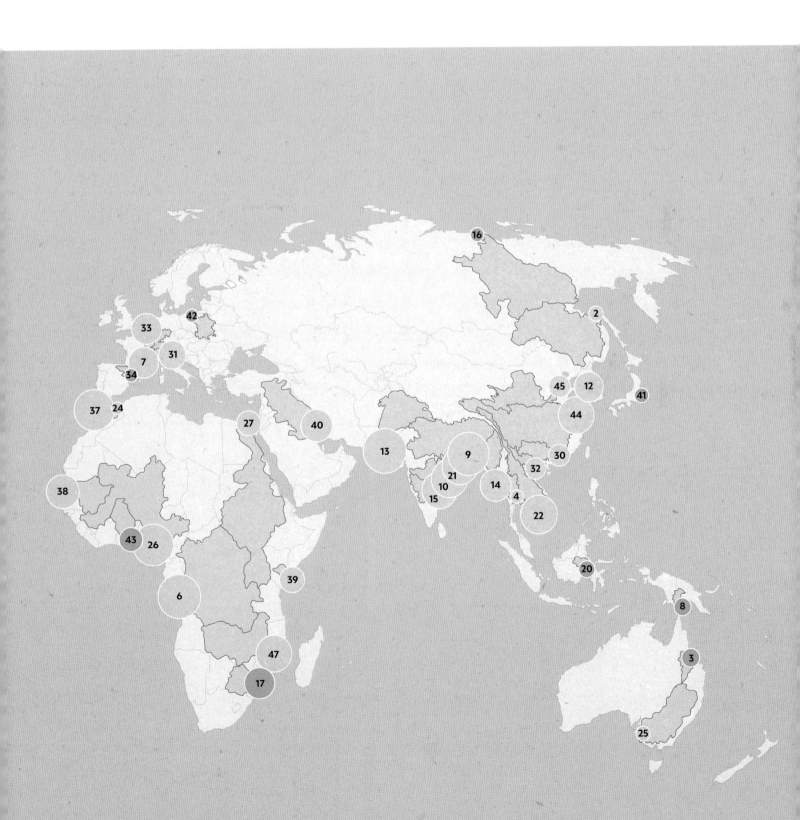

► **Troubled waters**

*Dianchi Lake in Yunnan
province, China, turns
green every year as a
result of algal blooms
stimulated by excessive
nutrient fluxes, mainly
from animal waste.*

Flooding

Model projections of future flood hazards show that anthropogenic climate change will drive substantial increases in both the frequency and severity of river and coastal flooding in the future. For example, based on sea-level projections for the year 2050, land that is currently home to 300 million people will fall below the elevation of the average annual flood. Similarly, the areas inundated by river flood events of a given probability are also expected to increase substantially as a result of climate change.

Algal blooms

Meanwhile, intensification of agriculture, coupled with increases in human population, will drive substantial increases in nitrate and phosphate emissions in the future, from about 200 million tonnes (220 million US tons) per year in 2020 to 250 million tonnes (275 million US tons) per year in 2050. These nitrate and phosphate emissions are likely to increase the frequency of harmful algal blooms. Ironically, this literal tendency for waterways to become "greener" poses a major threat of eutrophication (see page 235) to river and coastal ecosystems.

FUTURE FLOODING

Map showing the number of people affected by increases
in future river flooding in a scenario where the Earth heats
by 1.5 °C (2.7 °F) relative to the pre-industrial average,
as a result of human-induced climate change.

Population affected (% change) by global
rise in temperature of 1.5 °C (2.7 °F)

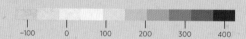

–100	0	100	200	300	400

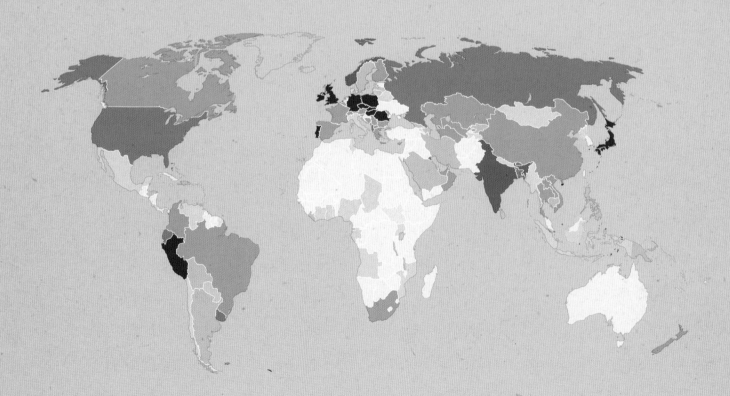

Future estuaries

At the interface between land and sea, estuaries are particularly vulnerable to increasing sea levels, water temperature, and pollution. How they respond to these pressures depends on their morphology and water balance, with many experiencing changes that are critical to the health of coastal ecosystems, economies, and communities.

Climate change affects natural and altered estuaries through increased temperatures, water volumes and sedimentation, more frequent flooding, habitat loss, shifts in species distributions, and changes in salinity. However, these impacts cannot be generalized as estuaries respond differently depending on their individual morphology and depth, the rate of relative sea-level rise, and changes in water flux. Estuaries are typically warming faster than their surrounding seas, particularly if they are shallow and have small water volumes and poor circulation. And those that are already under stress from human-induced changes in their catchments and along their shorelines are less resilient to climate change.

Saltwater intrusion

As sea levels rise, saltwater penetrates further inland, shifting the boundaries between estuaries and rivers, and leading to the salinization of aquifers and soils. Saltwater intrusion has been called the "silent killer," as it progressively creeps in from groundwater, threatening vegetation less tolerant to salts. Increased frequency of flooding during high tides, and groundwater abstraction for human consumption and irrigation, accelerate the problem. Low-lying areas around estuaries and coastal floodplains, which were once fertile agricultural fields, now face reduced yields and unviable soils. This is a growing problem worldwide, particularly on the East Coast of the United States and in Southeast Asia and Australia. Saltwater intrusion is worse in regions with dry climates, but less pronounced in estuaries where precipitation recharges aquifers and increases river discharge, keeping the saltwater closer to the sea.

Threatened freshwater ecosystems

Higher and more frequent tidal incursion expands the reach of salt-tolerant species into areas of freshwater habitats. As salinity increases, brackish and then marine species progressively dominate low-lying areas, and freshwater species are displaced further inland or to higher ground if conditions are favorable. In urbanized estuaries, coastal squeeze (see pages 240–41) can lead to the loss of freshwater and intertidal habitats due to inundation and salinization, as their inland shift is prevented by the presence of development and flood protection structures.

◥ **Nuisance flooding**
Water overtopping estuarine embankments during high spring tides is becoming more common as sea levels rise. This is so frequent in Venice, Italy, that elevated footpaths are deployed in tourist hotspots such as St. Mark's Square.

▶ **Marching mangroves**
Mangroves along tidal creeks of the Albert River in the Gulf of Carpentaria, Queensland, Australia. Due to sea-level rise, tidal creeks and mangroves are expanding inland into freshwater habitats in many catchments worldwide.

Future deltas

As human populations increase, the food and water resources of deltas are being stretched. Deltaic communities must adapt to living upon shifting—and sinking—landscapes that are continually plagued by floods, increased storminess, and salinity intrusion.

Exacerbating delta subsidence

The sediment supply to many deltas is projected to decline due to damming and sediment mining. Without new sediment, they will struggle in their battle against sea-level rise to maintain subaerial landscapes (see pages 262-67). Furthermore, increasing groundwater extraction for drinking water and agricultural purposes will exacerbate subsidence rates. This will result in future increases in flood risk to many delta communities, including megacities such as Cairo, Lagos, Kolkata, Dhaka, Ho Chi Minh City, Shanghai, Hong Kong, Bangkok, and Jakarta. While some communities are resorting to more expensive engineering practices to protect inhabited lands, maintaining these structures may become cost-prohibitive in the future.

RISK REDUCTION

Several approaches can be pursued to reduce future flood risk in deltas, from hard engineering structures, such as embankments and sediment infill, to more nature-based solutions, such as planting wetland vegetation to reduce the impact of storm surges and trap sediment.

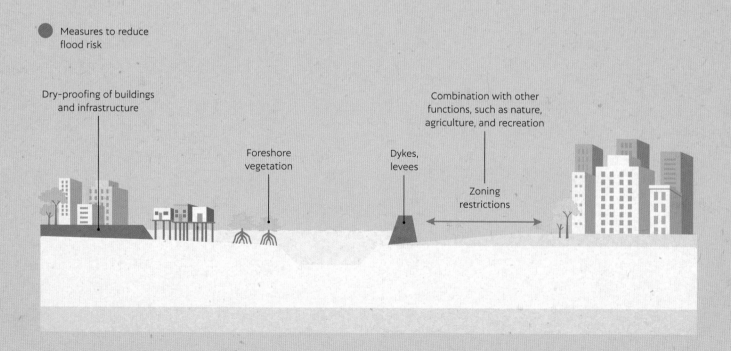

● Measures to reduce flood risk

Dry-proofing of buildings and infrastructure

Foreshore vegetation

Dykes, levees

Combination with other functions, such as nature, agriculture, and recreation

Zoning restrictions

Increased storminess and salinity

Shipboard measurements and weather buoys show that global sea-surface temperatures rose more than 0.5 °C (0.9 °F) in many of the world's oceans between 1982 and 2010. Our warming seas will produce more intense hurricanes, cyclones, and typhoons, with some studies predicting that storm intensities could increase by 12–20 percent. Stronger storms will result in greater rainfall and flooding, stronger winds, and larger storm surges, and many of these impacts will be felt hundreds of miles from where the storm moves inland. Ultimately, all of these impacts will increase the risk of economic damage and loss of life. Many coastal areas are already improving building codes and raising home elevations to protect against more intense storms, but these are costly adjustments.

Like estuaries, deltas are vulnerable to the salinity intrusion that accompanies sea-level rise, leading to species shifts as freshwater-tolerant plants and animals are replaced by more saline-tolerant lifeforms. In some locations, "ghost forests" preserve reminders of past environmental conditions. Salinity stress will be further evident in agricultural production. In the Mekong and Ganges-Brahmaputra-Meghna deltas, regions that once grew three rice crops each year now yield only one or two crops due to the high-salinity conditions generated during low river discharge. This reduced agricultural production is expected to become more severe in future.

Deltas in an Anthropocene bathtub?

As we have seen, the response of deltas to relative sea-level rise will potentially involve loss of land, changes in habitats, and migration of species, including humans. To assess the future shape of delta coastlines, we must account for the myriad factors determining delta morphology—those governing delta growth relative to those causing delta drowning.

◥ **Deadly storm**
Cyclone Nargis hit the Ayeyarwady delta in Myanmar in 2008, causing more than 140,000 casualties and the equivalent of an estimated US$12 billion in damage. As global storm intensity is predicted to increase, so will damage and loss of life.

Many maps adopt a simple "bathtub" approach to predict future coastline shape: elevation maps of existing delta morphology plot how the coastline will appear under different amounts of relative sea-level rise—a simple increase like the water surface rising as you fill a bathtub. However, the real situation is far more complex.

Previous chapters highlight how the amount of sediment supplied to a delta is key to its growth. Changes in precipitation, upstream damming, and sediment mining will impact this flux and will drive alterations in sediment transport. Besides these factors, the growth of marshlands and accumulation of organics in deltas can create a positive feedback, allowing greater accretion to combat sea-level rise. This imparts some natural resilience to deltas undergoing relative sea-level rise, but only up to a certain point.

Given these complexities and feedbacks, simple bathtub approaches do not predict realistic patterns of future delta shape or provide scenarios from which we can plan adaptable management strategies. To do so, we need to couple the drivers of change in a river basin with those at the coast and consider these under the spectrum of likely changes due to land-use change, dam construction and operation, alterations to sediment flux, growth of marshlands, and coastal processes.

FUTURE GLOBAL DELTA AREAS

Observed and predicted changes in delta area under one Shared Socioeconomic Pathway (see page 344) for greenhouse gas emissions.

Land loss (km²/year)

1   10

Land gain (km²/year)

Observed change in delta area 1985–2015

Projected change in delta area 2050–2100 (SSP2-4.5)

A FUTURE MEKONG DELTA?

The Mekong delta, Vietnam, seen here in a satellite image from 2007, covers 40,000 km² (15,444 square miles), is home to some 21 million people, and is one of the world's most vulnerable deltas. This vulnerability arises from a combination of sea-level rise, subsidence due to groundwater abstraction, and sediment starvation caused by upstream hydropower damming, embankment construction, and sediment mining.

Subsidence created by groundwater abstraction is critically important in dictating future patterns of inundation, as is the flux of sediment supplied to the delta, which is influenced strongly by the location and design of dams and embankments. Consideration of relative sea-level rise (see pages 266–67) and different scenarios of hydropower dam operation and upstream sediment mining allows more realistic estimates of future delta shape than simple bathtub models. The elevation map (below) shows the delta area likely inundated under different amounts of relative sea-level rise by 2100. This shows that, depending on the different scenarios of relative sea-level rise and sediment supply, between 7 percent and 90 percent of the delta may be below sea level by the end of the century. Integrated planning across river-to-delta environments (see diagram on page 375) is thus essential, particularly sediment management to resupply deltaic environments and increase resilience in the face of relative sea-level rise.

Future Mekong
Satellite image of the Mekong delta in 2007, together with a map illustrating land that is expected to fall below sea level for varying sediment supply and relative sea-level rise (rSLR) to 2100.

Below sea level for rSLR in meters

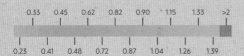

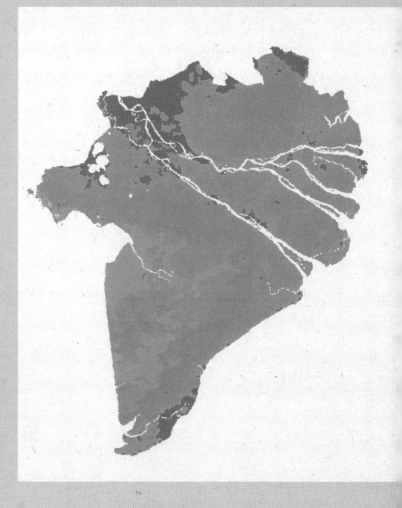

New methods to predict future changes

To manage a sustainable future for the world's rivers, estuaries, and deltas, it is imperative that we monitor these environments, and their changes, in response to human-induced stresses, and assess the progress of mitigation strategies and rehabilitation schemes. A range of innovative technologies and approaches is available to help in this grand challenge, allowing new understanding of these critical environments.

New data for a new future

After 16 years of planning, the international Surface Water and Ocean Topography (SWOT) satellite mission was launched on December 16, 2022. The initial three-year mission is led by NASA and the French space agency Centre national d'études spatiales (CNES), with contributions from the Canadian and UK space agencies. The SWOT satellite will provide an unrivaled quantitative view of the world's oceans and large waterbodies, including lakes and rivers more than 100 m (330 ft) wide, with overpasses every 21 days and covering 90 percent of the globe. The SWOT satellite uses a Ka-band (microwave frequencies in the range 26.5–40 GHz; here 35.75 GHz) radar

SWOT INTEROMETRY

SWOT INTERFEROMETRY

The SWOT satellite has two antennae mounted on a 10 m (30 ft) boom, collecting ground data across a 120 km-wide (75-mile) swath, and measuring water surface elevations across the Earth's oceans, estuaries, rivers, and lakes.

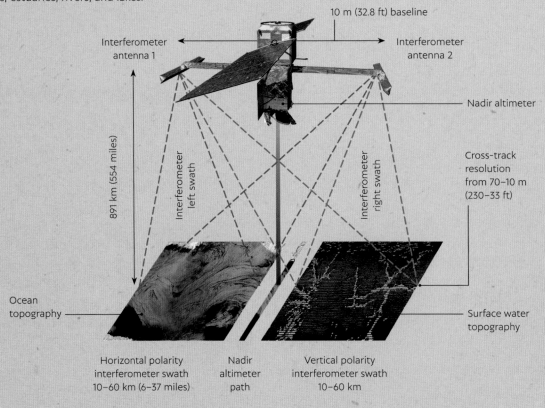

10 m (32.8 ft) baseline

Interferometer antenna 1

Interferometer antenna 2

Nadir altimeter

891 km (554 miles)

Interferometer left swath

Interferometer right swath

Cross-track resolution from 70–10 m (230–33 ft)

Ocean topography

Surface water topography

Horizontal polarity interferometer swath 10–60 km (6–37 miles)

Nadir altimeter path

Vertical polarity interferometer swath 10–60 km

interferometer (KaRIn) to measure water surface heights across a 120 km-wide (75-mile) swath, with a vertical resolution of about 1 cm (0.4 in). The KaRIn uses microwave radar and the precise position of the satellite to measure the distance between the antennae and target water surface. This high-precision, extensive coverage and short repeat frequency promise to revolutionize the monitoring of the watery surface of our Blue Planet.

Measuring flow from the air and space

Satellite measurements of water surface height, from SWOT and other remote sensing, can be used together with estimation of the water surface slope, channel width, and flow depth to measure the volume of water coursing through the world's rivers. This technique can be applied to extend our knowledge of water flow in many inaccessible river basins that can be monitored only remotely.

Techniques known as particle-imaging and particle-tracking velocimetry rely on detecting the movement of "tracers" on the water surface—such as froth, floating vegetation, or ice—to estimate surface velocity. If we know the exact position of these tracers at two time intervals—for example, through careful analysis of photographs taken from various platforms—we can use the distance moved and time gap between the images to calculate their velocity. Such photographs can be captured by cameras mounted on bridges or other infrastructure, attached to drones, or deployed on satellites. They provide a simple way to quantify the speed of water flows, which can aid in the estimation of water volumes and flow hazards.

Squeezing more out of water

As remote-sensing techniques and data analysis develop further, Earth observation satellites will increasingly be able to monitor other attributes of water in the world's rivers, estuaries, and deltas, such as temperature, sediment, pollutants, and organic matter. As an example, the International Initiative on Water Quality, launched by UNESCO in 2012, focuses on detection of five key indicators of water quality: turbidity and sediment concentration, chlorophyll-a, harmful algal blooms, organic absorption, and surface temperature. These indicators can be quantified remotely using the distinct spectral signatures in visible, near-infrared, and thermal infrared wavelengths from water surface reflectance that are produced by suspended solids, photosynthetic pigments, and organic matter in the water. Such data are key to providing information on water quality in urban areas, use of agricultural fertilizers, climate change, and dam and reservoir management. Indeed, achieving some Sustainable Development Goals, such as providing clean water and sanitation for all, improving water quality, reducing water pollution, and protecting and restoring water-related ecosystems, depends on such holistic water quality monitoring.

Using frozen water to measure liquid water
Tracking ice blocks using Planet cubesat imaging of the surface of the Amur River, Siberia, allows quantification of water surface velocity in meters per second (m/s), revealing regions of faster and slower water speed.

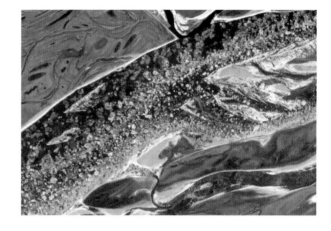

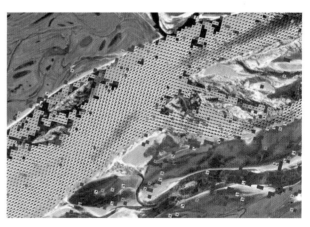

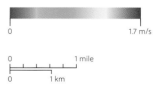

0   1.7 m/s

0   1 mile
0   1 km

THE COLOR OF THE WORLD'S RIVERS

Use of the reflectance data from 2.28 million Landsat images of the world's rivers over the period 1984–2022 allows the color of rivers to be assessed. Color is quantified using the dominant wavelength of the water surface reflectance in chromaticity color space, which characterizes how the human eye perceives color. Chromaticity transforms red-green-blue (RGB) values into human-perceived colors on the visible spectrum.

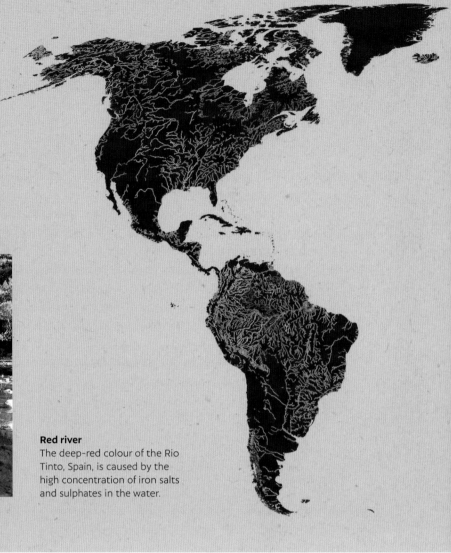

Red river
The deep-red colour of the Rio Tinto, Spain, is caused by the high concentration of iron salts and sulphates in the water.

December 22, 2020

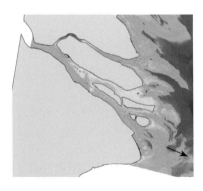

March 28, 2021

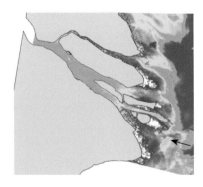

April 29, 2021

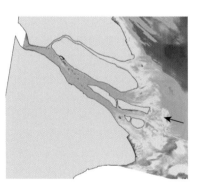

Chlorophyll concentration (µg/L)

0   1   2   3   4   5   6

Chlorophyll from space
Changes in the patterns of chlorophyll-a concentration over a period of 18 weeks at the mouth of the Chang Jiang (Yangtze) estuary, as revealed from the spectral signature of water surface reflectance in Landsat 8 imagery. Arrows show the local current direction.

0   100 miles
0   100 km

Drone swarms, remote boats, and robots

As drone and remote vehicle technologies increase in sophistication, decrease in cost, and thus become more available, so the deployment of boats, individually or in fleets, and swarms of aerial drones becomes feasible. Survey boats and autonomous underwater vehicles (AUVs), replete with multibeam echo sounders, GPS, and various flow and water quality sensors, can now be deployed remotely and operated from the other side of the world. These provide the ability to quantify the baseline conditions that characterize aquatic environments, thereby allowing assessment of change. Such vehicles can achieve this using multiple sensors and in conditions that may otherwise be too remote, too difficult, or too dangerous to access.

In addition to monitoring, remote vehicles, vessels, and drones can play an increasing role in direct environmental actions, such as investigating sites of hazardous natural and human-induced disasters, and aiding in environmental cleanup. Technologies that are being applied in the oceans, such the identification and cleanup of plastics and other waste on the ocean floor and surface, are also being used in shallower waters. As these technologies develop further, our ability to monitor the changing conditions present in rivers, estuaries, and deltas, and interact with these environments, will improve, promising a bright new era of monitoring and environmental management.

Numerical modeling

Computer models are widely used to predict how the flow of water, sediment, and nutrients, as well as critical properties such as water temperature, will change in rivers, estuaries, and deltas. But how do these models work, and how reliable are their predictions?

Models represent waterbodies using a series of discrete grid cells. The quantity of interest—such as water and sediment volume, or temperature—"stored" in each cell is updated based on the rate at which these quantities are transported into, or out of, each cell. Mathematical equations, based on the fundamental laws of physics, chemistry, and biology, are used to predict these local rates of transport at each cell.

HEATING UP

Changes in the annual mean temperature of the world's rivers between 1960 and 2014, simulated by a hydrological model.

Change in annual water temperature (Celsius per decade)

-0.8 -0.7 -0.6 -0.5 -0.4 -0.3 -0.2 -0.1   0   0.1   0.2   0.3   0.4   0.5   0.6   0.7   0.8

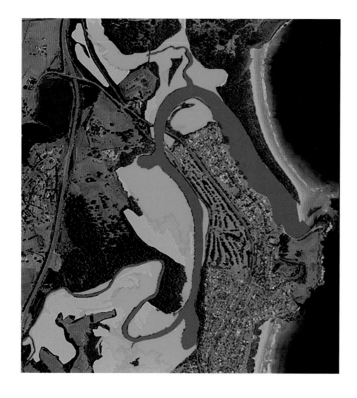

Model validation
Models are tested by comparing their predictions with observations. Here, predictions (right) of mangrove and saltmarsh distributions in the Minnamurra River estuary, Australia, are compared with real-world data (left) to test their accuracy.

- River
- Mangrove
- Saltmarsh
- *Casuarina*

In general, the smaller the size of the grid cells used, the more accurately these natural processes can be represented, but trade-offs are involved. Increasing grid cell resolution (making cells smaller) increases the total number in an area of interest (for example, at 10 m resolution it takes $100 \times 100 = 10{,}000$ grid cells to represent an area of 1 km², but at 1 m resolution it takes $1{,}000 \times 1000 = 1$ million cells), substantially increasing the computing power required. Importantly, increasing grid cell resolution also means that land surface properties (the input data for the model), such as the underlying terrain, must also be specified with increased precision.

Divergence from reality

Indeed, the main source of model uncertainty is typically not due to errors in the governing transport laws, but rather our planet's natural variability. The high natural variability in properties such as terrain and vegetation, which control the movement of water and sediment, can be challenging to represent at small spatial scales.

All models are idealizations of reality and thus their predictions will also diverge from reality, at least to a certain extent. The degree to which such model error is acceptable depends on the application of interest—for example, the error acceptable for flood hazard assessments for a nuclear power station is much less than that for agricultural land. Validation tests, whereby models are run for periods of time so that their predictions can be directly compared to actual observations, are used to quantify the confidence with which they can be used to predict future changes.

Assessing biodiversity

In all Earth surface environments, we require a detailed knowledge of the presence, abundance, and diversity of life as a prerequisite to establishing how ecosystems function and how they may be changing due to natural and human-induced stresses. Although such data form an essential baseline, they are often extremely difficult to obtain easily, resulting in incomplete pictures of the biological makeup of many rivers, estuaries, and deltas. However, new technologies are rapidly changing this situation and promise a novel way to monitor and manage bioenvironmental change.

Environmental DNA

A revolution in ecosystem assessment has come from the measurement of environmental DNA (eDNA) in water, sediment, and soils. Deoxyribonucleic acid (DNA) holds the blueprint of life, being the molecule that carries genetic information for the development and functioning of organisms, and each species and each individual has a unique and identifiable DNA profile. As organisms interact with their environment, they continuously shed DNA to their surroundings—for instance, from skin, scales, hair, fur, flesh, urine, and feces, and as their cells break down when they die. When cell membranes decay, DNA is yielded to the environment, where, depending on conditions, it may be preserved—for weeks in temperate waters, and up to hundreds of thousands of years in cold, dry permafrost conditions.

Methods developed to sample and sequence eDNA have proved extraordinarily successful at providing a robust, efficient, and cost-effective way to document species occurrences in a wide array of aquatic environments. No individuals need to be caught for eDNA biomonitoring, and thousands of organisms can often be identified from a small sample of water or sediment. Research has shown that river networks are conveyor belts of eDNA biodiversity information, from both aquatic and terrestrial biomes. Current projects are using eDNA profiling to map the complete picture of life in all the world's rivers.

Imaging fish using sound
Sonars and acoustic cameras provide nonintrusive methods of detecting and quantifying fish types and numbers. Here, an aggregation of Shortnose Sturgeon (*Acipenser brevirostrum*) swim near the bed of the Saint John River, New Brunswick, Canada. In late fall and winter, the Shortnose Sturgeon form dense winter aggregations where they remain almost completely stationary when water temperature drops below about 3 °C (37 °F).

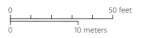

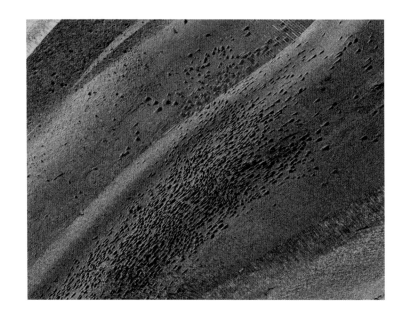

MONITORING LIFE

New technologies, including eDNA analysis, biotelemetry, and direct observations, are transforming how we can monitor organisms living in, on, and around water. Applications specific to eDNA sampling are labeled in italics.

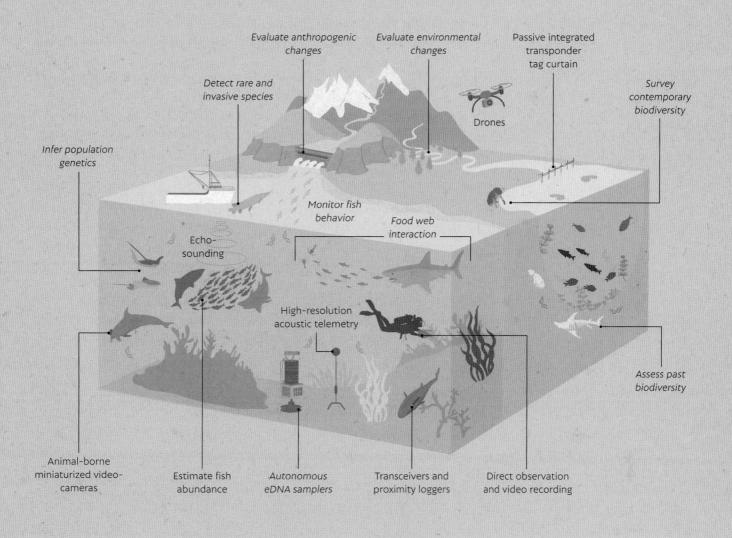

Counting numbers

Although eDNA provides a powerful method for determining the organisms that live in an environment, it cannot tell us their relative abundance. To do this, we need to rely on other methods. These have traditionally involved sampling and trapping, but new techniques are making this task easier, more efficient, and more complete. The use of drone and AUV surveys is becoming increasingly common, as is the use of acoustic technologies that examine the sound echoes reflected from organisms such as fish within the water column. These latter are becoming so refined that they can differentiate between fish species and allow quantification of fish size and number. Tracking organisms using biotelemetry has also become more feasible as transmitters have become both miniaturized and cheaper to produce, and GPS technology and internet communication make monitoring and data transfer possible from extremely remote locations.

Citizen science

The Italian Renaissance polymath Leonardo da Vinci once said, "Science is the observation of things possible, whether present or past." In citizen science, the public participates voluntarily in the scientific process, addressing real-world problems and helping advance knowledge.

Science often needs more eyes, ears, and perspectives than any one group of scientists can provide. Citizen science is an opportunity for collaboration between scientists and local volunteers who are curious, motivated, informed, and interested in making a difference. Participants typically collect, report, or analyze data. This could be done by taking pictures of local weather conditions (Weather Watchers in the United Kingdom), documenting local stream water levels (NASA-Terra), identifying or keeping track of local wildlife (Backyard Birdwatchers or iNaturalist), or even using smartphone sensors to monitor water and air quality (Microclimate apps).

A citizen science project can involve one person or millions of people collaborating toward a common goal. Anyone can participate, and a citizen does not need to possess special skills or knowledge to contribute. In a citizen science project, participants use a standard protocol to ensure data are high quality, and these data become available to the public as well as to scientists for analysis so that conclusions can be drawn from them. There are many opportunities to get involved and many areas of interest, from protecting endangered species to safeguarding water resources and recording environmental change.

▶ **Endangered species**
The Monarch Butterfly (Danaus plexippus) from the Americas and Australasia was identified recently as a threatened species due to population declines since the 1980s, mostly from habitat loss and climate change that has disrupted crucial stages in the species' life cycle. Citizen scientists help tag and track populations.

*Amphibians, such as this Emerald Glass Frog (*Espadarana prosoblepon*) in Panama, are seeing massive population declines as habitats are lost through deforestation and infectious diseases. Citizen scientists help track the occurrence and abundance of species around the world.*

The Great Nature Project

The Great Nature Project, sponsored by the National Geographic Society from 2013 to 2015, was a successful species inventory conducted by thousands of citizen scientists in more than 100 countries. Volunteers helped celebrate biodiversity by gathering over half a million images of plants, animals, and fungi. This program continues through the iNaturalist program, in which more than 6 million users have identified and tracked over 400,000 species to date!

Biodiversity can be observed simply through our own eyes, or by using tools such as binoculars, a net, and a magnifying lens, and keeping in mind that life can be found in all sorts of environments: under rocks, in the water, in the soil, up trees, or in the air. Such citizen science programs help scientists track the occurrence, abundance, movement, and changes of individual organisms and populations over time, providing fundamental information to help identify vulnerable or endangered species.

Managing rivers, estuaries, and deltas

Addressing future changes in habitats in response to dynamic climatic and environmental variables can be achieved only through the framework of integrated environmental management. But how are we to accomplish this?

▶ **Sinking deltas**

In many deltas, subsidence is the dominant driver of rising sea levels, as evident from the abandoned boat docks of the Ebro delta in Catalonia, Spain. Relative sea-level rise can be accelerated when fluvial sediments, which would otherwise be delivered to the delta, are trapped behind dams upstream or removed by sand-mining operations.

▶ **Europe's last wild river**

The Vjosa River, seen here in Qesarat, Albania, was saved by an effective campaign that prevented construction of a major dam, offering new hope for a future of more sustainable river management.

▶ **A city's river revitalized**

The Cheonggyecheon River in Seoul, South Korea, is the world's largest urban river restoration project to date.

Restoring connection from catchment to coast

A theme that has run, like a river, through this atlas is the need to ensure that water, sediment, and biota can flow unimpeded through rivers to their estuaries and deltas. This vital need for "connectivity" between catchment and coast points to the necessity for integrated river, estuary, and delta management plans. Such plans seek to retain pollutants and floodwaters "at source"—for example, by restoring straightened rivers to more natural sinuous courses, or by deliberately allowing floodplains in upstream regions to be inundated to slow the flow of water to more vulnerable locations downstream. The need for system-wide connectivity of sediment flows is also amply demonstrated by the problems of subsiding deltas.

Successful schemes

Examples of effective management plans include the restoration of sediment fluxes to the Mississippi delta (see page 368), dam removals (see page 52), and the growing recognition that we must protect rivers from fragmentation by damming. Another example is the Vjosa River in Albania, which was under threat from dam construction. A partnership between civil society, business, and government campaigned successfully to create the Vjosa Wild River National Park in March 2023, the first of its kind in Europe. The new park serves as an international model of effective water conservation.

With rapid urbanization, and with densely inhabited cities often being a locus of water insecurity and pollution, restoration of watercourses flowing through cities is vital. The Cheonggyecheon River runs through Seoul, South Korea, a city of 23 million people. Little more than a polluted drainage channel with a six-lane highway built over it in the 1970s, the river was restored in 2003–2005 to create an 8 km-long (5-mile) green corridor. The environmental benefits of the restoration are multiple: in addition to enhanced resilience to flooding and improvements to air quality, the number of bird species in the river corridor increased from 6 to 36, fish species from 4 to 25, and insect species from 15 to 192. Other cities in East Asia and North America are now studying the project to transfer its benefits for ecology, environmental quality, and urban sustainability.

Delta restoration

Although delta restoration is a huge challenge, scientists, coastal managers, and communities are coming together to brainstorm and implement feasible solutions to ameliorate the deterioration of millions of hectares of deltaic lands lost over the past century. These solutions often focus on re-establishing the natural flows of water and sediment to deltas that have been disrupted by environmental changes upstream.

Since the 1990s, Romania has reconnected river flow to interior lakes and reintroduced natural grazers to the Danube delta. Although recovery is slow, food production, tourism, and recreation, as well as habitat provision and biodiversity, are all improving, especially for the millions of waterfowl that use the delta on their yearly migration.

In the Shatt al-Arab delta, Iraq, river reconnections implemented since 2003 have restored about 50 percent of the marshland area, resupplying culturally significant resources such as reeds, water buffalo, and fish to the local Ma'dan communities. However, drought years remain a challenge, and additional national and international agreements on water and sediment supply are needed.

In the Yellow River Delta Natural Reserve, China, restoration of fresh water to degraded Common Reed (*Phragmites australis*) wetlands has greatly decreased salinization and improved stocks of wildlife. However, it takes time for restored wetlands to resemble their natural counterparts, so long-term monitoring and implementation are vital.

In the Ganges-Brahmaputra-Meghna delta, Bangladesh, restoring river and tidal flows to embanked regions has had mixed results: it has taken almost a decade for gains in land elevation to occur, and many households must be relocated in the process. Clearly, there is no one-size-fits-all solution to delta restoration, and negotiations that start with local grassroots efforts appear the most successful.

Restoration of the Mississippi delta

In 1897, American engineer Elmer Corthell (1840–1916) wrote that nineteenth-century levee construction along the Mississippi River would lead to short-term flood protection, but at the expense of long-term land loss that would become a challenge for future generations. Present-day conditions on the Mississippi delta confirm his prediction: more than 4,800 km² (1,850 square miles) of the delta have been lost since 1932 due to levee and dam construction, and through the extraction of groundwater and other subterranean fluids, which has exacerbated natural subsidence (see also pages 262–67). As sea levels rise globally, a future without action for the Mississippi delta is bleak.

After Hurricanes Katrina and Rita in 2005, which caused the migration of more than a million people, flooding of over 200,000 properties, and 3,000 deaths, the Louisiana Coastal Protection and Restoration Authority (CPRA) was created with the sole purpose of restoring the Mississippi delta and protecting its inhabitants. The CPRA seeks to achieve this through the implementation of the Louisiana Coastal Master Plan, which is updated every five years. The 2023 version of the plan illustrates how the Mississippi delta will change in terms of its landscape, natural resources, and future hurricane risk over the next 50 years, and identifies priority plans for restoration and risk reduction.

To date, more than 594 km (369 miles) of levee improvements, 116 km (72 miles) of barrier island and headland restoration, and 228 km² (88 square miles) of marshland creation have been achieved. Over the next 50 years, this US$50 billion project will enhance the connection between the coast and those who depend on it, and help create a more sustainable future for the delta.

FIXING THE FLOW

Harnessing the natural flows of water and sediment in deltas is key to restoring land loss due to rising seas, declining sediment loads, stronger storms, and land subsidence. Here, several possible solution scenarios are depicted, highlighting where they are being implemented around the world.

Channelization
(Danube delta)

Small-scale river diversions
(Mississippi Balize delta)

New/impacted channels

Coastal wetlands

Upland/natural levees

Waterbodies

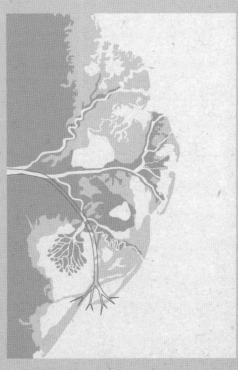

Large-scale river diversions
(Mississippi sediment diversions)

Lobe building
(Huang He delta)

Renewable energies

The flow of water in rivers, estuaries, and deltas offers significant potential for the generation of renewable energy. As we have seen (Chapter 5), rivers already provide massive global hydroelectricity generation, with potential for more, but the green credentials and impacts of damming, and especially megadams in tropical regions, raise many significant environmental concerns. New technologies are now allowing some of these detrimental effects to be mitigated.

▶ **Fish transit**

Fish ladder on the Elbe River, Germany, designed to allow fish passage around a dam. Such fish ladders have had variable degrees of success in allowing fish to bypass dams and mitigate river fragmentation.

▶ **Fish-safe turbines**

New blade designs allow safe fish passage through the turbines of some hydroelectric dams that have up to 30 m (100 ft) of water head.

▶ **Underwater power farms**

Underwater fields of hydrokinetic turbines may provide electricity generation close to sites of power consumption, and without many of the environmentally damaging impacts of conventional hydropower storage dams.

Fragmentation of river networks by dams has been a significant issue for fish migration, and many fish ladders or passages—which are designed to allow fish to bypass the dams—have proven relatively ineffective. However, new designs of turbines with thick, forward-slanted blades that have blunt leading edges can permit fish and eel passage without injury, partially alleviating some of these fragmentation issues, although significant concerns remain for large animals such as dolphins.

Hydrokinetic turbines

Smaller hydrokinetic turbines can be placed within a moving flow—in a large river, estuary, or the ocean—and use the kinetic energy of the fluid to generate electricity. These turbines generate electricity without requiring construction of a water-storage reservoir, and simply require flow of water over a certain depth, and at a sufficient rate and period, to be cost-effective. Although hydrokinetic turbines produce lower total energy yields than hydropower reservoirs, they are far less environmentally damaging as they do not interrupt the passage of sediment or organisms. They also have a longer total timespan of operation, since unlike hydropower reservoirs, they do not fill with sediment. Hydrokinetic turbines can also potentially be sited nearer the local communities that use the electricity they produce, rather than the electricity being exported and used far away. Many different designs of hydrokinetic turbines are under development, and submerged fields of turbines may provide a far less environmentally damaging way of harnessing the power of water in future.

The potential positive impact of such schemes is illustrated by the example of the Amazon Basin, where many hydropower dams are planned for construction over the next 20 years. It has been estimated that use of in-stream turbines could generate about 63 percent of the electricity planned through conventional hydropower, and at 50 percent of the cost. Such estimates suggest a global route toward power generation to meet energy needs, along with other renewable sources such as solar, wind, and thermal power, while also minimizing negative socioenvironmental impacts.

River management

While the effective, sustainable management of rivers, estuaries, and deltas is essential, it is complex to achieve in practice. This challenge is compounded when these environments cross geopolitical borders— some 40 percent of the world's population currently lives in transboundary river basins.

As the second-longest river in Europe, the Danube is a transboundary river that enjoys effective governance through the International Commission for the Protection of the Danube River (ICPDR). Established in 1998 and with 15 member nations, the ICPDR seeks to ensure sustainable water management, control pollution, and manage flood and ice hazards in the river. A significant recent success concerns efforts to protect the native sturgeons that represent a key part of the Danube's natural heritage. Through its Danube Sturgeon Task Force, the ICPDR has worked to protect fragile habitats, develop migration aids, and police the illegal fishing and caviar trade, helping to slow the dramatic decline in sturgeon numbers seen in recent decades. The importance, and impact, of transboundary cooperation is also shown by recovery of sturgeon populations in the Danube estuary.

CONFLICT RISK

Many countries in transboundary river basins are highly dependent on water from upstream, introducing risks of conflict. Upstream basin dependence is especially pronounced in parts of Africa, South America, and Central and Southeast Asia.

Upstream basin dependence

● Very high

● High

● Low

● Very low

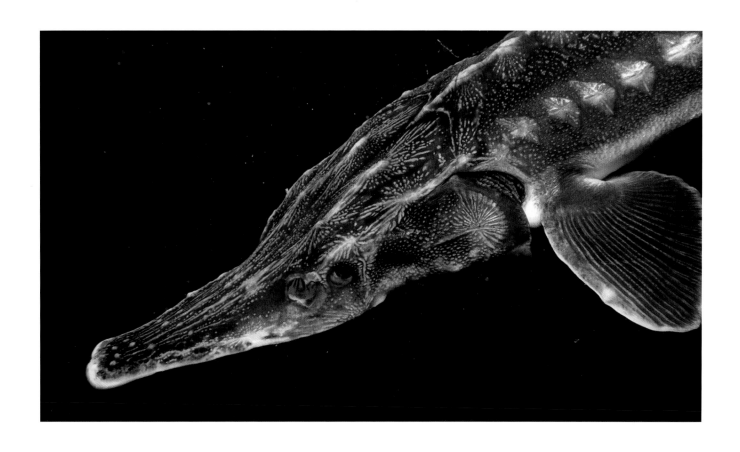

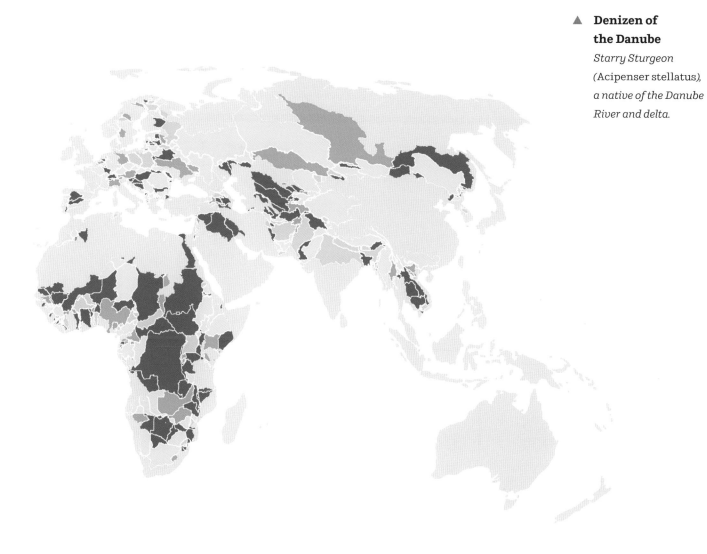

▲ **Denizen of
the Danube**

*Starry Sturgeon
(Acipenser stellatus),
a native of the Danube
River and delta.*

Management measures

Numerous river management measures have been developed, but the most important are those that focus on three key priorities. First, they should reconsider the need for further major dams, or seek to provide ways for sediment and fish to more effectively bypass dams. Second, they should prevent further loss of wetlands in order to provide better flood storage services for downstream deltas and estuaries. And third, they should focus on reducing groundwater abstraction to slow delta subsidence. A second tier of priority measures should focus on restoring lost wetlands and reducing the emissions of key pollutants, such as fertilizers, antibiotics, and human and animal waste.

Within the context of flood management, a combined approach is needed that both protects against the hazard (such as through the construction of dikes and levees, and greater use of natural flood management techniques), and also promotes reductions in exposure (for example, by limiting urban encroachment on floodplains and deltaplains). Finally, adapting to the challenges posed by rivers, estuaries, and deltas is as important as efforts to mitigate problems. Indeed, measures such as floodproofing buildings and critical infrastructure are among the most cost-effective approaches to reducing vulnerability to the threats posed in these environments.

▼ **Floating houses**

One way of making our homes more resilient to flooding is to allow them to float. Ijburg, a suburb of Amsterdam, Netherlands, is one area where homes simply rise and fall with changing water levels.

WHOLE-SYSTEM THINKING

A diverse suite of potential management measures is
required to improve the sustainability of rivers, estuaries,
and deltas under future environmental pressures

CLIMATE CHANGE
- Increasing temperature and heat
- Changing precipitation patterns
- Increasing drought and flood risks
- Increasing water temperature

**Reduce the risk of
transboundary conflict by:**
- Improving institutional
  resilience and transboundary
  collaboration
- Reducing water use/
  transboundary
  dependence

Forestation/deforestation
- Share the planet: improve
  protection of natural areas

Reduce flood risks by:
- Improving protection
  (dykes/levees)
- Waterproofing buildings
- Spatial zoning
- Integrating climate adaptation
- Nature-based solutions

Reduce water use by:
- Improving water use efficiency
  in agriculture, households,
  and industries
- Integrating climate adaptatio

**Reduce the impacts
of dams by:**
- Ecological hydropower
  construction
- Alternative sources of
  renewable energy

River water

Disturbed
water and
sediment flow

Sea-level
rise

Salinization

Delta

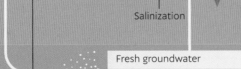

Fresh groundwater

Brackish groundwater

**Restore ecological
quality by:**
- Ecological dam construction
- Reducing water pollution
- Reducing water use/
  over-exploitation
- Nature-based solutions
  in flood protection
- Restoring wetlands

**Reduce water pollution
and nutrient emissions:**
- From agriculture:
  sustainable intensification
- By adding nutrient
  buffer zones
- From cities: improving
  sanitation and wastewater
  treatment

Nature-based solutions

Human actions have profoundly altered the Earth's surface, causing significant losses in natural habitats and biodiversity. We need to recognize the resilience of natural systems and their crucial contributions to our health, wellbeing, and economy, and give nature the space and conditions it needs for a sustainable future.

Nature-based solutions are initiatives that aim to maintain, improve, or restore nature and natural processes, so that people and the environment benefit from the many ecosystem services (see pages 125-27) they provide. Mangrove forests, for example, help mitigate climate change through carbon sequestration, reduce flood risk by dissipating wave energy, improve water quality by retaining sediments, are nursery grounds that support fisheries, and provide wood and other goods and services that support the livelihoods of millions of people worldwide.

SPACE FOR NATURE
Nature-based solutions range from the conservation of natural ecosystems to creating space in cities for blue (water) and green (vegetation) infrastructure.

Conservation

Restoration

Sustainable management

Green/blue infrastructure

Range of nature-based solutions

More natural (green/blue)

More artificial (gray)

Increased benefits from multiple ecosystem services

Reforestation

Forests are biodiversity hotspots with a vital role in climate regulation, air and water purification, and the formation and retention of soils. Deforestation is taking place at an alarming pace, mostly to clear areas for the production of food and fodder. Between 1990 and 2020, 4.2 million km^2 (1.6 million square miles) of forest were lost, an area equivalent to half of the size of Brazil. To halt and revert this loss, reforestation efforts are spreading worldwide, with an estimated 590,000 km^2 (228,000 square miles) of forest regenerated between 2000 and 2021. An area of this size can potentially store enough carbon to offset the current annual greenhouse gas emissions of the United States. The largest forest expansions have been in the Atlantic forest of Brazil and boreal forests of Mongolia and Canada.

Wherever possible, forest restoration involves simply stopping human pressures to allow natural regeneration. In some areas, human assistance is required to create conditions for natural vegetation to grow, such as removing invasive species or installing fences to reduce grazing pressure. Active replanting of native species and agroforestry management are needed in degraded areas where natural seeding and growth are limited.

▼ **Natural protection**
Mangrove reforestation in the Sundarbans, India. After the devastating floods caused by Cyclone Yaas in May 2021, the West Bengal government decided to plant 150 million mangrove saplings to help reduce the impact of flooding from future cyclones.

Blue carbon

Reducing emissions and removing greenhouse gases from the atmosphere are key to mitigating global warming. Blue carbon ecosystems, such as the mangroves and saltmarshes found in estuaries and deltas, are highly efficient at sequestering and storing carbon.

Plants remove more carbon from the atmosphere through photosynthesis than they release during their lifetime, and thus they sequester carbon. This carbon is stored in their biomass (leaves, stems, trunks, and root systems) and in soils, and this can represent significant storage in some types of vegetation, such as marshes and mangroves. Although vegetated areas act as carbon sinks if they take up more carbon than they release, the stored carbon is released back into the atmosphere when plants decompose or specific conditions arise, shifting the carbon budget from a sink to a source.

Coastal and marine ecosystems effectively sequester and store carbon, which in an allusion to the sea is known as blue carbon. Indeed, mangroves, saltmarshes, and seagrasses sequester and store more carbon per unit of area than terrestrial ecosystems. Therefore, reducing habitat loss and increasing the extent of these ecosystems through reforestation have been advocated as priorities for climate change mitigation. The larger their extent, the more carbon they can potentially take out of the atmosphere and store.

Blue carbon is a major component of carbon offsetting in countries that have extensive areas of mangroves, saltmarshes, and seagrasses. However, its potential contribution to mitigating climate change at a global scale has been questioned due to the limited geographic extent of these areas. There is also concern that carbon-offsetting strategies are based on estimates of carbon storage capacity that are still unreliable, because high variations in the rates of carbon burial through time and across different locations have not been considered adequately. Nevertheless, the widely recognized benefits provided by coastal ecosystems to coastal protection, biodiversity conservation, and supporting fisheries, warrant investment for their protection and restoration.

▶ **Measuring carbon**
Sediment core being taken to quantify rates of carbon sequestration and storage in a seagrass bed, Hobsons Bay, near Melbourne, Australia.

MANGROVE WINS

At a local scale, mangrove forests store more carbon per unit area than all other vegetated biomes in Brazil. Here, carbon quantity has been estimated based on the top 1 m (3 ft) of organic soil, together with the above- and below-ground biomass.

Brazilian vegetated biomes

- Biomass (above and below ground)
- Soil (top 1 m)

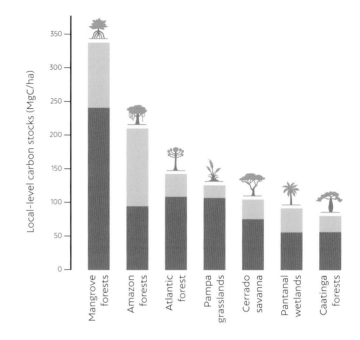

Local-level carbon stocks (MgC/ha)

Mangrove forests · Amazon forests · Atlantic forest · Pampa grasslands · Cerrado savanna · Pantanal wetlands · Caatinga forests

CARBON STORAGE AND FLUXES

Excluding the Earth's rocks, blue carbon is the Earth's largest carbon store, with only a minor part stored in coastal ecosystems. Blue carbon and terrestrial sinks remove over half the annual global carbon emissions from fossil fuel and land-use change.

Global carbon storage and average fluxes for the decade 2012–2021

↑↓ Carbon cycling GtC per year

◯ Stocks GtC

✦ Atmospheric increase

1 Gt = 1 billion tonnes

Budget imbalance = −0.3

Fossil carbon (9.6 ±0.5)

Land use change in carbon (1.2 ±0.7)

Terrestrial carbon sink (3.1 ±0.6)

130

130

Atmospheric CO_2 (+5.2) 875 GtC

Ocean carbon sink (2.9 ±0.4)

80

80

Vegetation 450 GtC

Rivers and lakes

Gas reserves 115 GtC

Oil reserves 230 GtC

Permafrost 1,400 GtC

Soils 1,700 GtC

Coal reserves 560 GtC

Coasts 10–45 GtC

Surface sediments 1,750 GtC

Organic carbon 700 GtC

Marine biota 3 GtC

Dissolved inorganic carbon 37,000 GtC

Blue carbon

Making room for water

Through history, many rivers have been straightened and canalized to facilitate agriculture or urban development. Without meanders, rivers lose aquatic habitat and water storage decreases, increasing the risk of downstream flooding. Many countries are now returning rivers to a more natural state to manage flood risk and enhance biodiversity.

▼ **Sponge cities**
Pond and green corridor in Qian'an, Hebei province, China, part of a green infrastructure and river restoration scheme. This has been designed to alleviate urban flooding and follows the "sponge cities" concept, a national pilot scheme started in 2015.

Creating space for rivers to regain a more natural form can bring multiple environmental and socioeconomic benefits. The presence of meanders increases habitat diversity and biodiversity, regulates water flows, improves water quality by reducing the dispersion of pollutants, and reduces the risk of downstream flooding. Nature-based solutions to restore the natural functions of rivers vary in scale and can be implemented in both rural and urban areas. For example, the Emscher River in Germany is 83 km (52 miles) long and since the late 1800s parts of it were channelized with concrete and used as an open wastewater canal. Restoration efforts since the 1990s dismantled the man-made channelization and have improved the flow of this urban river. In other examples worldwide, green infrastructure such as floodways and flood-retention ponds showcase the impressive ecological and socio-economic transformations that are possible.

A return to nature: room for the river

The name Netherlands means "low-lying land" and refers to the landscape formed by the Rhine and Meuse Rivers, with a delta that has been active for more than 7,500 years. Systematic peat draining and poldering of the deltaplain since 1000 CE have created a landscape that is, in many regions, more than 5 m (16 ft) below sea level (see pages 324–25). In 1953, a major coastal storm created a surge up to 5.6 m (18.3 ft) above the mean sea level of the North Sea, breaching embankments, producing extensive flooding, and causing more than 1,800 deaths. This catastrophic flood instigated a fortified armoring of the coastal zone—the Delta Works—to prevent future coastal flooding.

Today, the Netherlands is more at risk from riverine flooding due to increased precipitation. Major flooding occurred in the 1990s, and enormous efforts have been under way since 2007 to make "Room for the River," essentially dismantling hard engineering structures along the Rhine, Meuse, Waal, and IJssel Rivers to ameliorate flood risk. This may sound counterintuitive, but it is being accomplished by restoring the rivers' natural floodplain function—and hence the ability to store excess floodwater in places where flooding will cause least harm. Measures include creating water buffers, relocating levees, increasing the depth of side channels, and constructing flood bypasses. With more than 30 projects implemented to date, the "Room for the River" program is reducing risk for people and infrastructure, as well as improving the overall environmental quality of the rivers and adjacent wetland corridors. It gives the rivers room to breathe once again.

Making room for rivers in an anthropogenic delta
With more than 17 million people currently living on the Rhine-Meuse delta, much of which is below sea level, flood control is a major priority. The "Room for the River" program is enabling the rivers to inundate more of their natural floodplain, sacrificing agricultural land to protect towns and cities. This enlarged and reconnected floodplain along the Ijssel River allows more storage space for floodwaters, which reduces flooding elsewhere.

▲ **Wiggling with life**

The restored and newly meandering Swindale Beck, England.

Swindale success story

Swindale Beck, a stream in England's Lake District National Park, illustrates the benefits of integrating restoration efforts along a river catchment. The stream was straightened in the mid-nineteenth century to drain farmland. However, changes to the water flow adversely impacted salmon and trout spawning grounds, degraded soil quality, and increased water turbidity and downstream flooding. The land is now a conservation area owned by a water company and managed by the Royal Society for the Protection of Birds. To reduce flood risk and improve water quality and biodiversity, 40,000 trees have been planted to recover riparian woodland and 1 km (0.6 miles) of the watercourse has been restored.

In 2016, artificial embankments were removed to reconnect Swindale Beck with its floodplain and the meandering channel was restored, extending the length of the watercourse by 18 percent to slow the water flow. After restoration the channel developed naturally, creating gravel bars where salmon and trout spawned only a few months after the work was completed. Water and soil quality improved, which helped revive 1,000 ha (2,471 acres) of bogs and 15 ha (37 acres) of meadows. The old channel was covered with soil and planted with seeds from local meadows. In 2022, the project received the European River Prize for excellence in river management and conservation.

▲ **Butterflies revived**
*Restoration has created favourable wildflower meadows that have helped the return of Marsh Fritillaries (*Euphydryas aurinia*), which were extinct in the region.*

Rewilding

Rewilding aims to identify and enable nature to take its course and reverse environmental degradation, with little or no further input from humans.

▼ **Local attraction**

In May 2019, seven Water Buffaloes (Bubalus bubalis) were released on Ermakov Island on the Ukrainian side of the Danube delta, followed by a further ten in 2021. Buffaloes are ecosystem engineers as they disperse seeds and shape habitats by opening scrub and creating puddles and pools.

Rewilding returns to nature the core elements needed for natural processes to thrive. It focuses on reconnecting wilderness areas, removing artificial structures that constrain natural processes, and reintroducing species that will have a positive impact on biodiversity and ecosystem health. Apex predators can control population growth lower in the food chain, keystone species are essential for the healthy functioning of their ecosystem, and ecosystem engineers improve conditions and create habitats for other species. When natural processes re-establish habitat diversity, biodiversity flourishes and has a better chance of adjusting to changing climate.

Many rewilding projects are taking place in Europe, such as those supported by Rewilding Europe, an independent not-for-profit foundation. The Danube delta rewilding scheme crosses the boundaries of Ukraine, Romania, and Moldova to restore 40,000 ha (98,900 acres) of habitats in the largest wetland in Europe. Here, work with Rewilding Ukraine has seen the release of buffaloes and wild horses on Ermakov and Tataru Islands, removed dams, reinstated tidal flow into polders, and developed community-based ecotourism.

Beavers are back

The largest rodent in Europe, the Eurasian Beaver (*Castor fiber*), was widespread in the Continent until the twelfth century. By the twentieth century, the loss of riparian woodland habitat and overhunting of the animals for their meat, fur, and scent glands reduced numbers to 1,200 in eight isolated populations. From the 1920s in Scandinavia and later in the twentieth century elsewhere in Europe, Beavers were reintroduced to river catchments in more than 26 European countries, including those where they had become extinct. Helped by habitat restoration and protected status, the population increased from 150,000 in the early 1990s to 1.5 million in 2020, with half living in Russia and 40 percent across Norway, Sweden, Poland, Lithuania, and Latvia. As populations expand into the species' past range and mature, there will be an increasing need to monitor and manage population density and its effects on river catchments.

While the presence of Eurasian Beavers attracts visitors and saves money that would otherwise be spent on flood risk management, some conflicts with humans can arise. However, these can be resolved easily. Wrapping wire mesh around tree trunks or sandpainting them can prevent tree felling by beavers, and undesired floods near dams can be alleviated by installing a drain to allow some flow when water levels are high.

The photo here shows a Beaver building a dam in Kuikka Lake, Finland. Beavers are a keystone species, playing an important role in controlling floods. They dig burrows that improve the connection of the river with the floodplain, and their leaky dams slow water flow and enhance habitat diversity, which benefits species ranging from otters and fish to fungi.

Glossary

accommodation space The space available for sediment deposition.

alluvial fan A triangular-shaped deposit of sediment formed by a river, often at a break in slope.

anabranching A type of river planform with multiple interconnected channels separated by vegetated, semi-permanent alluvial islands.

anadromous Fish that are born in fresh water and spend portions of their life cycle in salt water, returning to fresh water to spawn.

anoxic An environment that does not contain dissolved oxygen.

avulsion The rapid abandonment of an existing channel via the occupation of an alternate pathway.

backwater zone The region of a river where the water is slowed and held back, for example by high tide.

barform Accumulation of sediment, forming an island whose size scales with the width and/or depth of the flow.

barrier beach A narrow, low-elevation strip of sediment that extends across a river mouth or estuary inlet and restricts tidal incursion.

barrier island An elongated island, generally sandy and narrow, parallel to the mainland and separated from it by a lagoon or bay.

baselevel The lowest elevation that a river flows into, referring mostly to inland lake elevation or sea level.

bathymetry The depth of water to the sediment bed.

bedload The sediment load that is carried by a flow, and is bounced or rolled in contact with the bed.

benthos Organisms that live on, in, or near the bottom of a waterbody.

bioaccumulation An increased concentration of a chemical (pollutant) in an organism over time.

biodiversity The variety of organisms found in a particular habitat.

biofilm A greenish-gold slimy coating formed by microorganisms, such as cyanobacteria and diatoms.

braided channel A river planform in which the main channel is divided into multiple smaller channels by barforms and river islands.

carbon sink/carbon source Carbon sinks absorb and store more carbon than they produce over an extended period; carbon sources release more carbon than they retain.

catadromous Fish born in salt water that migrate to fresh water as juveniles and grow into adults there, before migrating back to the ocean to spawn.

catchment The geographical area into which water from higher elevations flows (also termed "watershed" in the United States).

coastal plain Low-lying, flat-elevation land found in coastal areas, typically constructed by rivers and through coastal processes (waves, tides, storms).

coastal squeeze When inland migration of intertidal habitats is hindered by upstream human development or infrastructure.

continental shelf The flat, shallow ocean floor immediately seaward of the coastline, typically reaching depths of 100 m (300 ft).

crevasse splay A deposit of floodplain sediment formed when a river overflows its banks.

cyanobacteria Photosynthetic bacteria found in most aquatic environments, and including some of the oldest organisms on Earth. Although known as blue-green algae, they are not plants.

delta cycle The life cycle of a delta over time, starting with a new delta growing into a body of water, reaching a point of maximum growth, then senescing and retreating.

density current A current that moves due to gravity acting on differences in density between two fluids.

diapir A column of sediment that is formed when denser material lies on top of less dense sediment, causing the underlying material to flow upwards.

diatom A group of unicellular microalgae with a shell made of silica, found in the water or on the bottom of most aquatic environments.

diurnal tide Lunar tides that occur in some restricted ocean basins where there is only one high tide and one low tide each day.

drainage basin An alternative name for "catchment."

ebb tide The falling limb of the tide, when water is receding from high water levels (high tide) to low water levels (low tide).

ecosystem service The direct or indirect contribution that ecosystems provide for human wellbeing and quality of life.

embankment A wall or bank of earth or stone built to prevent flooding.

equinox When the lengths of day and night are equal, occurring once in spring and once in fall each year.

estuarine turbidity maximum The location of the highest concentration of suspended sediments in an estuary, usually at the saline seawater wedge near the estuary floor.

eutrophication The process by which a waterbody becomes enriched in nutrients, particularly nitrogen and phosphorus.

flocculation The aggregation of smaller, often clay, particles to form a larger grain.

flood pulse Seasonal flooding that dominates river ecological functioning via lateral exchange of water, nutrients, and organisms between channel and floodplain.

flood tide The rising limb of the tide, when water is rising from low water levels (low tide) to high water levels (high tide).

food chain/food web A series of organisms, each dependent on the next as a source of food.

halocline The position in the water column where salinity changes sharply with depth, reflecting a boundary between water layers.

halophyte Salt-tolerant plants, such as mangrove and saltmarsh species.

harmful algal bloom (HAB) An overgrowth of algae or cyanobacteria causing harm to the environment, species, or human health.

headcut erosion An erosion process whereby a vertical step in the river long profile migrates upstream.

homopycnal flow A flow (e.g., river) that has a density equal to that of the fluid it is entering (e.g., ocean). Hyperpycnal and hypopycnal refer, respectively, to when the flow is of greater or lesser density compared to the fluid it is entering.

hydrolysis A chemical reaction in which a substance is split into other compounds by reacting with water, constituting a major weathering process.

hydroperiod The amount of time that an intertidal or floodplain area is inundated by water.

hypersaline Where concentrations of salts in water (salinity) exceed 50 parts per thousand (ppt), or 50 g of salt per 1,000 g of water.

hypoxic An environment that possesses low concentrations of dissolved oxygen—in water, less than 2-3 parts per million (ppm).

insolation weathering Weathering due to repeated thermal expansion and contraction of a rock surface.

intertidal area The region regularly covered and uncovered by rising and falling tides, often colonized by species tolerant to wetting and drying and changes in salinity.

invasive species Non-native species that cause ecological harm.

Last Glacial Maximum (LGM) The time period approximately 20,000 years ago when Earth's climate was much colder and glaciers reached their maximum extent.

legacy pollutants Pollutants created through industry or manufacturing, and stored in the environment, which have long-term polluting effects.

lentic Standing waterbodies with no flow.

levee A ridge of sediment deposited alongside a river by overflowing water.

littoral drift The material (sand, gravel, shells) transported along the coast by a longshore current.

longshore current/littoral current A current that flows parallel to the shore in the surf zone when waves approach the coast at an angle.

lotic Waterbodies with flowing water.

mangrove Shrub or tree vegetation that dominates in tropical coastal regions and has adapted to varying tidal and salinity conditions.

megadam A very large hydropower dam, often referring to those more than 15 m (50 ft) in height and generating more than 400 MW of power.

mudflat A relatively flat, muddy, and unvegetated habitat that occurs between the low and high tide lines on sheltered shores.

natural capital The natural resources of a given area that have an economic value by providing a service to humans.

neap tide Tidal phenomenon that corresponds to smaller-than-average tidal ranges, occurring during the first- and third-quarter stages of the moon.

non-native species Species that are not indigenous to a region but cause no ecological harm.

oxbow lake A lake formed in an abandoned river meander.

permafrost Permanently frozen ground that remains below 0° C (32° F) for more than two years.

phytoplankton Planktonic plants, usually microscopic, vital to aquatic food chains.

plankton Aquatic organisms that live in saline or fresh water, drifting with the flow as they cannot swim.

plate tectonics Movement of the Earth's crustal plates, resulting in new ocean basins, mountain ranges, volcanic arcs, and deep-sea trenches.

point bar An arcuate-shaped deposit of sediment formed on the inside of a meander bend.

polder A region found typically alongside river channels and in deltas that has been embanked, or made into an island, to protect agricultural land or infrastructure from flooding.

pore space The empty space between particles in a sediment, soil, or rock, and that may be filled by gases, fluids, or mineral precipitation.

relative sea-level rise The combined result of global eustatic (worldwide) sea-level rise from meltwater of glaciers and ocean warming, plus subsidence or lowering of the land.

remote sensing The process of scanning the Earth's surface from a distance to obtain information about it.

ria A drowned river valley, often produced as a result of sea-level rise.

salinization Increase in the concentration of salts in soils or fresh water due to excessive evaporation, or concentration by salt water as a result of sea-level rise.

saltation The process of sediment transport involving the bouncing of sediment particles along the bed.

semi-diurnal tide Lunar tides that occur in most ocean basins where there are two high tides and two low tides each day.

slack water The quiescent periods of the tide, when water reaches its highest or lowest height and tidal currents diminish to no movement.

spring tide Tidal phenomenon occuring twice a month that corresponds to larger-than-average tidal ranges, occurring during the new and full stages of the moon.

storm surge The increase in water level in coastal areas associated with the passage of a storm due to a decrease in atmospheric pressure and/or increase in wind velocity.

subsidence Lowering of the land surface; common in deltas due to the natural compaction of sediment.

suspended load The amount of sediment that is carried within the body of the flow and away from the bed, supported by fluid turbulence.

thermokarst lake Lake produced by water infilling a depression in the land surface that has been created by thawing permafrost.

tidal bore Wave-like incursion of the highest (equinox) rising tides traveling long distances upstream in shallow, funnel-shaped, macrotidal estuaries.

tides Daily fluctuations in ocean water level due to gravitational forces from the moon and sun.

transpiration Release of water vapor by plants, mostly when the stomata in their leaves open to capture carbon monoxide during photosynthesis.

trophic level Position in a food chain occupied by organisms that have similar energy needs. Primary producers (plants) and top predators are at the lowest and highest levels, respectively.

turbidity An optical characteristic that is a measure of the relative clarity of a liquid.

UN Sustainable Development Goals Seventeen integrated targets to end poverty and protect the Earth, adopted by the United Nations in 2015.

water footprint The amount of water required to produce and supply the goods and services used by a particular person or group.

watershed The ridge of land that separates different catchments (i.e., the boundary of the catchment).

water stratification The presence of layers (strata in Latin), which in estuaries are caused by vertical variations in water density, usually with fresh water (less dense) at the surface and sea water (more dense) at the bottom.

wave Movement of water in an orbital motion, typically resulting from the disturbing forces of wind.

wave base Depth in the water column where wave orbital movement no longer occurs.

zooplankton Planktonic animals, typically microscopic, including jellyfish, some crustaceans, and fish larvae.

Resources

FURTHER READING

Chapter 1

Beer, A.-J. *The Flow: Rivers, Water and Wildness*. London: Bloomsbury Publishing, 2022.

Cagle, A. J., R. J. Wenke, and R. Redding, eds. *Kom El-Hisn (Ca. 2500–1900 bc): An Ancient Settlement in the Nile Delta*. Columbus, GA: Lockwood Press, 2016.

da Cunha, D. *The Invention of Rivers: Alexander's Eye and Ganga's Descent*. Pittsburg, PA: University of Pennsylvania Press, 2023.

Hourly History. *Indus Valley Civilization: A History from Beginning to End*. Hourly History, 2019.

Kenawi, M. *Alexandria's Hinterland: Archaeology of the Western Nile Delta, Egypt*. Oxford: Archaeopress Archaeology.

Smith, L. C. *Rivers of Power: How a Natural Force Raised Kingdoms, Destroyed Civilizations, and Shapes Our World*. New York, NY: Little, Brown, Spark/Hachette Book Group and London: Penguin Random House, 2020.

Wantzen, K. M., ed. *River Culture: Life as a Dance to the Rhythm of the Waters*. Paris: UNESCO Publishing, 2023, https://unesdoc.unesco.org/ark:/48223/pf0000382775

Chapter 2

Ouellet Dallaire, C., B. Lehner, R. Sayre, and M. Thieme. "A Multidisciplinary Framework to Derive Global River Reach Classifications at High Spatial Resolution." *Environmental Research Letters* 14 (2019): 024003, https://doi.org/10.1088/1748-9326/aad8e9

Park, E., and E. M. Latrubesse. "A Geomorphological Assessment of Wash-Load Sediment Fluxes and Floodplain Sediment Sinks Along the Lower Amazon River." *Geology* 47 (2019): 403–406, https://doi.org/10.1130/G45769.1

Potter, P. E., and W. K. Hamblin. "Big Rivers Worldwide." *Brigham Young University, Geology Studies* 46 (2006), https://geology.byu.edu/0000017d-0ff3-d1e7-a77d-aff386cf0001/pdf-icon-volume-48-2006-pdf

Wohl, E. *A World of Rivers*. Chicago, IL: University of Chicago Press, 2012.

Chapter 3

Ashworth, P. J., and J. Lewin. "How Do Big Rivers Come to Be Different?" *Earth-Science Reviews* 114 (2012): 84–107.

Gupta, A., ed. *Large Rivers: Geomorphology and Management*. 2nd ed. Oxford: Wiley Blackwell, 2022.

Chapter 4

Hoorn, C., F. P. Wesselingh, H. Ter Steege, et al. "Amazonia Through Time: Andean Uplift, Climate Change, Landscape Evolution, and Biodiversity". *Science* 330 (2010): 927–931.

Opperman, J. J., P. B. Moyle, E. W. Larsen, J. L. Florsheim, and A. D. Manfree. *Floodplains: Processes and Management for Ecosystem Services*. Oakland, CA: University of California Press, 2017.

Thorp, J. H., M. C. Thoms, and M. D. Delong. *The Riverine Ecosystem Synthesis: Toward Conceptual Cohesiveness in River Science*. Boston, MA: Academic Press, 2008.

Chapter 5

Best, J. "Anthropogenic Stresses on the World's Big Rivers." *Nature Geoscience* 12 (2019): 7–21, https://doi.org/10.1038/s41561-018-0262-x

Best, J., and S. E. Darby. "The Pace of Human-Induced Change in Large Rivers: Stresses, Resilience and Vulnerability to Extreme Events." *One Earth* 2 (2020): 510–514, https://doi.org/10.1016/j.oneear.2020.05.021

Grill, G., B. Lehner, M. Thieme, et al. "Mapping the World's Free-Flowing Rivers." *Nature* 569 (2019): 215–221, https://doi.org/10.1038/s41586-019-1111-9

PBL Netherlands Environmental Assessment Agency. "Geography of Future Water Challenges: Bending the Trend," 2023, 14 March, https://www.pbl.nl/en/publications/geography-of-future-water-challenges

Sabater, S., A. Elosegi, and R. Ludwig. *Multiple Stressors in River Ecosystems: Status, Impacts and Prospects for the Future*. Amsterdam: Elsevier.

UNEP-DHI and UNEP. *Transboundary River Basins: Status and Trends*. Nairobi: United Nations Environment Programme (UNEP), 2016, http://geftwap.org/publications/river-basins-technical-report

Zhang, A. T., and V. X. Gu. "Global Dam Tracker: A Database of More Than 35,000 Dams with Location, Catchment, and Attribute Information.". *Scientific Data* 10 (2023): 111, https://doi.org/10.1038/s41597-023-02008-2

Chapter 6

Dyer, K. R. *Estuaries: A Physical Introduction*. London: John Wiley, 1997.

Elliott, M., J. W. Day, R. Ramachandran, and E. Wolanski. "A Synthesis: What Is the Future for Coasts, Estuaries, Deltas and Other Transitional Habitats in 2050 and Beyond?" In *Coasts and Estuaries: The Future*, edited by Eric Wolanski, John W. Day, Michael Elliott, Ramesh Ramachandran, 1–28. Amsterdam: Elsevier, 2019, https://doi.org/10.1016/B978-0-12-814003-1.00001-0

Kraft, J. C., G. Rapp, H. Brükner, et al. "Results of the Struggle at Ancient Ephesus: Natural Processes 1, Human Intervention 0." *Geological Society, London, Special Publications* 352, no. 1 (2011): 27–36, https://doi.org/10.1144/SP352.3

Pinto P. J., and G. M. Kondolf. "Evolution of Two Urbanized Estuaries: Environmental Change, Legal Framework, and Implications for Sea-Level Rise Vulnerability." *Water* 8, no. 11 (2016): 535, https://doi.org/10.3390/w8110535

Chapter 7

Biguino, B., I. D. Haigh, J. M. Dias, and A. C. Brito. "Climate Change in Estuarine Systems: Patterns and Gaps Using a Meta-Analysis Approach." *Science of the Total Environment* 858, no. 1 (2023): 159742, https://doi.org/10.1016/j.scitotenv.2022.159742

Bonneton, P., A. G. Filippini, L. Arpaia, N. Bonneton, and M. Ricchiuto. "Conditions for Tidal Bore Formation in Convergent Alluvial Estuaries." *Estuarine, Coastal and Shelf Science* 172 (2016): 121–127, https://doi.org/10.1016/j.ecss.2016.01.019

Fichot, C. G., M. Tzortziou, and A. Mannino. "Remote Sensing of Dissolved Organic Carbon (DOC) Stocks, Fluxes and Transformations Along the Land-Ocean Aquatic Continuum: Advances, Challenges, and Opportunities." *Earth-Science Reviews* 242 (2023): 104446, https://doi.org/10.1016/j.earscirev.2023.104446

Mucci, A., G. Chaillo, and M. Jutras. "Why the St. Lawrence Estuary is Running Out of Breath." *The Conversation*, June 15, 2022, https://theconversation.com/why-the-st-lawrence-estuary-is-running-out-of-breath-184626

Postacchini, M., A. J. Manning, J. Calantoni, et al. "A Storm-Driven Turbidity Maximum in a Microtidal Estuary." *Estuarine, Coastal and Shelf Science* 288 (2023): 108350, https://doi.org/10.1016/j.ecss.2023.108350

Wolanski, E., and M. Elliott. *Estuarine Ecohydrology: An Introduction*. 2nd ed. London: Elsevier, 2015.

Chapter 8

Day, J. W. Jr, B. C. Crump, W. M. Kemp, and A. Yáñez-Arancibia, eds. *Estuarine Ecology*. 2nd ed. Hoboken, NJ: John Wiley & Sons, 2012.

Kwon, B. O., H. Kim, J. Noh, et al. "Spatiotemporal Variability in Microphytobenthic Primary Production Across Bare Intertidal Flat, Saltmarsh, and Mangrove Forest of Asia and Australia." *Marine Pollution Bulletin* 151 (2020): 110707, https://doi.org/10.1016/j.marpolbul.2019.110707

Li N., N. Tang, Z. Wang, and L. Zhang. "Response of Different Waterbird Guilds to Landscape Changes Along the Yellow Sea Coast: A Case Study." *Ecological Indicators* 142 (2022): 109298, https://doi.org/10.1016/j.ecolind.2022.109298

Narayan, S., M. W. Beck, P. Wilson, et al. "The Value of Coastal Wetlands for Flood Damage Reduction in the Northeastern USA." *Scientific Reports* 7 (2017): 9463, https://doi.org/10.1038/s41598-017-09269-z

Niella, Y., V. Raoult, T. Gaston, K. Goodman, et al. "Reliance of Young Sharks on Threatened Estuarine Habitats for Nutrition Implies Susceptibility to Climate Change." *Estuarine, Coastal and Shelf Science* 268 (2022): 107790, https://doi.org/10.1016/j.ecss.2022.107790

Whitfield, A. K., M. Elliott, and A. Basset, "Paradigms in Estuarine Ecology – A Review of the Remane Diagram with a Suggested Revised Model for Estuaries." *Estuarine, Coastal and Shelf Science* 97 (2012): 78–90, https://doi.org/10.1016/j.ecss.2011.11.026

Chapter 9

Bailey, S. A. "An Overview of Thirty Years of Research on Ballast Water as a Vector for Aquatic Invasive Species to Freshwater and Marine Environments." *Aquatic Ecosystem Health & Management* 18, no. 3 (2015): 1–8, https://doi.org/10.1080/14634988.2015.1027129

Cloern J. E., N. Knowles, L. R. Brown, et al. "Projected Evolution of California's San Francisco Bay-Delta-River System in a Century of Climate Change." *PLoS ONE* 6, no. 9 (2011): e24465, https://doi.org/10.1371/journal.pone.0024465

Esteves, L. S. *Managed Realignment: Is It a Viable Long-Term Coastal Management Strategy?* New York, NY: Springer, 2014.

Howie, A. H., and M. J. Bishop. "Contemporary Oyster Reef Restoration: Responding to a Changing World." *Frontiers in Ecology and Evolution* (2021): 9:689915, https://doi.org/10.3389/fevo.2021.689915

Little, S., J. P. Lewis, H. Pietkiewicz. "Defining Estuarine Squeeze: The Loss of Upper Estuarine Transitional Zones Against In-Channel Barriers Through Saline Intrusion." *Estuarine, Coastal and Shelf Science* 278 (2022): 108107, https://doi.org/10.1016/j.ecss.2022.108107

Warwick, R. M., J. R. Tweedley, and I. C. Potter. "Microtidal Estuaries Warrant Special Management Measures That Recognise Their Critical Vulnerability to Pollution and Climate Change." *Marine Pollution Bulletin* 135 (2018): 41-46, https://doi.org/10.1016/j.marpolbul.2018.06.062

Chapter 10

Galloway, W. E. "Process Framework for Describing the Morphologic and Stratigraphic Evolution of Deltaic Depositional Systems." In: *Delta: Models for Exploration*, 87-98. Tulsa, OK: American Association of Petroleum Geologists, 1975.

Nienhuis, J. H., A. D. Ashton, D. A. Edmonds, A. J. F Hoitink, A. J. Kettner, J. C. Rowland, and T. E. Törnqvist. "Global-Scale Human Impact on Delta Morphology Has Led to Net Land Area Gain." *Nature* 577, no. 7791 (2020): 514-518.

Paszkowski, A., S. Goodbred Jr, E. Borgomeo, M. S. A. Khan, and J. W. Hall. "Geomorphic Change in the Ganges-Brahmaputra-Meghna Delta." *Nature Reviews Earth & Environment* 2, no. 11 (2021): 763-780.

Zhang, Y., H. Huang, Y. Liu, and Y. Liu. "Self-Weight Consolidation and Compaction of Sediment in the Yellow River Delta, China." *Physical Geography* 39, no. 1 (2018): 84-98.

Chapter 11

Blum, M., J. Martin, K. Milliken, and M. Garvin. "Paleovalley Systems: Insights from Quaternary Analogs and Experiments." *Earth-Science Reviews* 116 (2013): 128-169.

Forte, A. M., and E. Cowgill. "Late Cenozoic Base-Level Variations of the Caspian Sea: A Review of Its History and Proposed Driving Mechanisms." *Palaeogeography, Palaeoclimatology, Palaeoecology* 386 (2013): 392-407.

Roberts, H. H. "Dynamic Changes of the Holocene Mississippi River Delta Plain: The Delta Cycle." *Journal of Coastal Research* 13, no. 3 (1997): 605-627.

Zavala, C., and S. X. Pan. "Hyperpycnal Flows and Hyperpycnites: Origin and Distinctive Characteristics." *Lithologic Reservoirs* 30, no. 1 (2018): 1-27.

Zhuang, Y., and T. R. Kidder. "Archaeology of the Anthropocene in the Yellow River Region, China, 8000-2000 cal. bp." *The Holocene* 24, no. 11 (2014): 1602-1623.

Chapter 12

Bailey, A. *Okavango: Africa's Wetland Wilderness*. Cape Town: Struik Publishers, 1998.

Lauria, V., I. Das, S. Hazra, et al. "Importance of Fisheries for Food Security Across Three Climate Change Vulnerable Deltas." *Science of the Total Environment* 640 (2018): 1566-1577.

Liebner, S., and C. U. Welte. "Roles of Thermokarst Lakes in a Warming World." *Trends in Microbiology* 28, no. 9 (2020): 769-779.

Overeem, I., J. H. Nienhuis, and A. Piliouras. "Ice-Dominated Arctic Deltas." *Nature Reviews Earth & Environment* 3, no. 4 (2022): 225-240.

Paola, C., R. R. Twilley, D. A. Edmonds, W. Kim, D. Mohrig, G. Parker, E. Viparelli, and V. R. Voller. "Natural Processes in Delta Restoration: Application to the Mississippi Delta." *Annual Review of Marine Science* 3 (2011): 67-91.

Spalding, M., M. Kainuma, and C. Collins. *World Atlas of Mangroves*. Abingdon: Routledge, 2010.

Chapter 13

Blum, M. D., and H. H. Roberts. "The Mississippi Delta Region: Past, Present, and Future." *Annual Review of Earth and Planetary Sciences* 40 (2012): 655-683.

Muehlmann, S. *Where the River Ends: Contested Indigeneity in the Mexican Colorado Delta*. Durham, NC: Duke University Press, 2013.

Syvitski, J. P. "Deltas at Risk." *Sustainability Science* 3 (2008): 23-32.

Tessler, Z. D., C. J. Vörösmarty, M. Grossberg, I. Gladkova, H. Aizenman, J. P. Syvitski, and E. Foufoula-Georgiou. "Profiling Risk and Sustainability in Coastal Deltas of the World." *Science* 349, no. 6248 (2015): 638-643.

Webb, E. L., N. R. Jachowski, J. Phelps, D. A. Friess, M. M. Than, and A. D. Ziegler. "Deforestation in the Ayeyarwady Delta and the Conservation Implications of an Internationally-Engaged Myanmar." *Global Environmental Change* 24 (2014): 321-333.

Chapter 14

Beiser, V. *The World in a Grain: The Story of Sand and How It Transformed Civilization*. New York, NY: Riverhead Books, 2018.

Best, J., P. Ashmore, and S. E. Darby. "Beyond Just Floodwater." *Nature Sustainability* 5 (2022): 811-813, https://doi.org/10.1038/s41893-022-00929-1

Brierley, G. J. *Finding the Voice of the River: Beyond Restoration and Management*. London: Palgrave Macmillan, 2019.

Coldren, G. A., J. A Langley, I. C. Feller, and S. K. Chapman. "Warming Accelerates Mangrove Expansion and Surface Elevation Gain in a Subtropical Wetland." *Journal of Ecology* 107, no. 1 (2018): 79-90, https://doi.org/10.1111/1365-2745.13049

Eyler, B. *Last Days of the Mighty Mekong*. London: Zed Books Ltd, 2019.

Giosan, L., J. Syvitski, S. Constantinescu, and J. Day. "Climate Change: Protect the World's Deltas." *Nature* 516, no. 7529 (2014): 31-33.

Hilmi, N., R. Chalmi, M. D. Sutherland, et al. "The Role of Blue Carbon in Climate Change Mitigation and Carbon Stock Conservation." *Frontiers in Climate* 3 (2021): 710546, https://doi.org/10.3389/fclim.2021.710546

Hirabayashi, Y., R. Mahendran, S. Koirala, et al. "Global Flood Risk Under Climate Change." *Nature Climate Change* 3 (2013): 816-821, https://doi.org/10.1038/nclimate1911

Nienhuis, P. H. *Environmental History of the Rhine-Meuse Delta: An Ecological Story on Evolving Human-Environmental Relations Coping with Climate Change and Sea-Level Rise*. New York, NY: Springer, 2008.

Sijmons, D., Y. Feddes, and E. Luiten. *Room for the River - Safe and Attractive Landscapes*. Wageningen: Blauwdruk, 2017.

Wolanski, E., J. W. Day, M. Elliott, and R. Ramesh, eds. *Coasts and Estuaries: The Future*. Amsterdam: Elsevier, 2019.

WEBSITES

Europe's Eyes on Earth
https://www.copernicus.eu/en

European Space Agency Sentinel Online
https://sentinels.copernicus.eu/web/sentinel/home

Global Surface Water Explorer
https://global-surface-water.appspot.com/map

Google Earth Engine
https://earthengine.google.com

Google Earth Engine Timelapse
https://earthengine.google.com/timelapse

Intergovernmental Panel on Climate Change
https://www.ipcc.ch

International Rivers
https://www.internationalrivers.org

IUCN (International Union for Conservation of Nature) Nature-based Solutions
https://www.iucn.org/our-work/nature-based-solutions

NASA Earth Observatory
https://earthobservatory.nasa.gov

NOAA (National Oceanic and Atmospheric Administration) Ocean Service
https://oceanservice.noaa.gov

PBL Netherlands Environmental Assessment Agency Rivers and Deltas
https://themasites.pbl.nl/future-water-challenges/river-basin-delta-tool

Planet Labs Gallery
https://www.planet.com/gallery/?utm_campaign=evr&utm_source=google&utm_medium=paid-search&utm_content=pros-leads-brdresponsivesearch-0923

Restore America's Estuaries
https://estuaries.org

Rewilding Europe
https://rewildingeurope.com

River Runner
https://river-runner-global.samlearner.com

Surface Water and Ocean Topography Mission
https://swot.jpl.nasa.gov

Transboundary Waters Assessment Programme (TWAP)
http://geftwap.org

UN Environment Programme
https://www.unep.org

USGS Earth Explorer
https://earthexplorer.usgs.gov

World Wildlife Fund
https://www.worldwildlife.org

Notes on contributors

Jim Best

Jim Best holds the Jack C. and Richard L. Threet Chair in Sedimentary Geology in the Department of Earth Science and Environmental Change at the University of Illinois at Urbana-Champaign, United States, where he is also professor of physical geography and holds affiliate appointments in the Department of Mechanical Science and Engineering, the Ven Te Chow Hydrosystems Laboratory, and the Center for Latin American and Caribbean Studies. After obtaining his BSc at the University of Leeds, United Kingdom, Jim conducted his PhD at Birkbeck College, University of London, before taking up a lectureship at the University of Hull, and then lectureship, readership, and personal chair appointments at the University of Leeds, before moving to Illinois in 2006.

Jim is an Earth scientist with research interests spanning experimental, field, and numerical investigations of Earth surface processes and both contemporary and ancient sedimentary environments. His research ranges from grain-scale dynamics through to investigation of the world's largest rivers, and examining timescales from turbulent vortices through to sedimentary environments hundreds of millions of years old. Jim has authored and co-authored more than 250 papers in journals and books, co-edited six books, and conducted field investigations of many large rivers, including the Amazon, Brahmaputra, Huang He (Yellow River), Meghna, Mekong, Mississippi, Paraná, and Paraguay. Jim was elected a fellow of the American Geophysical Union in 2015 for "pioneering the investigation of fluid flow and bedforms and field quantification of large rivers, their morphology, and flow structure," and was awarded the 2018 Jean-Baptiste Lamarck Medal by the European Geosciences Union in recognition of his "major contributions to our understanding of physical sedimentary processes and their products in the geological record."

Stephen E. Darby

Steve Darby is a professor in physical geography at the School of Geography and Environmental Sciences, University of Southampton, United Kingdom. He is a river scientist with more than 30 years' experience of researching the world's diverse river systems. His main interests span how changes in patterns of river erosion and sedimentation are driven by natural and human-induced processes, and the implications of these changes for the management of flooding and erosion risk. Steve has worked in a range of river environments in Australia, Asia, Europe, and North America. His academic career started in Italy in the Department of Civil Engineering at the University of Florence, before a period working for the United States Department of Agriculture at the National Sedimentation Laboratory in Oxford, Mississippi. He took up his current post at the University of Southampton in 1997. He received his PhD in fluvial geomorphology from the University of Nottingham. Steve has co-authored more than 100 publications in journals and books.

Luciana S. Esteves

Lu Esteves is an associate professor in physical geography at the Department of Life and Environmental Sciences, Bournemouth University, United Kingdom. She is a coastal scientist with more than 30 years' experience in research and teaching focused on coastal systems. Her main interests relate to quantifying and communicating coastal change driven by natural and human-induced processes, and the implications for the sustainable management of flooding and erosion risk. Interested in applied research, she has worked with coastal managers in Europe, Latin America, and Africa. Her academic career started in southern Brazil at the Universidade Federal do Rio Grande (FURG), where she worked for 14 years before moving to the United Kingdom in 2006. She studied oceanography at FURG, and received her MSc in marine geology from Florida Atlantic University in the United States and her PhD in geosciences (coastal and marine) from the Universidade Federal do Rio Grande do Sul (UFRGS) in Brazil. Lu has co-authored more than 60 publications in books and journals that specialize in coastal and marine systems, including *Estuaries and Coasts*; *Estuarine, Coastal and Shelf Science*; *Ocean and Coastal Management*; and the *Journal of Geophysical Research Oceans*.

Carol A. Wilson

Carol Wilson is an associate professor in sedimentology and ecogeomorphology in the Department of Geology and Geophysics at Louisiana State University (LSU). She is an expert in coastal deltaic and wetland studies, bringing together the intertwined processes of biology, geology, and hydrodynamics. After growing up on the United States Gulf Coast and attending the University of New Orleans and Tulane University, where she studied wetland loss in Louisiana surrounding the devastating events of Hurricane Katrina in 2005, Carol pursued a PhD at Boston University and post-doctoral studies at Vanderbilt University. Over the past 20 years, she and colleagues have carried out extensive research on coastal deltas and wetland systems around the world, including the East Coast and Gulf Coast of the United States, Canada, and Bangladesh, where her investigations quantify how these systems respond to sea-level rise, nutrients, sediment supply, salinity stress, hurricanes and storms, animal bioturbation, and human modification. Carol is a member of the LSU Coastal Studies Institute, president of the LSU Science Club, and persistent volunteer for the Cajun Navy, a post-emergency aid organization. She received the Sigma Phi Epsilon Certificate of Recognition for Teaching in 2016. Carol and her students have authored and peer-reviewed book chapters and scientific journal articles featured in *Geology*, *Nature Communications*, *Nature Climate Change*, *Proceedings of the National Academy of Sciences*, *Annual Review of Marine Science*, *Geomorphology*, *Earth Surface Processes and Landforms*, *Estuarine Coastal Shelf Science*, and *Estuaries and Coasts*.

Author acknowledgments

Jim Best is indebted to colleagues across many countries, including Argentina, Bangladesh, Brazil, Canada, Cambodia, China, Netherlands, the United Kingdom, and the United States, for their friendship and collaboration that have been at the core of much of his joint research on the world's large rivers and deltas. In addition, his gratitude goes to the wealth of supremely talented graduate students and post-doctoral fellows with whom he has been fortunate to work, and who have provided inspiration and insight into many aspects of Earth surface processes and sedimentary environments. He is indebted for funding over several decades provided by agencies including the National Science Foundation and Jack C. and Richard L. Threet Chair in Sedimentary Geology in the United States, and the Natural Environment Research Council, Royal Society of London, and Leverhulme Trust in the United Kingdom, which has enabled much of his research and permitted work on some of the world's most beautiful and inspirational rivers.

Stephen Darby thanks the numerous colleagues who have generously shared their knowledge and time, especially during fieldwork on the Ganges, Mekong, Chang Jiang (Yangtze), Indus, and Mississippi rivers. He is also grateful for support from the Natural Environment Research Council, the UK Global Challenges Research Fund, and the Royal Society.

Luciana Esteves would like to thank the invaluable contributions of researchers and practitioners worldwide who have generously shared their knowledge about estuaries and coasts over the years, as well as support from the Department of Life and Environmental Sciences at Bournemouth University. She is also grateful for the interest of students and funders, particularly Bournemouth University, the UK Global Challenges Research Fund, Newton Funds, the Natural Environment Research Council, the Arts and Humanities Research Council, and the European Union, as well as CAPES and CNPq, who fueled her enthusiasm for coastal research and teaching.

Carol Wilson would like to thank the Louisiana State University Department of Geology and Geophysics and Coastal Studies Institute for their support; the National Science Foundation, Office of Naval Research, the Louisiana Center of Excellence, the Bureau of Ocean Energy Management, and the World Bank for funding for research; and numerous students and colleagues from whom she has learned so much.

All four authors are deeply grateful to Kate Shanahan for the initial invitation to write this book; David Price-Goodfellow for his expert guidance, advice, and unending encouragement throughout the writing of the *Atlas*; Lindsey Johns for her design skills and converting our ideas into a beautiful book; and Susi Bailey for her detailed, constructive, and inspiring editing that greatly assisted in evolving the final text. Thank you so much to you all.

Index

Picture credits

The publisher would like to thank the following for permission to reproduce copyright material and for use of reference material for illustrations. All reasonable efforts have been made to contact copyright holders and to obtain their permission for the use of copyright material. The publisher apologizes for any errors or omissions and will gladly incorporate any corrections in future reprints if notified.

© Morgan Adler (https://www.morganadler.com) 149b

Adobe Stock /robertharding 14-15

Alamy Stock Photo /Abaca Press 233t; /aerial-photos. com 160-61; /AfriPics.com 87b; /Alberto Rigamonti 6-7; /Album 13; /All Canada Photos 87t; /Andrew Vaughan/The Canadian Press 164; /Art-Studio 384; / Arterra Picture Library 244-45; /Associated Press 135t; /Avalon.red 153b; /Bengal Picture Library 128-29; /Blue Planet Archive/Wolfgang Poelzer 210; /Cavan Images 2 & 374; /Cernan Elias 219; /Cynthia Lee 214; / David Wall 91b; /Denis-Huot/Nature Picture Library 104-105; /Design Pics Inc 53b & 78b; /dpa picture alliance 142, 158 & 237l; /Frans Lemmens 381 & 385; / GC Photo 286-87; /Genevieve Vallee 351b; /Glyn Genin 200; /Herbert Frei/mauritius images GmbH 373t; / history_docu_photo 22; /ImageBROKER.com GmbH & Co. KG 132 & 371tl; /Imaginechina Limited 143; /Jeremy Moeran 239; /Joe Klementovich/Cavan Images 239; / KAR Photography 289; /Kevin Schafer 118; /Li Linhai/ Xinhua 261; /lophius 367t; /Łukasz Szczepanski 90; / Malcolm Schuyl 108t; /Mark Pearson 353; /Martin Bertrand 135b; /Matjaz Corel 367c; /Michael Dietrich/ imageBROKER.com GmbH & Co. KG 206; /Mu Yu/ Xinhua 380; /Muhammad Mostafigur Rahman 290c; /Nature Picture Library 60, 109, 197, 208 & 307t; / NSF Photo 232; /Oleksandr Malovichko 66t; /Pacific Imagica 27b; /Rob Crandall 272; /Robert Wyatt 349; / Rodrigo Abd/Associated Press 150; /Rudi Sebastian/ imageBROKER.com GmbH & Co. KG 226; /Sandro Santioli/RealyEasyStar 63; /Simon Dack 198; /Sipa US 146; /Süddeutsche Zeitung Photo 133; /Thomas Hanahoe 213t; /tonymills 199; /Universal Images Group North America LLC 20 & 250; /Victor Paul Borg 218l; /Vijit Ghosh/SOPA Images/Sipa USA 377; /Yang Bin/Xinhua 314; /ZUMA Press, Inc. 342

American Geographical Society Library—Maps 23

Courtesy Dr Sam Andrews (Acadia University, Canada) and Dr Antóin O'Sullivan (University of New Brunswick, Canada) 362

Ashworth, P. J., Sambrook Smith, G. H., Best, J. L., Bridge, J. S., Lane, S. N., Lunt, I. A., Reesink, A.J.H., Simpson, C. J. and Thomas, R. E. "Evolution and sedimentology of a channel fill in the sandy braided South Saskatchewan River and its comparison to the deposits of an adjacent compound bar." *Sedimentology* 58 (2011): 1860-83. https://doi.org/10.1111/j.1365-3091.2011.01242.x 72-73

© Andy Ball/University of Southampton 37b & 38

Cleveland Public Library Photograph Collection 149t

Andy Coburn/Program for the Study of Developed Shorelines at Western Carolina University 52

Daniel E. Coe 44

Reprinted from Journal of Hydrology, 563, Cohen, S., Wan, T., Islam, M. T., and Syvitski, J.P.M., "Global river slope: A new geospatial data set and global-scale analysis." 1057-67. Copyright 2018, https://doi. org/10.1016/j.jhydrol.2018.06.066. With permission from Elsevier 82-83

David Rumsey Map Collection, David Rumsey Map Center, Stanford Libraries 98

© de Santana, C. D., Crampton, W.G.R., Dillman, C.B., et al. "Unexpected species diversity in electric eels with a description of the strongest living bioelectricity generator." *Nature Communications* 10 (2019): 4000. Fig 4. https://doi 10.1088/1748-9326/ac9197 120

Dreamstime /Alexey Kornylyev 225; /Florian Blümm 11; /Brayden Stanford 80-81

European Space Agency /contains modified Copernicus Sentinel data 2019 processed by ESA 4-5 & 191; /contains modified Copernicus

Sentinel data 2020 processed by ESA 99 & 274t; / Copernicus Sentinel-2 imagery 185, 297, 306 & 338br; / ENVISAT 355t

Photo courtesy Professor Chris Fielding, University of Connecticut 103b

Flickr /Andy Morffew 308; /Antonio Santa-Pau Ramírez 333; /Ashley Coates 230r; Bernard Dupont 337r; /Björn S.. 383; /Dennis 127b; /Jose A 358tl; /Sergei Gussev 318; /Valdiney Pimenta 290t; Rebecca Wynn/ USFWS/GPA Photo Archive/US Dept of State 293

Getty Images /Daniel Bosma 340-41; /Patricia Hamilton/larigan 84; /Streeter Lecka 166-67; /Lam Yik Fei /AsiaPac 235

Google Earth 77, 101b, & 192 (all 4)

Grasshopper Geography /Artwork by Robert Szucs 51

John Hammond/Rivers from Above (www. johnchammond.com) 96

© 2021 Haskins, J., Endris, C., Thomsen, A. S., Gerbl, F., Fountain, M. C., and Wasson, K. "UAV to inform restoration: A case study from a California tidal marsh." *Frontiers in Environmental Science* 9 (2021): 642906. https://doi.org/10.3389/fenvs.2021.642906 46

istockphoto /Aerial Essex 202-203; /BanarTABS 376ccl; /imaginima 371b; /plej92 376cr; /ASMR 184

Library of Congress /Geography and Map Division 21b; /Prints and Photographs Division 21t, 238, 265b, 323b & 328c

© Mulligan, M., van Soesbergen, A., & Sáenz, L. "GOODD, a global dataset of more than 38,000 georeferenced dams." *Scientific Data* 7 (31) (2020). Fig 1. https://doi.org/10.1038/s41597-020-0362-5 141 t&b

© Nardi, F., Annis, A., Di Baldassarre, G., et al. "GFPLAIN250m, a global high-resolution dataset of Earth's floodplains." *Scientific Data* 6 (180309) (2019). https://doi.org/10.1038/sdata.2018.309 75tl, tr, bl & br

NASA Earth Observatory /International Space Station Crew Earth Observations experiment and Image Science & Analysis Laboratory, Johnson Space Center 249c; /Jesse Allen and Robert Simmon, using Landsat data provided by the United States Geological Survey 248; /Johnson Space Center 33; /Joshua Stevens using Landsat data from the U.S. Geological Survey 16-17, 173, 234, 265t, 140l & 140r; /Landsat 8-OLI 180-81; /Lauren Dauphin, using Landsat data from the U.S. Geological Survey 243; /Lauren Dauphin, using Landsat data from the U.S. Geological Survey. Photo by Weiguang Teng 224; /Mike Taylor 91t; /US Geological Survey 54; /USGS EROS Data Center 277b

NASA/Goddard Space Flight Center 34, 168-69b & 174; /MITI/ERSDAC/JAROS and the US-Japan ASTER Science Team 92

NASA/International Space Station Crew Earth Observations experiment and Image Science & Analysis Laboratory, Johnson Space Center 249b

NASA/Jet Propulsion Laboratory 291t; /NGA 43

NASA/Johnson Space Center 246

Courtesy Natel Energy 371tr

Nature in Stock /D P Wilson/FLPA 205

Naval Intelligence Division, Geographical Handbook, *Iraq and the Persian Gulf*, September 1944, fig 162 323t

Jeffrey Neal, Laurence Hawker (2023): FABDEM V1-2. https://doi.org/10.5523/bris. s5hqmjcdj8yo2ibzi9b4ew3sn 50. Used with permission.

From Bruce Norman Bjornstad, *Ice Age Floodscapes of the Pacific Northwest*. Cham: Springer, 2021. 100t

Planet.com Planet Labs PBC (Planet.com) 9, 41b, 47t, 47b, 94, 114-15, 130, 137b, 156, 268-69, 335t & 335b

PXhere /Jong Myung Lim 367b

Rijkswaterstaat Archieven 126

RSPB /Lee Schofield, RSPB Site Manager, Haweswater 382

Science Photo Library /GEOEYE 170

Shutterstock /Adi Dharmawan 227t; /adwar 260r; / Agami Photo Agency 215 & 313bl; /Air Camargue 253t; /Alberto Loyo 193b; /Alex Couto 211; /Altrendo Images 78t; /amperespy44 260l; /Ana Dracaena 365; /Anetlanda 190; /Anirut Krisanakul 337l; /Anton_ Ivanov 48-49; /Bigc Studio 125; /Bob Hilscher 364; /C. Ray Shea 271t; /Calin Stan 313t; /Chalalai Atcha 327; / ChiccoDodiFC 351t; /corlaffra 611; /Damsea 204 & 376cl; /Danny Ye 121; /djavitch 113; /Doug McLean 229b; / EcoPrint 217; /Ed Metz 328b; /Elena Larina 187t; / Erni 230l; /Foto Para Ti 325br; /FrentaN 257b; /FTiare 220-21; /Gaston Piccinetti 301b; /gnomeandi 227b; / Grodza 301t; /guentermanaus 310; /Guillem Lopez Borras 172; /Halit Sadik 19t; /I. Noyan Yilmaz 378; / iliuta goean 332b; /Infinity T29 201; /ivSky 237r; /Jez Bennett 116; /jimcatlinphotography.com 189; /Johan Larson 257t; /John Brueske 112; /Joop Hoek 263t; /Lam Van Linh 339l; /lavizzara 255t; /Lee Yiu Tung 376ccr; / maphke 271b; /marekuliasz 117t; /Max Lindenthaler 307b; /Max R Miller 108b; /Michael G McKinne 299; /Michal Balada 277t; /Mihai_Andritoiu 188; /Mike Mareen 334t; /Mistervlad 10; /Monica Viora 313br; / mwesselsphotography 282; /MyVideoimage.com 263b; /ohrim 233b; /Ondrej Prosicky 278; /Pascale Gueret 296; /Peter Stuckings 326; /Photographer Lili 291c; /Photojulia 260c; /PradeepGaurs 148; /Quang nguyen vinh 124; /RLS Photo 193t; /Roberto Rizzi 110; / Rosamar 231; /rospoint 309t; /Rudmer Zwerver 114l & 242; /Ruud Morijn Photographer 331; /Sergey Bezgodov 216; /Sergey Uryadnikov 122; /Sergey Yeromenko 155; / slowmotiongli 336l & 336r; /Soumyajit Nandy 338bl; / Srinivas Piratla 61r; /StevenK 325bl; /T8 stock 218r; / Talukdar David 330; /Thomas Retterath 249t; /tony mills 196; /Troutnut 303; /ventdusud 280; /Viacheslav Lopatin 305; /Vietnam Stock Images 332t; /Viktar Malyshchyts 76; /Visual Collective 26; /Vladimir Melnik 127t; /Vladimir Wrangel 106bl; /xamnesiacx84 24; /zaferkizilkaya 316-17; /zuzabah Texture on coloured backgrounds throughout the book

© Susanne Sokolow 39

M.L.J.Stiassny, AMNH 86

Varun Swamy (Field Projects) San Diego Zoo Institute for Conservation Research 89t

U.S. Geological Survey /Benjamin Jones 304b; /Emily Roeder 64; /NASA 179 & 194; /NUSO 213b; /US Corona Spy satellite 41t

Washington Geological Survey/Department of Natural Resources /Photo by Dan Coe 100b

Wessex Environment Agency Paul Gainey, Environment Agency and Wildfowl & Wetlands Trust (WWT) 240

Wiki Commons /Marcus Cyron 298; /Ulf Mehlig 295tr

Illustration credits

Many figures in the book were redrawn, adapted or modified from the following sources:

18: Adapted from https://commons.princeton.edu/mg/ wp-content/uploads/2017/04/Ancient_Civilizations_ of_the_Old_World_3500_to_after_600_BCE.jpg • 25: Adapted from USGS figure: https://labs.waterdata.usgs. gov/visualizations/water-cycle/index.html#/ • 30/31, 345, 352, 372-73 & 375: Adapted from Ligtvoet, W., et al. *The Geography of Future Water Challenges; Bending the Trend*, The Hague: PBL Netherlands Environmental Assessment Agency, 2023 • 35m: Adapted from Garzanti, E., et al. "Congo River sand and the equatorial quartz factory." *Earth Science Reviews* 197 (2019): 102918 • 35b: Adapted from Babonneau, N., et al. "Sedimentary architecture in meanders of a submarine channel: detailed study of the present Congo turbidite channel (Zaiango Project)." *Journal of Sedimentary Research* 80 (10) (2010): 852-66 • 37t: Redrawn from FAO. *The State of World Fisheries and Aquaculture 2020. Sustainability in Action*. Rome: Food and Agriculture Organization of the United Nations, 2020 • 40: Redrawn from Nagel, G. W., Darby, S. E., and Leyland, J. "The use of satellite remote sensing for exploring river meander migration." *Earth-Science Reviews* 247 (2023): 104607 • 42: Modified from Langhorst, T., and Pavelsky, T. "Global observations of riverbank erosion and accretion from Landsat imagery", *JGR Earth Surface* 128 (2023): e2022JF006774 • 45: Redrawn from Johnson, K., et al. "Rapid mapping of ultrafine fault zone topography with structure from motion." *Geosphere* 10 (5) (2014): 969-86 • 53: Adapted from https://www.economist.com/the-americas/2014/06/06/salmon-en-route • 55: Redrawn from Hoorn, C., et al. "Amazonia through time: Andean uplift, climate change, landscape evolution, and biodiversity." *Science*. 2010 Nov 12; 330(6006): 927-31 • 56-57: Redrawn by Camille Ouellet Dallaire and from: Dallaire, C. O., et al. *Environmental Research Letters* 14 024003 (2019) • 58l: Redrawn from Irwanto, D. *Sundaland: Tracing the Cradle of Civilizations*. West Java: Indonesia Hydro Media. 2019 • 58r: Redrawn from Hanebuth, T.J.J., Voris, H. K., Yokoyama, Y., Saito, Y., and Okuno, J. "Formation and fate of sedimentary depocentres on Southeast Asia's Sunda Shelf over the past sea-level cycle and biogeographic implications." *Earth-Science Reviews* 104, Issues 1-3 (2011): 92-110 • 59: Adapted from Pazzaglia, F. J., "Fluvial terraces." In *Treatise on Geomorphology*, ed. J. Shroder and E. Wohl vol. 9, *Fluvial Geomorphology*, 379-412. San Diego, CA, 2013 • 62: Adapted from Marshak, S. *Earth: Portrait of a Planet*. New York: W.W. Norton & Co., 2005 • 65: Adapted from https://ww2.mathworks.cn/company/newsletters/articles/analyzing-and-visualizing-flows-in-rivers-and-lakes-with-matlab.html; and Jackson, P. R., et al. "Velocity mapping in the Lower Congo River: a first look at the unique bathymetry and hydrodynamics of Bulu Reach." 2009. https://pubs.usgs.gov/publication/70158956#:~:text=Results%20show%20that%20the%20flow,channel%20flow%20structures%20are%20absent • 66-67: Modified from Cohen, S., Kettner, A. J., and Syvitski, J.P.M. "Global suspended sediment and water discharge dynamics between 1960 and 2010: Continental trends and intra-basin sensitivity." *Global and Planetary Change* 115 (2014), 44-58 • 68-69: Courtesy of Edward Park and modified from Park, E., and Latrubesse, E. M. "A geomorphological assessment of wash-load sediment fluxes and floodplain sediment sinks along the lower Amazon River." *Geology* 47 (5) (2019): 403-6 • 71: Adapted from McClain, M. E., and Naiman, R. J. "Andean influences on the biogeochemistry and ecology of the Amazon River." *BioScience* 58, (4) (2008): 325-38 • 79: Adapted from Junk, W. J., Bayley, P. B., and Sparks, R. E. "The flood pulse concept in river-floodplain systems." *Canadian Journal of Fisheries and Aquatic Science* 106 (1989): 110-27 • 85: Redrawn from Chen, S-A., et al. "Aridity is expressed in river topography globally." *Nature* 573 (2019): 573-77 • 88: Modified from Strick, R.J.P., et al. "Quantification of bedform dynamics and bedload sediment flux in sandy braided rivers from airborne and satellite imagery." *Earth Surface Processes and Landforms* 44 (2019): 953-72 • 93: Redrawn based on Nicholas, A. P. "Morphodynamic diversity of the world's largest rivers." *Geology* 41 (4) (2013): 475-78 • 95 & 97: Redrawn from Sylvester, Z., Durkin, P. R., Hubbard, S. M., and Mohrig D. "Autogenic translation and counter point bar deposition in meandering rivers." *Geological Society of America Bulletin* 133 (2021): 2439-56 • 101t: Redrawn from Bjornstad, B. N. *Ice Age Floodscapes of the Pacific Northwest*. Cham: Springer International Publishing, 2021 • 103tl and m: Redrawn from Ghinassi, M., et al. "Plan-form evolution of ancient meandering rivers reconstructed from longitudinal outcrop sections." *Sedimentology* 61 (2014): 952-77 • 103tr: Modified from Strick, R.J.P., Ashworth, P. J., Awcock, G, and Lewin, J. "Morphology and spacing of river meander scrolls." *Geomorphology* 310 (2018): 57-68 • 106-107: Courtesy of Pedro Val, and modified from Val, P., Lyons, N. J., Gasparini, N., Willenbring, J. K., and Albert, J. S. "Landscape evolution as a diversification driver in freshwater fishes." *Frontiers in Ecology and Evolution* 9 (2022): 788328 • 111: Adapted from Vannote, R. L., et al. "The river continuum concept." *Canadian Journal of Fisheries and Aquatic Sciences* 37 (1) (1980): 130-37 • 117b: Redrawn from Lytle, D. A., et al. "Linking river flow regimes to riparian plant guilds: a community-wide modeling approach." *Ecological Applications* 27 (4) (2017): 1027-1377 • 119: Courtesy Tacio Bicudo and modified from Bicudo, T. C., et al. "Andean tectonics and mantle dynamics as a pervasive influence on Amazonian ecosystem." *Scientific Reports* 9 (2019): 16879 • 123: Redrawn from Takemoto, H., Kawamoto, Y., and Furuichi, T. "How did bonobos come to range south of the Congo river? Reconsideration of the divergence of *Pan paniscus* from other *Pan* populations." *Evolutionary Anthropology* 24 (5) (2015): 170-84 • 131t: Modified from Andreadis, K. M., et al. "Urbanizing the floodplain: global changes of imperviousness in flood-prone areas." *Environmental Research Letters* 17 (2022): 104024 • 131b: Modified from Varis, O., Taka, M., and Tortajada, C. "Global human exposure to urban riverine floods and storms." *River* 1 (2022): 80-90 • 136: Courtesy of Gustavo Naumann and modified from https://doi.org/10.1002/2017GL076521 • 137t: Modified from Hirabayashi, Y., et al. "Global exposure to flooding from the new CMIP6 climate model projections." *Scientific Reports* 11 (2021): 3740 • 138: Modified from Moragoda, N., and Cohen, S. "Climate-induced trends in global riverine water discharge and suspended sediment dynamics in the 21st century." *Global and Planetary Change* 191 (2020): 103199 • 139: Redrawn from Dongfeng Li, et al. "Exceptional increases in fluvial sediment fluxes in a warmer and wetter High Mountain Asia." *Science* 374 (2021): 599-603 • 144: Redrawn from Best, J. "Anthropogenic stresses on the world's big rivers." *Nature Geoscience* 12 (2019): 7-21 • 145: Redrawn from Serrano, A., et al. "Virtual water flows in the EU27: A consumption-based approach." *Journal of Industrial Ecology* 20 (2016): 547-58 • 147: Redrawn from Knox, R. L., Wohl, E. E., and Morrison, R. R. "Levees don't protect, they disconnect: A critical review of how artificial levees impact floodplain functions." *Science of the Total Environment* 837 (2022): 155773 • 151: Adapted from Sergeant, C. J., et al. "Risks of mining to salmonid-bearing watersheds." *Science Advances* 8 (2022) • 153t: Courtesy of Günther Grill and modified from Grill, G., et al. "Mapping the world's free-flowing rivers." *Nature* 569 (2019): 215-221 • 154: Redrawn from O'Neill Jr., C. R., and Dextrase, A. "The introduction and spread of the zebra mussel in North America." *Proceedings of the Fourth International Zebra Mussel Conference, Madison, Wisconsin* (1994): 433-46 • 157: Redrawn from Sanders, B. F., et al. "Large and inequitable flood risks in Los Angeles, California." *Nature Sustainability* 6 (2023): 47-57 • 159: Redrawn from Best, J., and Darby, S. E. "The pace of human-induced change in large rivers: Stresses, resilience, and vulnerability to extreme events." *One Earth* 2 (2022): 510-14 • 162-63: Adapted from https://www.coastalwiki.org/wiki/File:Figure1_3_COLOR.png • 165: Adapted from https://www.bayoffundy.com/about/highest-tides/ • 169t: Adapted from Dalrymple, R. W., Zaitlin, B., and Boyd, R. R. "Estuarine facies models; conceptual basis and stratigraphic implications." *Journal of Sedimentary Research* 62 (6) (1992): 1130-46 • 171: Adapted from Defontaine, S., et al. "Microplastics in a salt-wedge estuary: Vertical structure and tidal dynamics." *Marine Pollution Bulletin* 160 (2020): 111688 • 175: Some elements adapted from The Open University, *Waves, Tides and Shallow-water Processes*. Oxford: Pergamon Press, 1993 • 209: Created based on data from The IUCN Red List of Threatened Species. Version 2022-2 • 222-23: Created using data from FAO. *The State of Food and Agriculture 2022*. Rome: Food and Agriculture Organization of the United Nations, 2022 • 228: Adapted from Molnar, J. L., et al. "Assessing the global threat of invasive species to marine biodiversity." *Frontiers in Ecology and the Environment* 6 (2008): 485-492 • 229: Adapted from https://www.grida.no/resources/7191. Credit: Hugo Ahlenius, UNEP/GRID-Arendal • 241: Redrawn from Esteves, L. S. *Managed Realignment: A Viable Long-term Coastal Management Strategy?* New York: Springer, 2014 • 252-53: Adapted from Nienhuis, J. H., et al. "Global-scale human impact on delta morphology has led to net land area gain." *Nature* 577 (2020): 514-18 • 255b: Adapted from Zavala, C., and Pan, S.X. "Hyperpycnal flows and hyperpycnites: Origin and distinctive characteristics." *Lithologic Reservoirs* 30, (1) (2018): 1-27 • 256: Adapted from Paszkowski, A., et al. "Geomorphic change in the Ganges-Brahmaputra-Meghna delta." *Nature Reviews Earth & Environment* 2, (11) (2021): 763-80 • 259: Adapted from Yi Zhang, et al. "Self-weight consolidation and compaction of sediment in the Yellow River Delta, China." *Physical Geography* 39, (1) (2018): 84-98 • 264: Redrawn courtesy of Professor Jeff Hanor, Louisiana State University, USA • 275 & 281: Redrawn using information from Blum, M. D., and Roberts, H.H. "The Mississippi delta region: Past, present, and future." *Annual Review of Earth and Planetary Sciences* 40 (2012): 655-83 • 279: Adapted from https://www.eurekalert.org/multimedia/754469 courtesy of the *Journal of Archaeological and Anthropological Sciences* • 283: Adapted from Van Wagoner, J. C., Mitchum, R. M., Campion, K. M., and Rahmanian, V. D. "Siliciclastic sequence stratigraphy in well logs, cores, and outcrops: Concepts for high-resolution correlation of time and facies." *AAPG Methods in Exploration Series*, 7. Tulsa: American Association of Petroleum Geologists, 1990 • 285: Adapted from Kroonenberg, S. B., et al. "Two deltas, two basins, one river, one sea: The modern Volga Delta as an analogue of the Neogene Productive Series, South Caspian Basin." In *River Deltas—Concepts, Models and Examples* ed. Liviu Giosan and Janok P. Bhattacharya, *SEPM* 83 (2005): 231-56 • 292: Adapted from Paola, C., et al. "Natural processes in delta restoration: Application to the Mississippi Delta." *Annual Review of Marine Science* 3 (2011): 67-91 • 294: Adapted from a report by the National Oceanic and Atmospheric Administration. 295l: Adapted from Daniel Cole/Alamy Stock Vector • 304t: Adapted from In 't Zandt, M. H., Liebner, S., and Welte, C. U. "Roles of thermokarst lakes in a warming world." *Trends in Microbiology* 28 (9) (2020): 769-79 • 309b: Redrawn from data from Maryland Department of Natural Resources • 311t: Redrawn from data in FAO. *Ecosystems and Human Well-being: Synthesis*. Washington: Island Press, 2005 • 311b: Redrawn from Lauria, V., et al. "Importance of fisheries for food security across three climate change vulnerable deltas." *Science of the Total Environment* 640-641 (2018): 1566-77 • 319: Modified from Webb, E. L., et al. "Deforestation in the Ayeyarwady Delta and the conservation implications of an internationally-engaged Myanmar." *Global Environmental Change* 24 (2014): 321-33 • 320-21: Redrawn from Tessler, Z. D., et al. "Profiling risk and sustainability in coastal deltas of the world." *Science* 349, (6248) (2015): 638-43 • 334b: Redrawn from map of United States Geological Survey • 339r: Redrawn from https://www.aljazeera.com/wp-content/uploads/2020/04/a3b55db85ea04d35a8e6f61745f06d8f_18.jpeg?quality=80 • 343: Modified from Gutiérrez, J. M., et al. *Climate Change 2021: The Physical Science Basis. Contribution of Working Group I to the Sixth Assessment Report of the Intergovernmental Panel on Climate Change*. Cambridge: Cambridge University Press. In press. Interactive atlas available from http://interactive-atlas.ipcc.ch/ • 344t and 344b: Redrawn from https://sealevel.nasa.gov/ipcc-ar6-sea-level-projection-tool?type=global&info=true and https://climatedata.ca/resource/understanding-shared-socio-economic-pathways-ssps/ • 346-47: Redrawn from Dunn, F. E., et al. "Projections of declining fluvial sediment delivery to major deltas worldwide in response to climate change and anthropogenic stress." *Environmental Research Letters* 14 (8) (2019): 084034 • 348: Redrawn from Alfieri, L., et al. "Global projections of river flood risk in a warmer world." *Earth's Future* 5 (2016): 171-82 • 354: Redrawn from Nienhuis, J. H., et al. "River deltas and sea-level rise." *Annual Review of Earth and Planetary Sciences* 51 (2023): 79-104 • 355b: Modified from Schmitt, R.J.P., et al. "Strategic basin and delta planning increases the resilience of the Mekong Delta under future uncertainty." *PNAS* 118 (36) (2021): e2026127118 • 356: Modified from https://www.earthdata.nasa.gov/learn/articles/swot-calibration-validation • 357: Modified from Kääb, A,. Altena, B., and Mascaro, J. "River-ice and water velocities using the Planet optical cubesat constellation." *Hydrology and Earth System Sciences* 23 (10) (2019) • 358-59: Modified from image courtesy of John Gardner, University of Pittsburgh, USA • 358b: Modified from Juan Bu, et al. "Monitoring the Chl-a distribution details in the Yangtze River mouth using satellite remote sensing." *Water* 14 (8) (2022): 1295 • 360: Modified from Wanders, N., et al. "High-resolution global water temperature modeling." *Water Resources Research* 55 (4) (2019): 2760-78 • 361: Modified from Kumbier, K., et al. "An eco-morphodynamic modelling approach to estuarine hydrodynamics and wetlands in response to sea-level rise." *Frontiers in Marine Science* 9 (2022): 860910 • 363: Adapted from Meng Yao. "Fishing for fish environmental DNA: Ecological applications, methodological considerations, surveying designs, and ways forward." *Molecular Ecology* 31 (20) (2022): 5132-64; and Villegas-Ríos, D., Jacoby, D.M.P. and Mourier, J. "Social networks and the conservation of fish." *Communications Biology* 5, 178 (2022) • 369: Adapted from Giosan, L., et al. "Climate change: Protect the world's deltas." *Nature* 516 (7529) (2014): 31-33 • 379t: Redrawn from Rovai, A. S., et al. "Brazilian mangroves: blue carbon hotspots of national and global relevance to natural climate solutions." *Frontiers in Forests and Global Change* 4 (2021): 787533 • 379b: Adapted from https://doi.org/10.5194/essd-14-4811-2022